T0142179

New Era for Robust Speech Recognition

Shinji Watanabe • Marc Delcroix • Florian Metze •
John R. Hershey
Editors

New Era for Robust Speech Recognition

Exploiting Deep Learning

 Springer

Editors

Shinji Watanabe
Mitsubishi Electric Research Laboratories
  (MERL)
Cambridge
Massachusetts, USA

Marc Delcroix
NTT Communication Science Laboratories
NTT Corporation
Kyoto, Japan

Florian Metze
Language Technologies Institute
Carnegie Mellon University
Pittsburgh
Pennsylvania, USA

John R. Hershey
Mitsubishi Electric Research Laboratories
  (MERL)
Cambridge
Massachusetts, USA

ISBN 978-3-319-87849-2   ISBN 978-3-319-64680-0   (eBook)
DOI 10.1007/978-3-319-64680-0

Printed on acid-free paper

This Springer imprint is published by Springer Nature
The registered company is Springer International Publishing AG
The registered company address is: Gewerbestrasse 11, 6330 Cham, Switzerland

This book is dedicated to the memory of Yajie Miao († 2016), who tragically passed away during the preparation of this book.

Preface

The field of automatic speech recognition has evolved greatly since the introduction of deep learning, which began only about 5 years ago. In particular, as more and more products using speech recognition are being deployed, there is a crucial need for increased noise robustness, which is well served by deep learning methods. This book covers the state of the art in noise robustness for deep-neural-network-based speech recognition with a focus on applications to distant speech. Some of the main actors in the areas of front-end and back-end research on noise-robust speech recognition research gathered in Seattle for the 2015 Jelinek Speech and Language Summer Workshop. They significantly advanced the state of the art by tightly integrating these two areas together for the first time. This book compiles their insights, and presents detailed descriptions of some of the key technologies in the field, including speech enhancement, neural-network-based noise reduction, robust features, acoustic-model adaptation, training data augmentation, novel network architectures, and training criteria. The presentation of these technologies is augmented with descriptions of some of the most important benchmark tools and datasets that are instrumental for research in the field, and a presentation of recent research activities at some of the leading institutions in the area of noise-robust speech recognition.

This book is intended for researchers and practitioners working in the field of automatic speech recognition with an interest in improving noise robustness. This book will also be of interest to graduate students in electrical engineering or computer science, who will find it a useful guide to this field of research.

Cambridge, MA, USA   Shinji Watanabe
Kyoto, Japan   Marc Delcroix
Pittsburgh, PA, USA   Florian Metze
Cambridge, MA, USA   John R. Hershey

Acknowledgments

Much of the work reported in this book was initiated during the 2015 Jelinek Memorial Summer Workshop on Speech and Language Technologies (JSALT) at the University of Washington, Seattle, and was supported by Johns Hopkins University via NSF Grant No. IIS 1005411, and gifts from Google, Microsoft Research, Amazon, Mitsubishi Electric, and MERL. We would like to thank these sponsor companies and JSALT organizing committee members, in particular Professor Les Atlas at University of Washington, Professor Sanjeev Khudanpur at Johns Hopkins University, Professor Mari Ostendorf at University of Washington, and Dr. Geoffrey Zweig at Microsoft Research. Thank you for providing us the opportunity to work on these exciting research topics.

We also would like to thank Dr. Dong Yu and Dr. Mike Seltzer at Microsoft Research for their valuable comments on the initial design of the book.

Finally, the editors would like especially to thank all the authors for their hard work and their invaluable contributions.

Contents

Acronyms

AM	Acoustic model; acoustic modeling; amplitude modulation
ASGD	Asynchronous stochastic gradient descent
ASR	Automatic speech recognition
BLSTM	Bidirectional long short-term memory
BMMI	Boosted maximum mutual information
BPTT	Back-propagation through time
BRIR	Binaural room impulse response
BSS	Blind source separation
BSV	Bottleneck speaker vector
CAT	Cluster adaptive training
CLDNN	Convolutional long short-term memory deep neural network
CLP	Complex linear projection
CMLLR	Constrained maximum likelihood linear regression
CMN	Cepstral mean normalization
CMVN	Cepstral mean/variance normalization
CNN	Convolutional neural network
CNTK	Computational Network Toolkit
CTC	Connectionist temporal classification
DLSTM	Deep long short-term memory
DNN	Deep neural network
DOC	Damped oscillator coefficient
DS beamformer	Delay-and-sum beamformer
EM algorithm	Expectation maximization algorithm
FBANK	Log mel filterbank
FDLP	Frequency-domain linear prediction
fDLR	Feature-based discriminative linear regression
FHL	Factorized hidden layer
fMLLR	Feature space maximum likelihood linear regression
GCC-PHAT	Generalized cross-correlation phase transform
GFB	Gammatone filterbank
GLSTM	Grid long short-term memory

GMM	Gaussian mixture model
GPU	Graphics processing unit
HLSTM	Highway long short-term memory
HMM	Hidden Markov model
IAF	Ideal amplitude filter
ICA	Independent component analysis
ILD	Interaural level difference
IPD	Interaural phase difference
IRM	Ideal ratio mask
ITD	Interaural time difference
JTL	Joint-task learning
KL	Kullback–Leibler
KLD	Kullback–Leibler divergence
LCMV beamformer	Linearly constrained minimum variance beamformer
LDA	Linear discriminant analysis
LHN	Linear hidden network
LHUC	Learning hidden unit contribution
LIMABEAM	Likelihood-maximizing beamforming
LIN	Linear input network
LM	Language model
LON	Linear output network
LSTM	Long short-term memory
LSTMP	Long short-term memory with projection layer
LVCSR	Large-vocabulary continuous speech recognition
MAP	Maximum a posteriori
MBR	Minimum Bayes risk
MCGMM	Multichannel Gaussian mixture model
MCWF	Multichannel Wiener filter
MC-WSJ-AV	Multichannel Wall Street Journal audio visual corpus
MESSL	Model-based expectation maximization source separation and localization
MFCC	Mel-frequency cepstral coefficient
MLLR	Maximum likelihood linear regression
MLP	Multilayer perceptron
MMeDuSA	Modulation of medium-duration speech amplitudes
MMI	Maximum mutual information
MMSE	Minimum mean squared error
MPE	Minimum phone error
MSE	Mean squared error
MTL	Multitask learning
MVDR	Minimum variance distortionless response
NAB	Neural network adaptive beamforming
NaT	Noise-aware training
NMC	Normalized modulation coefficient
NMF	Nonnegative matrix factorization

NXT	NITE XML Toolkit
OMLSA	Optimally modified log spectral amplitude
PAC-RNN	Prediction–adaptation–correction recurrent neural network
PDF	Probability distribution function
PESQ	Perceptual evaluation of speech quality
PLP	Perceptual linear prediction
PNCC	Power-normalized cepstral coefficients
RASTA	RelAtive SpecTrA
RIR	Room impulse response
RLSTM	Residual long short-term memory
RNN	Recurrent neural network
RNNLM	Recurrent neural network language model
SAT	Speaker-adaptive training
SaT	Speaker-aware training
SDR	Source-to-distortion ratio
SER	Sentence error rate
sMBR	State-level minimum Bayes risk
SNR	Signal-to-noise ratio
SS	Spectral subtraction
STFT	Short-time Fourier transformation
STOI	Short-time objective intelligibility
SVD	Singular value decomposition
SWBD	Switchboard
TDNN	Time delay neural network
TDOA	Time difference of arrival
TRAPS	TempoRAl PatternS
TTS	Text-to-speech
UBM	Universal background model
VAD	Voice activity detection
VTLN	Vocal tract length normalization
VTS	Vector Taylor series
WDAS	Weighted delay-and-sum
WER	Word error rate
WFST	Weighted finite state transducer
WPE	Weighted prediction error
WSJ	Wall Street Journal

Part I
Introduction

Chapter 1
Preliminaries

Shinji Watanabe, Marc Delcroix, Florian Metze, and John R. Hershey

Abstract Robust automatic speech recognition (ASR) technologies have greatly evolved due to the emergence of deep learning. This chapter introduces the general background of robustness issues of deep neural-network-based ASR. It provides an overview of robust ASR research including a brief history of several studies before the deep learning era, basic formulations of ASR, signal processing, and neural networks. This chapter also introduces common notations for variables and equations, which are extended in the later chapters to deal with more advanced topics. Finally, the chapter provides an overview of the book structure by summarizing the contributions of the individual chapters and associates them with the different components of a robust ASR system.

1.1 Introduction

1.1.1 Motivation

Automatic speech recognition (ASR) is an essential human interface technology to convey human intentions to machines through human voices. The technology is well defined by solving a problem of converting voice signals captured by microphones to the corresponding texts. ASR has recently been developed and deployed in various applications with great success, including voice search, intelligent personal assistance, and car navigation, with the help of emergent deep learning technologies. Nevertheless, ASR applications are still limited due to the so-called *lack of robustness* against noise, room environments, languages, speakers, speaking styles,

S. Watanabe (✉) • J.R. Hershey
Mitsubishi Electric Research Laboratories (MERL), Cambridge, MA, USA
e-mail: shinjiw@ieee.org

M. Delcroix
NTT Communication Science Laboratories, NTT Corporation, 2-4, Hikaridai, Seika-cho, Kyoto, Japan

F. Metze
Carnegie Mellon University, 5000 Forbes Ave, Pittsburgh, PA, USA

and so on. Although by using large corpora covering many acoustic conditions (noise, speakers, etc.) and powerful deep learning techniques, the robustness of ASR systems can be improved, there still remains room for further improvement by using dedicated techniques.

For example, distant ASR is a scenario where the speakers and microphones are far from each other, and this scenario introduces difficult robustness issues caused by noise, impulse response, and microphone configuration variations. Actually, several distant-ASR benchmarks, including REVERB, CHiME, and AMI [1, 5, 15], show the drastic degradation of ASR performance in this scenario. In the AMI benchmark, the evaluation set captured by a close-talk microphone scores 21.5% word error rate (WER), while that captured by a distant microphone scores 32.7% WER.[1] WER is a common metric for measuring the ASR performance by using the edit distance (Levenshtein distance), and when the error rate goes over 30%, it's very hard to use ASR for speech interface applications. Recently, many researchers have tackled robustness issues through individual and company research activities, common benchmark challenges, and community-driven research projects, and they show significant improvement on these scenarios. The 2015 Jelinek Summer Workshop on Speech and Language Technology (JSALT)[2] is one of the above activities, and over 20 researchers in the field gathered to solve the various aspects of robustness issues in ASR, including the distant-ASR scenario. The idea of this book started from our discussions during the JSALT workshop and was extended to include contributions from some of the main actors in the field.

This book introduces the recent progress in ASR by focusing on the issues related to robustness, and provides state-of-the-art techniques described by leading researchers on this topic. It covers all aspects of recent ASR studies, including data and software resources and product-level applications, in addition to technology developments.

1.1.2   Before the Deep Learning Era

Robustness issues in ASR have been studied for long time. The primary focus of the robustness issues in ASR is speaker variations. The use of statistical methods and a large amount of training corpora have enabled us to realize speaker-independent ASR systems with sufficient accuracies [18]. In addition, speaker adaptation and normalization techniques have further mitigated the robustness issues due to speaker variations [7, 8, 12, 19, 20]. By following this trend, many researchers have extended their research directions to the other robustness issues mainly due to noise, speaking style, and environments [21, 23]. Remarkably, many methodologies developed in

---

[1]The WERs refer to the Kaldi AMI recipe, November 15, 2016. https://github.com/kaldi-asr/kaldi/blob/master/egs/ami/s5b.

[2]http://www.clsp.jhu.edu/workshops/15-workshop/.

those directions were tightly integrated with traditional Gaussian-based acoustic models in ASR, and these have been changed after deep learning techniques were introduced [13]. Although this book mainly focuses on novel robustness techniques developed during this deep learning era, this section briefly reviews traditional robustness techniques before the emergence of deep learning.

The conventional robust ASR techniques depend highly on an acoustic model based on a Gaussian mixture model (GMM) and mel-frequency cepstral coefficients (MFCCs). The techniques can be categorized into *feature space* and *model space* approaches.

1.1.2.1 Feature Space Approaches

The most basic feature space approach is feature normalization based on cepstral mean/variance normalization (CMVN). Cepstral mean normalization corresponds to suppressing short-term convolutional distortions in the time domain by extracting the bias components in the log spectral domain. This has the effect of reducing some speaker variability and channel distortions. In addition, feature space maximum likelihood linear regression (fMLLR) [10] is another feature space approach that transforms MFCC features, where the transformation matrices are estimated by using the maximum likelihood criterion with GMM-based acoustic models. These feature transformation and normalization techniques were developed for GMM-based acoustic models, but can easily be incorporated into deep learning techniques, and still exist as feature extraction or preprocessing modules of many deep-neural-network (DNN)-based ASR systems.

Other feature space approaches target suppressing the noise components in noisy speech signals. They are referred to as noise reduction or speech enhancement techniques. Spectral subtraction and the Wiener filter [2, 4, 9] are some of the most famous signal-processing techniques applied to noise-robust ASR. These approaches estimate noise components at run time and subtract these components from noisy speech signals in the spectral domain. The enhanced speech signals are then converted to MFCC features for the back-end ASR processing. Other successful approaches consist of feature compensation techniques that use the noise signal statistics obtained in the MFCC domain [11, 22], and suppress the noise components in that domain. Since MFCC features have nonlinearity due to the logarithmic operation, the additiveness properties of the speech and noise signals in the time and short-time Fourier transformation (STFT) domains are not preserved anymore. Therefore, to remove the noise signal components in the MFCC feature domain, we need an approximation based on Taylor series. Vector-Taylor-series (VTS)-based noise compensation techniques were developed for such a purpose [22]. Although these noise reduction techniques showed significant improvements on GMM-based ASR systems, their effect is limited for DNN-based ASR systems. One possible reason for this limited performance gain is the powerful representation learning ability of DNNs, which may already include the above suppression and compensation functions in its nonlinear feature transformation. Therefore, limited gains can be obtained by the direct application of noise suppression and compensation techniques to DNN-based ASR systems.

Note that the feature space techniques described above use only single-channel signals. Some of the recent advanced technologies introduced in the later chapters fully use multichannel signals to develop robust ASR systems.

1.1.2.2 Model Space Approaches

The major model space approaches are based on model adaptation techniques developed for speaker adaptation. Maximum a posteriori (MAP) adaptation estimates GMM parameters by including a regularization through the prior distribution of generic GMM parameters [12, 19]. The MAP estimation of an acoustic model is performed efficiently based on the exponential-family property of Gaussian distributions. Maximum likelihood linear regression (MLLR) estimates the affine transformation matrices shared among several Gaussians based on the maximum likelihood estimation [7, 20]. Similarly to the MAP estimation, MLLR is also performed efficiently due to the closed-form solution based on a Gaussian distribution. In addition, uncertainty-decoding techniques have been developed for noise-robust ASR [6, 16]. Uncertainty decoding represents the feature uncertainties coming from noise suppression techniques with a Gaussian distribution, and integrates out the feature distribution with GMM-based acoustic models to include feature uncertainties in the acoustic models. The above model space approaches rely heavily on having a GMM-based ASR back end, and it is difficult to apply these approaches directly to DNNs.[3]

As a summary, several legacy feature space robust techniques are still applied to DNN acoustic models, while model space techniques have to be replaced with DNN-specific techniques. In addition, many novel speech enhancement front ends have been developed and evaluated recently with DNN-based ASR systems.

This book introduces various robust ASR techniques, which were newly developed or revisited by considering their integration with deep-learning-based ASR.

1.2   Basic Formulation and Notations

This section first provides general mathematical notations and specific notations for typical problems used in this book, where we follow the notation conventions used in textbooks in the field [3, 14, 24, 25]. We also provide basic formulations of speech recognition, neural networks, and signal processing, which are omitted in the following chapters dealing with advanced topics. However, since the book covers a

---

[3]However, these concepts have inspired related techniques for DNN-based acoustic models, such as DNN parameter regularization based on the L2 norm and Kullback–Leibler (KL) divergence, that can be regarded as a variant of MAP adaptation in the context of DNNs.

wide range of topics in speech and language processing, these notations sometimes conflict across different problems (e.g., *a* is used as a state transition in the hidden marked model (HMM), and a preactivation in the DNN). Also, some representations in the following chapters do not strictly follow the notations defined here, but follow the notation conventions for their specific problems.

1.2.1   General Notations (Tables 1.1 and 1.2)

Table 1.1 lists the notations used to describe a set of variables. We use a blackboard bold font or Fraktur font to represent a set of variables, which is used to define a domain of a variable.

Table 1.2 lists the notations for scalar, vector, and matrix variables. Scalar variables are represented by an italic font, and uppercase letters often represent constant values. This book uses lowercase and uppercase bold fonts for vector and matrix variables, respectively, and does not distinguish the use of upright and italic. Sequences and tensors are represented by uppercase letters. Note that this conflicts with the notation for a constant-value scalar. However, this is the case in many

Table 1.1 Sets of variables

$\mathbb{R}$ or $\mathfrak{R}$	Real number
$\mathbb{R}_{>0}$ or $\mathfrak{R}_{>0}$	Positive real number
$\mathbb{R}^D$ or $\mathfrak{R}^D$	D-dimensional real number
$\mathbb{C}$	Complex number
$\mathbb{C}^D$	D-dimensional complex number

Table 1.2 Variables

a, ϕ	Scalar
A	Scalar (for constant value)
$\mathbf{a}, a, \boldsymbol{\phi}$	Vector
$\mathbf{A}, A, \boldsymbol{\Phi}$	Matrix
A	Sequence, tensor
$\mathscr{A}, \Phi$	Set

Table 1.3 Matrix and vector operations

$[a]_d$	dth element of the vector (i.e., $[a]_d = a_d$)
I_D	$D \times D$ identity matrix
A^T	Transpose
$A^\dagger$ or A^H	Conjugate (Hermitian) transpose
$A \circ B$ or $A \otimes B$	Elementwise multiplication
$\mathrm{diag}(a)$	Diagonal matrix that uses vector a as diagonal elements

books and scientific papers, and this book also simply follows this convention. To avoid confusion, sequences and tensors are often explicitly defined. For example, an N-length vector sequence with a dimension D is defined by an element with the domain definition, i.e., $X \triangleq \{x_n \in \mathbb{R}^D | n = 1, \ldots, N\}$, when we first introduce it. This sequence can also be represented by a matrix, i.e., $X \in \mathbb{R}^{D \times N}$. Since there is no calligraphic style for Greek letters, the book uses uppercase letters of the Greek alphabet for a set, instead. When we first introduce vector and matrix variables, they can be defined with a domain definition as in Table 1.1, i.e., $a \in \mathbb{R}^D$ for a D-dimensional vector and $A \in \mathbb{R}^{N \times M}$ for an $N \times M$-dimensional matrix. However, when the domain definition is trivial or already defined, a sequence can simply be defined as $X \triangleq \{x_1, \ldots, x_N\}$ and $X \triangleq \{x_n\}_{n=1}^N$.

1.2.2 Matrix and Vector Operations (Table 1.3)

Table 1.3 lists the matrix and vector operations which are mainly used in signal processing and neural networks.

1.2.3 Probability Distribution Functions (Table 1.4)

Table 1.4 lists the probability distribution functions (PDFs) which are typically used in signal processing and speech recognition. A generic PDF is represented by using either $p(\cdot)$ or $P(\cdot)$. For a Gaussian distribution, we use a calligraphic style $\mathcal{N}(\cdot)$, following convention. Here, we provide the actual equation forms of the real-value

Table 1.4 Probability distribution functions

$p(\cdot), P(\cdot)$	Generic PDF
$\mathcal{N}(\cdot|\boldsymbol{\mu}, \boldsymbol{\Sigma})$	Real-value Gaussian (or normal distribution)
$\mathcal{N}_{\mathbb{C}}(\cdot|\boldsymbol{\mu}, \boldsymbol{\Sigma})$	Complex-value Gaussian

and complex-value Gaussian distributions as follows:

$$\mathcal{N}(\boldsymbol{x}|\boldsymbol{\mu}, \boldsymbol{\Sigma}) \triangleq (2\pi)^{-D/2}|\boldsymbol{\Sigma}|^{-1/2} \exp\left(-\frac{1}{2}(\boldsymbol{x} - \boldsymbol{\mu})^\top \boldsymbol{\Sigma}^{-1}(\boldsymbol{x} - \boldsymbol{\mu})\right), \tag{1.1}$$

$$\mathcal{N}_{\mathbb{C}}(\boldsymbol{x}|\boldsymbol{\mu}, \boldsymbol{\Sigma}) \triangleq (\pi)^{-D}|\boldsymbol{\Sigma}|^{-1} \exp\left(-(\boldsymbol{x} - \boldsymbol{\mu})^\dagger \boldsymbol{\Sigma}^{-1}(\boldsymbol{x} - \boldsymbol{\mu})\right). \tag{1.2}$$

Here, $\boldsymbol{\mu}$ and $\boldsymbol{\Sigma}$ are Gaussian mean vector and covariance matrix parameters, respectively. We use Greek letters for the parameters of the distribution functions.

1.2.3.1 Expectation

With the PDF $p(\cdot)$, we can define the expectation of a function $f(x)$ with respect to x as follows:

$$\mathbb{E}_{p(x)}[f(x)] \triangleq \begin{cases} \int_x f(x)p(x)\,dx & \text{for } x \in \mathbb{R}, \\ \sum_x f(x)p(x) & \text{for } x \in \mathbb{Z}. \end{cases}$$

$\mathbb{E}_{p(x)}[f(x)]$ can be used for either continuous or discrete variables for x.

1.2.3.2 Kullback–Leibler Divergence

The Kullback–Leibler divergence (KLD) [17] for continuous- and discrete-variable PDFs is defined as follows:

$$D_{\mathrm{KL}}(p(x)||p'(x)) \triangleq \begin{cases} \int p(x)\log\dfrac{p(x)}{p'(x)}\,dx & \text{for } x \in \mathbb{R}, \\ \sum_x p(x)\log\dfrac{p(x)}{p'(x)} & \text{for } x \in \mathbb{Z}. \end{cases}$$

The KLD is used as a cost function for measuring PDFs $p(x)$ and $p'(x)$ close to each other.

1.2.4 Signal Processing

Table 1.5 summarizes the variables used for signal processing. With these notations, the microphone signals can be expressed in the time domain as

$$y_j[n] = \sum_{i=1}^{I} \sum_{l=0}^{L-1} h_{ij}[n]x_i[n-l] + u_j[n] \tag{1.3}$$

$$= \sum_{i=1}^{I} h_i[n] * x_i + u_j[n], \tag{1.4}$$

where I is the total number of sources, L is the length of the room impulse response, and $*$ denotes the convolution operation. The observed speech signal can be expressed as $y[n]$, when considering a single-microphone scenario. Similarly,

Table 1.5 Signal processing

$x[n] \in \mathbb{R}$	Time domain signal at sample n
$X(t,f) \in \mathbb{C}$	Frequency domain coefficient at frame t and frequency bin f
$\hat{x}[n] \in \mathbb{R}$	Estimate of signal $x[n]$
$x_i[n] \in \mathbb{R}$	The ith source signal
$X_i(t,f) \in \mathbb{C}$	Frequency domain coefficient at frame t and frequency bin f of $x_i[n]$
$y_j[n] \in \mathbb{R}$	The observed speech signal in the time domain at sample n for microphone j
$Y_j(t,f) \in \mathbb{C}$	Frequency domain coefficient at frame t and frequency bin f of $y_j[n]$
$u_j[n] \in \mathbb{R}$	The noise signal in the time domain at sample n for microphone j
$U_i(t,f) \in \mathbb{C}$	Frequency domain coefficient at frame t and frequency bin f of $u_i[n]$
$h_{ij}[n] \in \mathbb{R}$	The room impulse response in the time domain at sample n from source i to microphone j
$H_{ij}(m,f) \in \mathbb{C}$	Frequency domain coefficient at frame t and frequency bin f of the room impulse response h_{ij}
$*$	The convolution operation

when considering only a single source, $x_i[n]$ can be simplified to $x[n]$, which refers then to clean speech. In this case the notation for the room impulse response becomes h_j, or $h[n]$ if there is only a single microphone.

In the frequency domain, (1.4) can be approximated as

$$Y_j(t,f) \approx \sum_{i=1}^{I} \sum_{m=1}^{M} H_{ij}(m,f) X_i(t-m,f) + U_j(t,f). \tag{1.5}$$

The following chapters often deal with speech enhancement, which estimates the target source signals $x_i[n]$ or $X_i(t,f)$ from microphone signals $y_j[n]$ or $Y_j(t,f)$. The enhanced speech is then notated as $\hat{x}[n]$ or $\hat{X}(t,f)$. Equations (1.4) and (1.5) are fundamental equations to describe the mathematical relationship between microphone and target signals.

1.2.5 Automatic Speech Recognition

This section introduces the fundamental formulation of automatic speech recognition and related technologies. The related notations are provided in Table 1.6. Based on Bayes decision theory, ASR is formulated as follows:

$$\hat{W} = \arg \max_{W \in \mathscr{W}} p(W|O), \tag{1.6}$$

where W and O are word and speech feature sequences, respectively. Thus, one major problem of ASR is to obtain the posterior distribution $p(W|O)$. In general, it is very difficult to directly consider the posterior distribution having an input

Table 1.6 Automatic speech recognition

$o_t \in \mathbb{R}^D$	D-dimensional speech feature vector at frame t
$w_n \in \mathscr{V}$	Word at nth position in an utterance with vocabulary $\mathscr{V}$
$O \triangleq \{o_t \| t = 1, \ldots, T\}$	T-length sequence of speech feature vectors
$W \triangleq \{w_n \| n = 1, \ldots, N\}$	N-length word sequence
$\hat{W}$	Estimated word sequence
$\mathscr{W}$	Set of all possible word sequences

sequence composed of continuous vectors and an output sequence composed of discrete symbols at the same time.[4] Instead of dealing with $p(W|O)$ directly, it is rewritten with the Bayes theorem to separately consider the likelihood function $p(O|W)$ and the prior distribution $p(W)$, as follows:

$$\hat{W} = \arg \max_{W \in \mathscr{W}} p(O|W)p(W). \tag{1.7}$$

$p(O|W)$ and $p(W)$ are called the acoustic model and language model, respectively. The following sections mainly deal with the acoustic model $p(O|W)$.

1.2.6 Hidden Markov Model

Although the likelihood function $p(O|W)$ is still difficult to handle, with the probabilistic chain rule and conditional-independence assumptions, $p(O|W)$ is factorized as follows:

$$p(O|W) = \sum_{S \in \mathscr{S}} p(O|S)p(S|W) \tag{1.8}$$

$$= \sum_{S \in \mathscr{S}} \prod_{t=1}^{T} p(\boldsymbol{o}_t|s_t)p(s_t|s_{t-1}, W), \tag{1.9}$$

where we introduce the HMM state sequence $S = \{s_t | t = 1, \ldots, T\}$. The likelihood function is factorized with S, and represented by the summation over all possible state sequences. The notation used in the HMM is listed in Table 1.7. $p(\boldsymbol{o}_t|s_t)$ is an acoustic likelihood function at frame t, and $p(s_t|s_{t-1}, W)$ is an HMM state transition probability given a word sequence W. The HMM state transition probability is usually defined for each phoneme or context-dependent phoneme, and the conversion from word to phoneme sequences is performed by a hand-crafted pronunciation dictionary. The following explanation omits the dependence on W for simplicity, i.e., $p(s_t|s_{t-1}, W) \rightarrow p(s_t|s_{t-1})$.

Note that $p(\boldsymbol{o}_t|s_t = j)$ is a frame-level likelihood function at frame t in state j. It is obtained from either the GMM likelihood in an HMM-GMM system or the pseudo-likelihood in an HMM-DNN hybrid system. The next section describes the frame-level likelihood function in an HMM-GMM system.

---

[4]This problem is discussed in Chap. 13.

Table 1.7 Hidden Markov model

$s_t \in \{1, \dots, J\}$	HMM state variable at frame t (the number of distinct HMM states is J)
$S = \{s_t | t = 1, \dots, T\}$	T-length sequence of HMM states
$\mathscr{S}$	Set of all possible state sequences
$a_j \in \mathbb{R}_{\geq 0}$	Initial weight for state $s_1 = j$
$a_{ij} \in \mathbb{R}_{\geq 0}$	Transition weight from $s_{t-1} = i$ to $s_{t-1} = j$
$p(o_t | j)$	Likelihood given state $s_t = j$

Table 1.8 Gaussian mixture model

k	Mixture component index
K	Number of components
$w_k \in \mathbb{R}_{\geq 0}$	Weight parameter at k
$\mu_k \in \mathbb{R}^D$	Mean vector parameter at k
$\Sigma_k \in \mathbb{R}^{D \times D}$	Covariance matrix parameter at k

1.2.7 Gaussian Mixture Model

For a D-dimensional feature vector $o_t \in \mathbb{R}^D$ at frame t, the likelihood of the GMM is represented as follows:

$$p(o_t | j) = \sum_{k=1}^{K} w_{jk} \mathcal{N}(o_t | \mu_{jk}, \Sigma_{jk}). \qquad (1.10)$$

The likelihood is represented by a weighted summation of K Gaussian distributions. The variables used in the GMM are summarized in Table 1.8, where the HMM state index j is omitted for simplicity. The GMM was used as a standard acoustic likelihood function, since its parameters, together with the HMM parameters, are efficiently estimated by using the expectation and maximization algorithm. However, it often fails to model high-dimensional features due to the curse of dimensionality. Also, the discriminative ability of the GMM is not enough even

Table 1.9 Neural network

$a_t^l \in \mathbb{R}^{D^l}$	D^l-dimensional preactivation vector at frame t and layer l
$h_t^l \in [0, 1]^{D^l}$	D^l-dimensional activation vector at frame t and layer l
$W^l \in \mathbb{R}^{D^l \times D^{l-1}}$	lth-layer transformation matrix
$b^l \in \mathbb{R}^{D^l}$	lth-layer bias vector
sigmoid(x)	Elementwise sigmoid function $1/(1 + e^{-x_d})$ for $d = 1, \ldots, D$
softmax(x)	Elementwise softmax function $e^{x_d}/(\sum_d e^{x_d})$ for $d = 1, \ldots, D$

with discriminative training, and the GMM has therefore been replaced with a neural network.

1.2.8 Neural Network

The alternative representation of $p(o_t|j)$ is obtained from a neural network. Using the Bayes theorem, $p(o_t|j)$ is represented by the frame-level posterior PDF $p(j|o_t)$ as follows:

$$p(o_t|j) = \frac{p(j|o_t)p(o_t)}{p(j)}, \tag{1.11}$$

where $p(o_t)$ and $p(j)$ are prior distributions of feature vector o and HMM state j, respectively. The $p(o_t|j)$ obtained via (1.11) is called the pseudo-likelihood.

A standard feedforward network provides the frame-level posterior PDF as follows:

$$p(j|o_t) = \left[\text{softmax}(a_t^L)\right]_j, \tag{1.12}$$

where a_t^L is called a preactivation vector in layer L at frame t and softmax() is a softmax function. All the notations used in a neural network are listed in Table 1.9. The preactivation a_t^L is recursively computed by affine transformations and nonlinear operations in $L - 1$ layers as follows:

$$\begin{aligned} a_t^l &= W^l h_t^{l-1} + b^l \\ h_t^l &= \text{sigmoid}(a_t^l) \end{aligned} \qquad \text{for } l = 1, \ldots, L. \tag{1.13}$$

Here we provide a sigmoid network with a sigmoid activation function, but it can be replaced with the other nonlinear activation functions. h_t^l is an activation vector at frame t in layer l. h_t^0 is defined as the original observation vector o_t. Particularly when we consider a large number (more than one) of hidden layers, the network is called a deep neural network. An acoustic model using an HMM with this pseudo-likelihood obtained by DNN is called a hybrid DNN-HMM system. The hybrid DNN-HMM system significantly outperforms the conventional GMM-HMM in various tasks due to its strong discriminative abilities [13]. In addition to the above feedforward network, there are powerful neural network architectures including recurrent and convolutional neural networks, which are explained in Chaps. 5, 7, 11, and 16. Also, the DNN described in this section does not have sequence-level discriminative ability, and the sequence-level discriminative training of DNNs is discussed in Chap. 12.

1.3 Book Organization

This book is organized in four parts, described below.

In *Part I: Introduction*, we introduce some preliminaries and review briefly the history of ASR, introduce the basics of ASR, and summarize the robustness issues of current ASR systems.

Part II: Approaches to Robust Automatic Speech recognition consists of 11 chapters each reviewing some key technologies for robust ASR. Figure 1.1 is a schematic diagram of a typical robust ASR system. In the diagram, we have included references to the chapters of Part II of the book to illustrate what part of the system each chapter deals with.

- Chapters 2 and 3 introduce various techniques for multichannel speech enhancement. These chapters focus on generative-model-based multichannel approaches

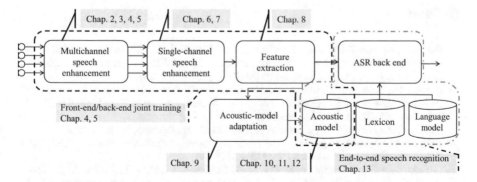

Fig. 1.1 Schematic diagram of a robust ASR system with the reference to chapters discussing the key technologies

and review some classical techniques such as linear-prediction-based dereverberation and beamforming.

- Chapters 4 and 5 deal with neural-network-based beamforming for noise reduction. These chapters also discuss the joint training of the front-end multichannel speech enhancement with the acoustic model using an acoustic-model training criterion.
- Chapters 6 and 7 discuss single-channel speech enhancement methods exploiting deep learning.
- Chapter 8 introduces recent work on the design of noise-robust features for ASR.
- Chapter 9 reviews the key approaches to adaptation of the acoustic model to speakers or environments.
- Chapters 10–12 deal with several aspects of acoustic modeling. Chapter 10 introduce approaches to generating multicondition training data and to training data augmentation. Chapter 11 reviews advanced recurrent network architectures for acoustic modeling. Chapter 11 introduces sequence-training approaches for acoustic models.
- Chapter 13 reviews recent efforts in creating end-to-end ASR systems, including connectionist temporal classification and encoder–decoder approaches. It also introduces the EESEN framework for developing end-to-end ASR systems.

Part III: Resources reviews some important tasks for robust ASR such as the CHiME challenge tasks discussed in Chap. 14, the REVERB challenge task discussed in Chap. 15, and the AMI meeting corpus discussed in Chap. 16. We also review some important toolkits for ASR, speech enhancement, deep learning, and end-to-end ASR in Chap. 17.

Part IV: Applications concludes the book by presenting recent activities in research and development for creating novel speech applications at some of the key industrial players. This includes contributions from Google in Chap. 18, Microsoft in Chap. 19, and Mitsubishi Electric in Chap. 20.

References

1. Barker, J., Marxer, R., Vincent, E., Watanabe, S.: The third "CHiME" speech separation and recognition challenge: dataset, task and baselines. In: 2015 IEEE Workshop on Automatic Speech Recognition and Understanding (ASRU), pp. 504–511 (2015)
2. Berouti, M., Schwartz, R., Makhoul, J.: Enhancement of speech corrupted by acoustic noise. In: IEEE International Conference on Acoustics, Speech, and Signal Processing. ICASSP'79, vol. 4, pp. 208–211. IEEE, New York (1979)
3. Bishop, C.M.: Pattern Recognition and Machine Learning. Springer, Berlin (2006)
4. Boll, S.: Suppression of acoustic noise in speech using spectral subtraction. IEEE Trans. Acoust. Speech Signal Process. 27(2), 113–120 (1979)
5. Carletta, J., Ashby, S., Bourban, S., Flynn, M., Guillemot, M., Hain, T., Kadlec, J., Karaiskos, V., Kraaij, W., Kronenthal, M., et al.: The AMI meeting corpus: a pre-announcement. In: International Workshop on Machine Learning for Multimodal Interaction, pp. 28–39. Springer, Berlin (2005)

6. Deng, L., Droppo, J., Acero, A.: Dynamic compensation of HMM variances using the feature enhancement uncertainty computed from a parametric model of speech distortion. IEEE Trans. Speech Audio Process. **13**(3), 412–421 (2005)
7. Digalakis, V.V., Rtischev, D., Neumeyer, L.G.: Speaker adaptation using constrained estimation of Gaussian mixtures. IEEE Trans. Speech Audio Process. **3**(5), 357–366 (1995)
8. Eide, E., Gish, H.: A parametric approach to vocal tract length normalization. In: IEEE International Conference on Acoustics, Speech, and Signal Processing, ICASSP 96, vol. 1, pp. 346–348. IEEE, New York (1996)
9. ETSI: Speech processing, transmission and quality aspects (STQ); distributed speech recognition; advanced front-end feature extraction algorithm; compression algorithms. ETSI ES 202, 050 (2002)
10. Gales, M.J.: Maximum likelihood linear transformations for HMM-based speech recognition. Comput. Speech Lang. **12**(2), 75–98 (1998)
11. Gales, M.J., Young, S.J.: Robust continuous speech recognition using parallel model combination. IEEE Trans. Speech Audio Process. **4**(5), 352–359 (1996)
12. Gauvain, J.L., Lee, C.H.: Maximum a posteriori estimation for multivariate Gaussian mixture observations of Markov chains. IEEE Trans. Speech Audio Process. **2**(2), 291–298 (1994)
13. Hinton, G., Deng, L., Yu, D., Dahl, G.E., Mohamed, A.R., Jaitly, N., Senior, A., Vanhoucke, V., Nguyen, P., Sainath, T.N., et al.: Deep neural networks for acoustic modeling in speech recognition: the shared views of four research groups. IEEE Signal Process. Mag. **29**(6), 82–97 (2012)
14. Huang, X., Acero, A., Hon, H.W.: Spoken Language Processing: A Guide to Theory, Algorithm, and System Development. Prentice Hall, Englewood Cliffs, NJ (2001)
15. Kinoshita, K., Delcroix, M., Yoshioka, T., Nakatani, T., Sehr, A., Kellermann, W., Maas, R.: The REVERB challenge: a common evaluation framework for dereverberation and recognition of reverberant speech. In: 2013 IEEE Workshop on Applications of Signal Processing to Audio and Acoustics, pp. 1–4. IEEE, New York (2013)
16. Kolossa, D., Haeb-Umbach, R.: Robust Speech Recognition of Uncertain or Missing Data: Theory and Applications. Springer Science & Business Media, Berlin (2011)
17. Kullback, S., Leibler, R.A.: On information and sufficiency. Ann. Math. Stat. **22**(1), 79–86 (1951)
18. Lee, K.F., Hon, H.W.: Large-vocabulary speaker-independent continuous speech recognition using HMM. In: IEEE International Conference on Acoustics, Speech, and Signal Processing. ICASSP 88, pp. 123–126. IEEE, New York (1988)
19. Lee, C.H., Lin, C.H., Juang, B.H.: A study on speaker adaptation of the parameters of continuous density hidden Markov models. IEEE Trans. Signal Process. **39**(4), 806–814 (1991)
20. Leggetter, C.J., Woodland, P.C.: Maximum likelihood linear regression for speaker adaptation of continuous density hidden Markov models. Comput. Speech Lang. **9**(2), 171–185 (1995)
21. Li, J., Deng, L., Gong, Y., Haeb-Umbach, R.: An overview of noise-robust automatic speech recognition. IEEE/ACM Trans. Audio Speech Lang. Process. **22**(4), 745–777 (2014)
22. Moreno, P.J., Raj, B., Stern, R.M.: A vector Taylor series approach for environment-independent speech recognition. In: IEEE International Conference on Acoustics, Speech, and Signal Processing. ICASSP 96, vol. 2, pp. 733–736. IEEE, New York (1996)
23. Virtanen, T., Singh, R., Raj, B.: Techniques for Noise Robustness in Automatic Speech Recognition. Wiley, New York (2012)
24. Watanabe, S., Chien, J.T.: Bayesian Speech and Language Processing. Cambridge University Press, Cambridge (2015)
25. Yu, D., Deng, L.: Automatic Speech Recognition. Springer, Berlin (2012)

Part II
Approaches to Robust Automatic Speech Recognition

Chapter 2
Multichannel Speech Enhancement Approaches to DNN-Based Far-Field Speech Recognition

Marc Delcroix, Takuya Yoshioka, Nobutaka Ito, Atsunori Ogawa, Keisuke Kinoshita, Masakiyo Fujimoto, Takuya Higuchi, Shoko Araki, and Tomohiro Nakatani

Abstract In this chapter we review some promising speech enhancement front-end techniques for handling noise and reverberation. We focus on signal-processing-based multichannel approaches and describe beamforming-based noise reduction and linear-prediction-based dereverberation. We demonstrate the potential of these approaches by introducing two systems that achieved top performance on the recent REVERB and CHiME-3 benchmarks.

2.1 Introduction

Recently, far-field automatic speech recognition (ASR) using devices mounted with a microphone array has received increased interest from both industry (see Chaps. 18–20) and academia [5, 10, 17, 26]. Speech signals recorded with distant microphones are corrupted by noise and reverberation, which severely affect recognition performance. Therefore, it is essential to make ASR systems robust to such acoustic distortions if we are to achieve robust distant ASR.

Current state-of-the-art ASR systems achieve noise robustness by employing deep-neural-network (DNN) based acoustic models and exploiting a large amount of training data captured under various noise and reverberation conditions. Furthermore, reducing noise or reverberation prior to recognition using a multimicrophone speech enhancement front end has been shown to improve the performance of state-of-the-art ASR back ends [5, 17, 27, 33].

M. Delcroix (✉) • T. Yoshioka • N. Ito • A. Ogawa • K. Kinoshita • M. Fujimoto • T. Higuchi • S. Araki • T. Nakatani
NTT Communication Science Laboratories, NTT Corporation, 2-4, Hikaridai, Seika-cho, Kyoto, Japan
e-mail: marc.delcroix@lab.ntt.co.jp

© Springer International Publishing AG 2017   21
S. Watanabe et al. (eds.), *New Era for Robust Speech Recognition*,
DOI 10.1007/978-3-319-64680-0_2

2.1.1 Categories of Speech Enhancement

Extensive research has been undertaken on speech enhancement algorithms designed to reduce noise and reverberation from microphone signals, including single-channel and multichannel approaches. Most approaches originally targeted acoustic applications, but some are also effective when used as an ASR front end.

Speech enhancement techniques can be classified into linear- and nonlinear-processing-based approaches. Linear-processing approaches enhance speech using a linear filter that is constant across the entire signal or long signal segments. Examples of linear-processing-based speech enhancement approaches include beamforming [37] and linear-prediction-based dereverberation [31]. Nonlinear-processing approaches include nonlinear filtering such as spectral subtraction [7], nonnegative matrix factorization (NMF) [38], neural-network-based speech enhancement [42], and frame-by-frame linear filtering such as Wiener filtering. Note that most single-channel speech enhancement techniques rely on nonlinear processing.

Nonlinear-processing-based speech enhancement has been shown to reduce noise significantly. However, most approaches also tend to introduce distortions that have a great impact on ASR performance.[1] In contrast, linear-processing-based approaches tend to introduce fewer distortions into the processed speech. For example, multichannel linear-filtering-based speech enhancement approaches have been shown to be particularly effective for ASR. In this chapter, we review some of these approaches, including linear-prediction-based speech dereverberation and beamforming. We focus here on batch-processing approaches, although we provide references to extensions to online processing for the benefit of interested readers.

2.1.2 Problem Formulation

We deal with a scenario where speech is recorded with a distant microphone array composed of J microphones. A microphone signal consists of the summation of the target speech signal with different source signals such as interfering speakers and

---

[1] We should mention the notable exception of neural-network-based speech enhancement, which may be jointly optimized with the ASR back end and has been shown to improve ASR performance [15, 32, 41, 42]. Neural-network-based enhancement is also discussed in Chaps. 4, 5, and 7.

noise. The jth microphone signal $y_j[n]$ at time sample n can be written as

$$y_j[n] = \sum_{l=0}^{L_h-1} h_j[l]x[n-l] + u_j[n] \tag{2.1}$$

$$= h_j[n] * x[n] + u_j[n] \tag{2.2}$$

$$= o_j[n] + u_j[n], \tag{2.3}$$

where $h_j[n]$ is the room impulse response between the target speaker and microphone j, $x[n]$ is the target speech signal, $u_j[n]$ is the noise signal at microphone j, $o_j[n] = h_j[n] * x[n]$ is the target speech source image at microphone j, L_h is the length of the room impulse responses, and $*$ represents the convolution operation.

In a general configuration, there may be several active speakers that cause the speech signals to overlap. However, in the following, we focus on the recognition of a single target speaker. Accordingly, we consider all other potential sources as interferences and include them in the noise term $u_j[n]$. Issues related to multispeaker situations, such as meeting recognition, are discussed in Chap. 16.

The source image at a microphone is delayed compared with the source image at a reference microphone by a time delay value that is given by the difference between the propagation times from the source to the respective microphones. Moreover, in most living environments, sounds are reflected by walls and objects in rooms and, consequently, the source image will usually be reverberant. The room impulse response models the multipath propagation of the sound between the sources and the microphones, including the relative propagation delays. Accordingly, the source image includes both the relative delays and reverberation.

We can approximate (2.3) in the short-term Fourier transform (STFT) domain as [46]

$$Y_j(t,f) \approx \sum_{m=0}^{M-1} H_j(m,f)X(t-m,f) + U_j(t,f) \tag{2.4}$$

$$= O_j(t,f) + U_j(t,f), \tag{2.5}$$

where $Y_j(t,f)$, $H_j(m,f)$, $X(t,f)$, $U_j(t,f)$, and $O_j(t,f)$ are the STFT at frame t and frequency bin f of the microphone signal $y_j[n]$, the room impulse response $h_j[n]$, the target speech signal $x[n]$, the noise signal $u_j[n]$, and the target source image $o_j[n]$, respectively. M is the length of the room impulse response in the STFT domain.

We further introduce a vector representation of the signals as

$$\mathbf{y}_{t,f} = \sum_{m=0}^{M-1} \mathbf{h}_{m,f}X(t-m,f) + \mathbf{u}_{t,f} \tag{2.6}$$

$$= \mathbf{o}_{t,f} + \mathbf{u}_{t,f}, \tag{2.7}$$

where $\mathbf{y}_{t,f} = [Y_1(t,f), \ldots, Y_J(t,f)]^T$, $\mathbf{h}_{m,f} = [H_1(m,f), \ldots, H_J(m,f)]^T$, $\mathbf{u}_{t,f} = [U_1(t,f), \ldots, U_J(t,f)]^T$, $\mathbf{o}_{t,f} = [O_1(t,f), \ldots, O_J(t,f)]^T$, and $()^T$ is the transpose operation; $|\cdot|$, $\cdot^*$; and $()^H$ denote the modulus, the complex conjugate, and the conjugate transpose or Hermitian transpose, respectively. In the following, we process each frequency f independently. Note that $\mathbf{h}_{0,f} = [H_1(0,f), \ldots, H_J(0,f)]^T$ is also referred to as the steering vector in the context of beamforming, as it contains information about the direction of the source included in the relative delays.

Speech enhancement aims at recovering the target speech signal $x[n]$, while suppressing noise and reverberation. This processing can be done blindly, meaning that it relies only on the observed noisy signals, $\mathbf{y}_{t,f}$. In the remainder of this chapter we review some of the main approaches that can be used to reduce reverberation and noise. The order of the discussion follows the processing flow of the distant ASR systems we describe in Sect. 2.4. Accordingly, we start by reviewing speech dereverberation in Sect. 2.2 and focus on linear-prediction-based multichannel dereverberation with the weighted prediction error (WPE) algorithm. We then introduce beamforming in Sect. 2.3 and review some of the major types of beamformer that have been used as ASR front ends. In Sect. 2.4 we describe two distant ASR systems that employ dereverberation and beamforming in their front ends, and demonstrate the impact of these techniques on recognition performance. Finally, Sect. 2.5 concludes the chapter and discusses future research directions.

2.2 Dereverberation

This section reviews the problem of speech dereverberation and briefly describes some of the existing approaches. We then review in more detail a linear-prediction-based approach that uses the WPE algorithm.

2.2.1 Problem Description

To simplify the derivations, let us consider a case where the observed signal is corrupted only by reverberation, and noise can be neglected. In such a case, the microphone signal becomes

$$Y_j(t,f) \approx \sum_{m=0}^{M-1} H_j(m,f)X(t-m,f). \tag{2.8}$$

Fig. 2.1 Room impulse measured in a meeting room illustrating the three components of a room impulse response, i.e., the direct path, early reflections, and late reverberation

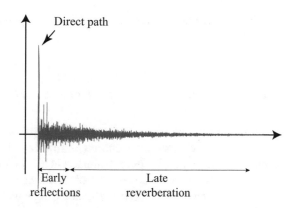

Neglecting noise is of course a strong hypothesis that is not usually valid. Dereverberation methods need to be robust to noise to be used in practice. The approaches we discuss in this section have been shown to be effective even in noisy conditions.

Figure 2.1 shows an example of a room impulse response recorded in a meeting room. The length of the room impulse response is related to the reverberation time of the room (RT60), which ranges from 200 to 1000 ms in typical offices and living environments. We can divide a room impulse response into three parts, the direct path, early reflections, and late reverberation [28]. The early reflections consist of the reflections that arrive at the microphone within about 50 ms of the direct path. The late reverberation consists of all the remaining reflections.

Early reflections may improve speech intelligibility for human listeners[8]. Moreover, early reflections do not pose severe problems for ASR as they can be partially mitigated with utterance-level feature mean normalization when dealing with ASR. Indeed, since early reflections can be represented as the convolution of a short impulse response, they can be reduced by mean normalization in the log spectrum domain[23]. Late reverberation is known to seriously affect the audible quality, speech intelligibility, and performance of ASR systems. Therefore, speech dereverberation usually focuses on suppressing late reverberation but may preserve the early reflections in the dereverberated speech. Accordingly, we denote the target signal of the dereverberation process as

$$D_j(t,f) = \sum_{m=0}^{\delta} H_j(m,f)X(t-m,f),\qquad(2.9)$$

where δ corresponds to the number of time frames associated with the duration of the early reflection.

It is important to note that reverberation consists of a long filtering operation, and thus has different characteristics from additive noise. Therefore, dereverberation requires specific techniques that are different from the approaches designed for noise reduction.

2.2.2 Overview of Existing Dereverberation Approaches

Several approaches have been developed for dealing with reverberation.[46] provides a review of some speech dereverberation techniques. Some of these approaches have recently been evaluated in the REVERB challenge task[27].

A widely used approach consists of modeling a room impulse response as white noise modulated by an exponentially decaying envelope in the time domain as [29]

$$h[n] = e[n]e^{-\Delta n}, \tag{2.10}$$

where $e[n]$ is a zero-mean white noise sequence and $\Delta = -3 \ln(10)/RT_{60}$. With this model, we can obtain an estimate of the power spectrum of the late reverberation, Φ_{Late}, as

$$\Phi_{\text{Late}} = e^{-2\Delta\delta_t}|Y(t-\delta,f)|^2, \tag{2.11}$$

where δ_t is a delay set at 50 ms. Note here that δ represents the delay in terms of taps in the STFT domain, whereas δ_t is the corresponding delay value in seconds. Given this late-reverberation model, dereverberation can be achieved by subtracting the power spectrum of the late reverberation from the power spectrum of the observed signal. This approach only requires the estimation of a single parameter, i.e., the reverberation time. It is based on a simple room reverberation model, which does not allow precise dereverberation, but has been shown to improve ASR performance [36]. However, this approach uses spectral subtraction, which is a form of nonlinear processing and may thus introduce distortions. Moreover, it is a single-channel approach that cannot exploit multiple microphone signals even if they are available.

Neural-network-based enhancement is another approach that has been used for dereverberation and has been shown to be effective as a front end for ASR in reverberant conditions. In this approach, a neural network is trained to predict clean speech from an observed reverberant speech signal, using a parallel corpus of clean and reverberant speech [40]. This approach is not specific to dereverberation, and similar neural networks have been used for noise reduction. Neural-network-based enhancement is discussed in more detail in Chaps. 4, 5, and 7.

Finally, linear-prediction-based speech dereverberation has been shown to be particularly effective as a front end to a DNN-based ASR system [12, 21, 47, 48]. We review this approach in more detail below.

2.2.3 Linear-Prediction-Based Dereverberation

We can rewrite (2.8) using an autoregressive model, which leads to the multichannel linear prediction expression [31, 43]

$$Y_1(t,f) = D_1(t,f) + \sum_{j=1}^{J} \bar{\mathbf{g}}_{j,f}^H \bar{\mathbf{y}}_{j,t-\delta,f} \tag{2.12}$$

$$= D_1(t,f) + \bar{\mathbf{g}}_f^H \bar{\mathbf{y}}_{t-\delta,f}, \tag{2.13}$$

where we use microphone 1 as the reference microphone. Here we use vector operations to express the convolution operation, and define the vectors as

$$\bar{\mathbf{y}}_{j,t,f} = [Y_j(t,f) \ldots Y_j(t-L,f)]^T,$$
$$\bar{\mathbf{g}}_{j,f} = [G_j(1,f) \ldots G_j(L,f)]^T,$$
$$\bar{\mathbf{g}}_f = [\bar{\mathbf{g}}_{1,f}^T, \ldots, \bar{\mathbf{g}}_{J,f}^T]^T,$$
$$\bar{\mathbf{y}}_{t,f} = [\bar{\mathbf{y}}_{1,t,f}^T, \ldots, \bar{\mathbf{y}}_{J,t,f}^T]^T.$$

Note that we use the notation $\bar{\mathbf{y}}_{t,f}$ to emphasize the difference from $\mathbf{y}_{t,f}$, which contains a set of microphone signal observations for the time frame t, as shown in (2.7).

The second term in (2.13), i.e., $\bar{\mathbf{g}}_f^H \bar{\mathbf{y}}_{t-\delta,f}$, corresponds to the late reverberation. Therefore, if we know the prediction filters $\bar{\mathbf{g}}_f$, the dereverberated signal can be obtained as

$$D_1(t,f) = Y_1(t,f) - \bar{\mathbf{g}}_f^H \bar{\mathbf{y}}_{t-\delta,f}. \tag{2.14}$$

Conventional linear prediction assumes that the target signal or prediction residual follows a stationary Gaussian distribution and does not include the prediction delay δ [18]. The prediction filter is then obtained using maximum likelihood estimation. However, employing conventional linear prediction for speech dereverberation destroys the time structure of speech, because the power density of speech may change greatly from one time frame to another, and thus cannot be well modeled as a stationary Gaussian signal. Moreover, linear prediction also causes excessive whitening because it equalizes the short-term generative process of speech [25]. Therefore, when using conventional linear prediction for speech dereverberation, the characteristics of the dereverberated speech signal are severely modified and ASR performance degrades.

These issues can be addressed by introducing a model of speech that takes better account of the dynamics of the speech signal, and by including the prediction delay δ to prevent overwhitening. Several models have been investigated [24, 30]. Here, we model the target signal as a Gaussian with a zero mean and a time-varying variance

$\phi_D(t,f)$. The variance $\phi_D(t,f)$ corresponds to the short-time power spectrum of the target speech. With this model, the distribution of the target signal $D_1(t,f)$ can be written as

$$
\begin{aligned}
p(D_1(t,f); \phi_D(t,f)) &= \mathcal{N}_C(D_1(t,f); 0, \phi_D(t,f)) \\
&= \frac{1}{\pi \phi_D(t,f)} e^{-|D_1(t,f)|^2/\phi_D(t,f)} \\
&= \frac{1}{\pi \phi_D(t,f)} e^{-|Y_1(t,f) - \bar{\mathbf{g}}_f^H \bar{\mathbf{y}}_{t-\delta,f}|^2/\phi_D(t,f)},
\end{aligned}
\tag{2.15}
$$

where $\mathcal{N}_C()$ represents a complex Gaussian distribution. Let $\Theta = \{\phi_D(t,f), \bar{\mathbf{g}}_f\}$ be the set of unknown parameters. We estimate the parameters by maximizing the log-likelihood function, defined as

$$
\begin{aligned}
\mathcal{L}(\Theta) &= \sum_t \log\left(p(D_1(t,f); \Theta)\right) \\
&= -\sum_t \left(\log(\pi \phi_D(t,f)) + \frac{|Y_1(t,f) - \bar{\mathbf{g}}_f^H \bar{\mathbf{y}}_{t-\delta,f}|^2}{\phi_D(t,f)} \right).
\end{aligned}
\tag{2.16}
$$

Equation (2.16) cannot be maximized analytically. Instead, we perform recursive optimization in two steps.

1. First we optimize $\mathcal{L}(\Theta)$ with respect to $\bar{\mathbf{g}}_f$ for a fixed $\phi_D(t,f)$. This is solved by taking the derivative of $\mathcal{L}(\Theta)$ with respect to $\bar{\mathbf{g}}_f$ and equating it to zero as

$$
\frac{\partial \mathcal{L}(\Theta)}{\partial \bar{\mathbf{g}}_f} = \sum_t \frac{2\bar{\mathbf{y}}_{t-\delta,f} Y_1^*(t,f) - 2\bar{\mathbf{y}}_{t-\delta,f} \bar{\mathbf{y}}_{t-\delta,f}^H \bar{\mathbf{g}}_f}{\phi_D(t,f)} = 0.
\tag{2.17}
$$

Solving (2.17) leads to the expressions for the prediction filters

$$
\bar{\mathbf{g}}_f = \bar{\mathbf{R}}_f^{-1} \bar{\mathbf{r}}_{f,\delta},
\tag{2.18}
$$

$$
\bar{\mathbf{R}}_f = \sum_t \frac{\bar{\mathbf{y}}_{t-\delta,f} \bar{\mathbf{y}}_{t-\delta,f}^H}{\phi_D(t,f)},
\tag{2.19}
$$

$$
\bar{\mathbf{r}}_{f,\delta} = \sum_t \frac{\bar{\mathbf{y}}_{t-\delta,f} Y_1^*(t,f)}{\phi_D(t,f)},
\tag{2.20}
$$

where $\bar{\mathbf{R}}_f$ is the covariance matrix of the microphone signals and $\bar{\mathbf{r}}_{f,\delta}$ is the covariance vector computed with a delay δ.

2. Then we optimize $\mathscr{L}(\Theta)$ with respect to $\phi_D(t,f)$. Taking the derivative of $\mathscr{L}(\Theta)$ with respect to $\phi_D(t,f)$ gives

$$\frac{\partial \mathscr{L}(\Theta)}{\partial \phi_D(t,f)} = -\frac{1}{\phi_D(t,f)} + \frac{|Y_1(t,f) - \bar{\mathbf{g}}_f^H \bar{\mathbf{y}}_{t-\delta,f}|^2}{\phi_D^2(t,f)}. \tag{2.21}$$

Equating (2.21) to zero leads to the following expression for the variance, $\phi_D(t,f)$:

$$\phi_D(t,f) = |Y_1(t,f) - \bar{\mathbf{g}}_f^H \bar{\mathbf{y}}_{t-\delta,f}|^2. \tag{2.22}$$

Equations (2.18)–(2.20) are very similar to those for conventional multichannel linear prediction except for the delay δ and the normalization by the variance $\phi_D(t,f)$ [18]. Because of this normalization, this algorithm is referred to as WPE. Note that the normalization tends to emphasize the contribution of time frames where $\phi_D(t,f)$ is small, i.e., time frames that may be dominated by reverberation.

It is possible to show that WPE induces little distortion in the processed signal [31]. Moreover, the algorithm can be extended to online processing [44].

Note that although in the discussion we assumed no additive noise, the WPE algorithm has been shown to perform well in the presence of noise [12, 45, 48] and to be effective for meeting recognition [21, 47]. The WPE algorithm can also be used for single-channel recordings [47]. Moreover, the WPE algorithm has the interesting property of shortening room impulse responses while preserving early reflections. This means that the spatial information contained in the observed signals is preserved after dereverberation. Consequently, multichannel noise reduction techniques that exploit spatial information, such as beamforming or clustering-based approaches, can be employed after dereverberation to reduce noise. We discuss beamforming in the next section.

2.3 Beamforming

In addition to dereverberation, it is also important to reduce noise prior to recognition. Beamforming is a class of multichannel SE approaches that is very effective for noise reduction. A beamformer is designed to capture sound coming from the target speaker direction, while reducing interfering sounds coming from other directions. This is realized by steering the beam of the beamformer in a target direction, which is realized by filtering the microphone signals with linear filters. The output of a beamformer, $\hat{x}[n]$, can be expressed as

$$\hat{x}[n] = \sum_{j=1}^{J} w_j[n] * y_j[n], \tag{2.23}$$

where $w_j[n]$ is the filter associated with microphone j. In the STFT domain, (2.23) becomes

$$\hat{X}(t,f) = \sum_{j=1}^{J} W_j^*(f)Y_j(t,f) \tag{2.24}$$

$$= \mathbf{w}_f^H \mathbf{y}_{t,f} \tag{2.25}$$

$$= \mathbf{w}_f^H \mathbf{o}_{t,f} + \mathbf{w}_f^H \mathbf{u}_{t,f}, \tag{2.26}$$

where $\mathbf{w}_f = [W_1(f), \dots, W_J(f)]^T$ is a vector containing the beamforming filter coefficients in the STFT domain. In general, the filter $\mathbf{w}_f$ of a beamformer is obtained by assuming that early reflections can be covered within the analysis frame of the STFT and that late reverberations are not correlated with the target speech and can thus be included in the noise term. Consequently, the expression for the source image can be simplified to

$$\mathbf{o}_{t,f} = \mathbf{h}_f X(t,f), \tag{2.27}$$

where $\mathbf{h}_f \triangleq \mathbf{h}_{0,f} = [H_1(0,f), \dots, H_J(0,f)]^T$ is the steering vector. A beamformer aims at reducing the noise term of its output, $\mathbf{w}_f^H \mathbf{u}_{t,f}$.

2.3.1 Types of Beamformers

There has been a lot of research on beamforming and many different types of beamformers have been developed. It is not our intention to provide extensive coverage of existing beamformers, but rather to focus on some of the approaches that have recently been used for distant ASR. These include the delay-and-sum (DS) beamformer [13, 33, 37], the max-SNR beamformer [19, 37, 39], the minimum variance distortionless response (MVDR) beamformer [16, 22, 48], and the multichannel Wiener filter (MCWF) [14, 34]. First we derive the expressions for the filters for these beamformers in terms of key quantities such as the steering vectors and/or spatial correlation matrices of the signals. We then elaborate on how to estimate these quantities in Sect. 2.3.2.

2.3.1.1 Delay-and-Sum Beamformer

The DS beamformer is the simplest possible beamformer, which functions by averaging the microphone signals after time-aligning them such that the target speech signal is synchronized among all microphones [37]. If we assume plane wave propagation of the speech signal (i.e., far field) without reverberation (i.e., free field), the room impulse responses reduce to propagation delays and the microphone

Fig. 2.2 Schematic diagram illustrating the TDOA at microphone j when microphone 1 is taken as the reference microphone, i.e., $\Delta\tau_{1,1} = 0$

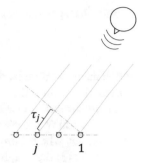

signals can be expressed as

$$\mathbf{y}_{t,f} = \mathbf{h}_f X(t,f) + \mathbf{u}_{t,f} \tag{2.28}$$

$$\approx \left[e^{-2\pi i f \Delta\tau_{r,1}}, \dots, e^{-2\pi i f \Delta\tau_{r,J}} \right]^T X(t,f) + \mathbf{u}_{t,f}, \tag{2.29}$$

where $\mathbf{h}_f = \left[e^{-2\pi i f \Delta\tau_{r,1}}, \dots, e^{-2\pi i f \Delta\tau_{r,J}} \right]^T$ is referred to as a steering vector in the context of beamforming. Under far-field and free-field conditions, the steering vector is entirely defined by the time differences of arrival (TDOAs), $\Delta\tau_{r,j}$, of the microphone signals. The TDOAs, $\Delta\tau_{r,j}$ represent the time difference of arrival between microphone j and a reference microphone r, as illustrated in Fig. 2.2. TDOAs can be estimated from the cross-correlation of the microphone signals as explained in Sect. 2.3.2.1.

The DS beamformer simply aligns the different microphone signals in time, such that the signals in the target direction sum constructively, and the interferences sum destructively. This is accomplished by setting the filter coefficients as

$$\mathbf{w}_f^{\mathrm{DS}} = \frac{1}{J}\mathbf{h}_f = \left[\frac{e^{-2\pi i f \Delta\tau_{r,1}}}{J} \cdots \frac{e^{-2\pi i f \Delta\tau_{r,J}}}{J} \right], \tag{2.30}$$

which corresponds to advancing the jth microphone signal by $\Delta\tau_{r,j}$ taps so that all signals can be synchronized. The enhanced speech is thus obtained as

$$\hat{X}(t,f) = (\mathbf{w}_f^{\mathrm{DS}})^H \mathbf{y}_{t,f} \tag{2.31}$$

$$= \frac{1}{J} \sum_{j=1}^{J} e^{2\pi i f \Delta\tau_{r,j}} Y_j(t,f) \tag{2.32}$$

or, in the time domain,

$$\hat{x}[n] = \frac{1}{J} \sum_{j=1}^{J} y_j[n + \Delta\tau_{r,j}]. \tag{2.33}$$

Note that, given the expression for the filter, we have

$$(\mathbf{w}_f^{DS})^H \mathbf{h}_f = 1, \tag{2.34}$$

which means that under the above assumptions, the target speech can be recovered without distortions at the output of the DS beamformer. This is further referred to as the *distortionless constraint*.

A variation of the conventional DS beamformer is the weighted delay-and-sum beamformer, which introduces different weights for each microphone signal. For example, in [2] the weights are related to a measure of the quality of the microphone signals derived from the cross-correlation among channels. Channels that have a low cross-correlation with a reference channel are considered harmful to the beamformer and are given a weight of zero, while the other microphones are given uniform weights. In addition, the different microphone signals can be weighted to account for the different signal powers. This approach is implemented in the BeamformIt toolkit [1], which has been successfully used for different distant ASR tasks [13, 33].

2.3.1.2 Minimum Variance Distortionless Response Beamformer

The MVDR beamformer is designed to minimize the noise at the output of the beamformer while imposing a distortionless constraint on the target speech signal [16]. The filter can thus be obtained by solving the following optimization problem:

$$\mathbf{w}_f^{MVDR} = \underset{\mathbf{w}_f}{\arg\min} \, E\{|\mathbf{w}_f^H \mathbf{u}_{t,f}|^2\}, \tag{2.35}$$

$$\text{subject to } \mathbf{w}_f^H \mathbf{h}_f = 1,$$

where $E\{|\mathbf{w}_f^H \mathbf{u}_{t,f}|^2\} = \mathbf{w}_f^H \mathbf{R}_{u,f} \mathbf{w}_f$ is the power spectrum density of the output noise signal and $\mathbf{R}_{u,f} = E\{\mathbf{u}_{t,f} \mathbf{u}_{t,f}^H\}$ is the spatial correlation matrix of the noise signals. If we assume that the target speech and noise signals are uncorrelated, we can express the power spectrum density of the output noise as

$$E\{|\mathbf{w}_f^H \mathbf{u}_{t,f}|^2\} = \mathbf{w}_f^H \mathbf{R}_{y,f} \mathbf{w}_f - \mathbf{w}_f^H \mathbf{R}_{o,f} \mathbf{w}_f \tag{2.36}$$

$$= \mathbf{w}_f^H \mathbf{R}_{y,f} \mathbf{w}_f - \mathbf{w}_f^H \mathbf{h}_f \Phi_X \mathbf{w}_f \mathbf{h}_f^H \tag{2.37}$$

$$= \mathbf{w}_f^H \mathbf{R}_{y,f} \mathbf{w}_f - \Phi_X, \tag{2.38}$$

where Φ_X is the power spectral density of the target speech and $\mathbf{R}_{y,f} = E\{|\mathbf{y}_{t,f}|^2\}$ and $\mathbf{R}_{o,f} = E\{|\mathbf{o}_{t,f}|^2\}$ are the spatial correlation matrices of the microphone signals and the source images, respectively. The distortionless constraint means that the second term does not depend on $\mathbf{w}_f$. Consequently, the optimization problem of

(2.36) can be reformulated as

$$\mathbf{w}_f^{\mathrm{MVDR}} = \operatorname*{argmin}_{\mathbf{w}_f} \mathbf{w}_f^H \mathbf{R}_{y,f} \mathbf{w}_f, \tag{2.39}$$

$$\text{subject to } \mathbf{w}_f^H \mathbf{h}_f = 1,$$

Solving this optimization problem gives us the following expression for the MVDR filters [16]:

$$\mathbf{w}_f^{\mathrm{MVDR}} = \frac{\mathbf{R}_{y,f}^{-1} \mathbf{h}_f}{\mathbf{h}_f^H \mathbf{R}_{y,f}^{-1} \mathbf{h}_f}. \tag{2.40}$$

Note that, to compute the filters, we first need to estimate the steering vector $\mathbf{h}_f$ and the spatial correlation matrix of the microphone signals. This is discussed in Sect. 2.3.2.2.

The fact that the MVDR beamformer optimizes noise reduction while imposing the distortionless constraint on the processed speech makes it particularly attractive for ASR applications where the acoustic model is usually sensitive to distortions. Accordingly, the MVDR beamformer has been shown to significantly improve ASR performance for many tasks [12, 22, 48].

2.3.1.3 Max-SNR Beamformer

The maximum-signal-to noise-ratio (max-SNR) beamformer or generalized eigenvalue beamformer [3, 37, 39] is an alternative to the MVDR, and its direct purpose is to optimize the output SNR without imposing a distortionless constraint. The filters of the max-SNR beamformer can be obtained directly from the spatial correlation matrices of the noise and microphone signals and do not require prior knowledge or estimation of the steering vectors. This may be an advantage since steering-vector estimation is prone to errors when noise and reverberation are severe.

To derive the max-SNR beamformer, we first introduce the power spectral density of the beamformer output, $\Phi_{\hat{X},f}$, which is defined as

$$\Phi_{\hat{X},f} = E\{|\hat{X}(t)|^2\}$$

$$= \mathbf{w}_f^H \mathbf{R}_{o,f} \mathbf{w}_f + \mathbf{w}_f^H \mathbf{R}_{u,f} \mathbf{w}_f, \tag{2.41}$$

assuming that the target speech source image, $\mathbf{o}_{t,f}$, and noise signals, $\mathbf{u}_{t,f}$, are independent, i.e., $\mathbf{R}_{y,f} = \mathbf{R}_{o,f} + \mathbf{R}_{u,f}$.

The filters of the max-SNR beamformer are obtained by maximizing the SNR as

$$\mathbf{w}_f^{\mathrm{maxSNR}} = \operatorname*{argmax}_{\mathbf{w}_f} \frac{\mathbf{w}_f^H \mathbf{R}_{o,f} \mathbf{w}_f}{\mathbf{w}_f^H \mathbf{R}_{u,f} \mathbf{w}_f}, \tag{2.42}$$

where $\mathbf{w}_f^H \mathbf{R}_{o,f} \mathbf{w}_f / \mathbf{w}_f^H \mathbf{R}_{u,f} \mathbf{w}_f$ represents the SNR at the output of the beamformer. Solving Eq. (2.42) leads to the following relation:

$$\mathbf{R}_{o,f} \mathbf{w}_f = \lambda \mathbf{R}_{u,f} \mathbf{w}_f, \tag{2.43}$$

where λ is an eigenvalue. Equation (2.43) is equivalent to a generalized eigenvalue problem, which can be solved by multiplying both sides by $\mathbf{R}_{u,f}^{-1}$. Consequently, the filters of the max-SNR beamformer can be obtained as the principal eigenvector of $\mathbf{R}_{u,f}^{-1} \mathbf{R}_{o,f}$,

$$\mathbf{w}_f^{\text{maxSNR}} = \mathscr{P}(\mathbf{R}_{u,f}^{-1} \mathbf{R}_{o,f}), \tag{2.44}$$

where $\mathscr{P}(\mathbf{A})$ is the principal eigenvector of matrix $\mathbf{A}$, which is the eigenvector associated with the maximum eigenvalue of $\mathbf{A}$, where the eigenvalues λ and eigenvectors $\mathbf{x}$ are obtained by solving $(\mathbf{A} - \lambda \mathbf{I})\mathbf{x} = 0$, where $\mathbf{I}$ is an identity matrix. Note that from the definition of the eigenvectors, we have $\mathscr{P}(\mathbf{A} + \mathbf{I}) = \mathscr{P}(\mathbf{A})$. Therefore, given the assumption of the independence of the speech source image and noise signals, we can easily see that $\mathscr{P}(\mathbf{R}_{u,f}^{-1} \mathbf{R}_{o,f}) = \mathscr{P}(\mathbf{R}_{u,f}^{-1} \mathbf{R}_{y,f})$. Consequently, (2.44) can be equivalently expressed using the microphone signal spatial correlation matrix as

$$\mathbf{w}_f^{\text{maxSNR}} = \mathscr{P}(\mathbf{R}_{u,f}^{-1} \mathbf{R}_{y,f}). \tag{2.45}$$

Equation (2.45) reveals that the max-SNR beamformer does not require knowledge of the steering vector $\mathbf{h}_f$, but rather it requires the spatial correlation matrix of the microphone signal and that of the noise signals. Since noise is not observed directly, the latter matrix needs to be estimated from the observed signals.

The max-SNR directly optimizes the SNR without imposing any distortionless constraint on the output signal. Therefore, although this beamformer may be optimal in terms of noise reduction, it can cause distortions in the output enhanced speech that may affect recognition performance. It has been proposed that postfiltering be used to impose a distortionless constraint. However, it is reported that such postfiltering may not always improve ASR performance [19].

2.3.1.4 Multichannel Wiener Filter

The MCWF [14] can be considered a type of beamformer since it realizes the multichannel filtering of microphone signals to reduce noise. Here we focus on a linear-processing-based MCWF, where the filters are constant over time as in the other discussions in this chapter. The MCWF is designed to preserve spatial information, represented by the steering vectors, at its output. Therefore, the MCWF output consists of multichannel source image signals. The multichannel output signals $\hat{\mathbf{o}}_{t,f}$ are given as

$$\hat{\mathbf{o}}_{t,f} = \mathbf{W}_f^H \mathbf{y}_{t,f}, \tag{2.46}$$

where $\mathbf{W}_f$ is a filter matrix of size $J \times J$.

The filter matrix is derived by minimizing the mean square error:

$$\mathbf{W}_f^{\text{MCWF}} = \underset{\mathbf{W}_f}{\operatorname{argmin}} E\{|\mathscr{E}|^2\}, \tag{2.47}$$

where the error signal, $\mathscr{E}$, is defined as the difference between the output signals and the source images as

$$\mathscr{E} = \mathbf{o}_{t,f} - \hat{\mathbf{o}}_{t,f}. \tag{2.48}$$

Solving (2.47) leads to the following expression for the filter matrix:

$$\mathbf{W}_f^{\text{MCWF}} = \mathbf{R}_{y,f}^{-1} \mathbf{R}_{y,o,f}, \tag{2.49}$$

where $\mathbf{R}_{y,o,f} = E\{\mathbf{y}_{t,f} \mathbf{o}_{t,f}^H\}$ is the cross-correlation matrix between signals $\mathbf{y}_{t,f}$ and $\mathbf{o}_{t,f}$. Note that if we assume that the target speech source image $\mathbf{o}_{t,f}$ and the noise signals, $\mathbf{u}_{t,f}$ are uncorrelated, $\mathbf{R}_{y,o,f} = \mathbf{R}_{o,f}$ and the filter matrix becomes

$$\mathbf{W}_f^{\text{MCWF}} = \mathbf{R}_{y,f}^{-1} \mathbf{R}_{o,f}. \tag{2.50}$$

The MCWF can thus be derived from the observed and source image spatial correlation matrices and does not require estimation of the steering vectors. Note that since the MCWF preserves spatial cues, it could be used as a preprocessor for a subsequent multichannel processing step [34].

2.3.2 Parameter Estimation

Table 2.1 summarizes the different types of beamformers, the expressions for the filters, their characteristics (distortionless, multichannel output), and the key quantities needed to compute their filters. The DS beamformer requires TDOA estimation ($\Delta \tau_{r,1}, \ldots, \Delta \tau_{r,J}$); the other beamformers require estimation of the spatial correlation matrices ($\mathbf{R}_{y,f}$, $\mathbf{R}_{u,f}$, and $\mathbf{R}_{o,f}$). In addition, the MVDR beamformer also needs to estimate the steering vector ($\mathbf{h}_f$), although it can be derived from $\mathbf{R}_{y,f}$ and $\mathbf{R}_{u,f}$. In this section we briefly review the main approaches to computing these quantities and describe in more detail a time–frequency-mask-based approach for spatial correlation matrix estimation.

Table 2.1 Classification of beamformers

Beamformer type	Filter expression	Distortionless constraint	Multichannel output	Key quantities
Delay-and-sum	$\mathbf{w}_f^{DS} = \frac{1}{J}\mathbf{h}_f = \left[\frac{e^{-2\pi f \Delta \tau_{r,1}}}{J} \cdots \frac{e^{-2\pi f \Delta \tau_{r,J}}}{J}\right]$	Yes	No	$\Delta \tau_{r,1}, \ldots, \Delta \tau_{r,J}$
MVDR	$\mathbf{w}_f^{MVDR} = \frac{\mathbf{R}_{y,f}^{-1}\mathbf{h}_f}{\mathbf{h}_f^H \mathbf{R}_{y,f}^{-1}\mathbf{h}_f}$	Yes	No	$\mathbf{R}_{y,f}, \mathbf{h}_f$
Max-SNR	$\mathbf{w}_f^{maxSNR} = \mathscr{P}(\mathbf{R}_{u,f}^{-1}\mathbf{R}_{y,f})$	No	No	$\mathbf{R}_{u,f}, \mathbf{R}_{y,f}$
Multichannel WF	$\mathbf{W}_f^{MCWF} = \mathbf{R}_{y,f}^{-1}\mathbf{R}_{o,f}$	No	Yes	$\mathbf{R}_{y,f}, \mathbf{R}_{o,f}$

2.3.2.1 TDOA Estimation

There has been a lot of research on TDOA estimation from microphone signals [9, 11]. One common approach assumes that the cross-correlation between the microphone signals is maximum when the signals are aligned. TDOAs are thus obtained by finding the positions of the peaks in the cross-correlation between the microphone signals. The conventional cross-correlation is sensitive to noise and reverberation, which cause spurious peaks that may lead to TDOA estimation errors. Therefore, generalized cross-correlation phase transform (GCC-PHAT) coefficients are usually preferred [9]. These coefficients are defined as

$$\psi_{k,l}(d) = \text{IFFT}\left\{\frac{X_k(f)X_l^*(f)}{|X_k(f)||X_l(f)|}\right\}_d, \tag{2.51}$$

where $\text{IFFT}\{\}_d$ is the dth coefficient of the inverse Fourier transform, $X_k(f)$ and $X_l(f)$ are the Fourier transforms of microphone signals l and k, respectively. The GCC-PHAT coefficients are computed for a relatively long time segment.

TDOAs are obtained from the GCC-PHAT coefficients as

$$\Delta \tau_{k,l} = \underset{d}{\text{argmax}}\ \psi_{k,l}(d), \tag{2.52}$$

and the corresponding steering vector is then $\mathbf{h}_f = [e^{-2\pi f \Delta \tau_{r,1}} \ldots e^{-2\pi f \Delta \tau_{r,J}}]$, where r is the index of a microphone taken as a reference for the TDOA calculation.

Even when using GCC-PHAT coefficients, TDOAs remain sensitive to noise and reverberation. For example, using estimated TDOAs for the steering vectors employed to design an MVDR beamformer demonstrated poor performance for the real recordings in the CHiME-3 task [5].

It is possible to further improve the TDOA estimation by filtering out TDOA estimates for regions of the signals where noise is dominant and by tracking TDOAs across segments using a Viterbi search. These improvements are implemented in the BeamformIt package [1, 2], which uses these TDOA estimates to perform weighted-DS beamforming. These refinements are critical and may account for the fact that BeamformIt has steadily improved in performance for numerous tasks [13, 33].

Another approach used to increase the robustness to noise and reverberation is to estimate the steering vector directly from the signal correlation matrices instead of relying on the error-prone TDOA estimation step.

2.3.2.2 Steering-Vector Estimation

It is possible to estimate the steering vector of the source image $\mathbf{h}_f$ as the principal eigenvector of the source image spatial correlation matrix $\mathbf{R}_{o,f}$ as

$$\mathbf{h}_f = a_f \mathscr{P}(\mathbf{R}_{o,f}), \tag{2.53}$$

where a_f is a scalar complex coefficient that represents the frequency-dependent gain of $\mathbf{h}_f$. Intuitively, assuming a far-field condition without reverberation, we can assume that $\mathbf{R}_{o,f} = \Phi_X \mathbf{h}_f \mathbf{h}_f^H$ is rank 1, where Φ_X is a scalar representing the power spectral density of the target speech. Therefore, there is a unique eigenvalue that can be obtained by solving $(\mathbf{R}_{o,f} - \lambda \mathbf{I})\mathbf{v} = 0$. Consequently, the principal eigenvector is equal to the steering vector, and the principal eigenvalue, $\tilde{\lambda}$, is equal to the power spectral density of the target speech, Φ_X, if the steering vectors and the eigenvectors are normalized, i.e., $\tilde{\lambda} = \Phi_X$. We can set the value of a_f by assuming that the norm of $\mathbf{h}_f$ equals the number of microphones J.

Note that the rank-1 assumption is not necessary. For example, if the source images are corrupted with white noise uncorrelated across the microphone signals, the spatial correlation matrix becomes

$$\tilde{\mathbf{R}}_o = \mathbf{R}_{o,f} + \sigma_N \mathbf{I}, \tag{2.54}$$

where σ_N represents the noise power spectrum. In this case, since $\mathscr{P}(\mathbf{A} + \mathbf{I}) = \mathscr{P}(\mathbf{A})$, the principal eigenvector is still equal to the steering vector, and the principal eigenvalue becomes $\tilde{\lambda} = \Phi_X - \sigma_N$. However, if noise becomes dominant, $\sigma_N > \Phi_X$, the principal eigenvalue is negative and the steering-vector estimation may become inaccurate.

It is possible to derive an alternative expression for the steering vector. Assuming that $\mathbf{R}_{o,f}$ is rank 1 and that $\mathbf{R}_{u,f}$ is full rank, the steering vectors can also be obtained as [39]

$$\mathbf{h}_f = b_f \mathbf{R}_{u,f} \mathscr{P}(\mathbf{R}_{u,f}^{-1} \mathbf{R}_{y,f}), \tag{2.55}$$

where b_f can be chosen so that the norm of $\mathbf{h}_f$ equals the number of microphones J.

In practice, both approaches to computing the steering vectors require the spatial correlation matrix of the source images, which are unseen and must be estimated. Assuming that speech is uncorrelated with noise, we can estimate the source image spatial correlation matrix as

$$\mathbf{R}_{o,f} = \mathbf{R}_{y,f} - \mathbf{R}_{u,f}. \tag{2.56}$$

Therefore, we only need to estimate the microphone signals and noise spatial correlation matrices to derive the steering vectors. If noise is stationary, $\mathbf{R}_{u,f}$ can be estimated during speech-absent periods, e.g. from the first frames of the signal. If noise is nonstationary, we need to estimate the noise from the observed microphone signals.

In the following subsection, we discuss how time–frequency-masking-based speech enhancement schemes can be used for that purpose.

2.3.2.3 Time–Frequency-Masking-Based Spatial Correlation Matrix Estimation

Principle

Since the microphone signals are directly observed, the spatial correlation matrix of the microphone signals can be computed as

$$\mathbf{R}_{y,f} = \sum_{t=1}^{T} \mathbf{y}_{t,f} \mathbf{y}_{t,f}^{H}. \tag{2.57}$$

However, the noise spatial correlation matrix must be estimated because it is not observed directly. From the sparseness assumption of speech signals, the target speech will only be active in some time–frequency bins of the observed signal and some time–frequency bins will be dominated by noise. Let us assume that we know a time–frequency mask, $\Omega(t,f)$, which represents the probability that the (t,f) frequency bin consists solely of noise. By knowing such a time–frequency mask, we can estimate the spatial correlation matrix of the noise by averaging $\mathbf{y}_{t,f} \mathbf{y}_{t,f}^{H}$ as

$$\mathbf{R}_{u,f} = \frac{\sum_{t=1}^{T} \Omega(t,f) \mathbf{y}_{t,f} \mathbf{y}_{t,f}^{H}}{\sum_{t=1}^{T} \Omega(t,f)}. \tag{2.58}$$

Finally, assuming that the speech and noise signals are independent, the spatial correlation matrix of the target source image can be computed as

$$\mathbf{R}_{o,f} = \mathbf{R}_{y,f} - \mathbf{R}_{u,f}. \tag{2.59}$$

Therefore, if we can compute the time–frequency masks, $\Omega(t,f)$, we can estimate all the spatial correlation matrices needed to compute the steering vector and the

Fig. 2.3 Schematic diagram of beamformer using time–frequency masking for estimating spatial correlation matrices

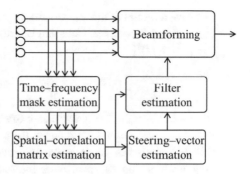

beamformer filters. Figure 2.3 is a schematic diagram of a beamformer using time–frequency masking to estimate a noise spatial correlation matrix. As shown in the figure, first a time–frequency mask is estimated. The mask is then used to estimate the spatial correlation matrices, which are subsequently employed to estimate the steering vectors and compute the beamformer filter.

There has been a lot of work on time–frequency mask estimation for speech enhancement. In the context of robust ASR, neural-network-based approaches [19] and clustering-based approaches [20, 35, 48] to mask estimation have recently gained interest. Related approaches are also discussed in Chaps. 3 and 4. In this chapter, to illustrate mask-based spatial-statistic estimation, we elaborate on a clustering-based approach, which relies on a complex Gaussian mixture model (CGMM) of the sources.

Modeling Sources with Complex Gaussian Mixture Model

Using the sparseness property of speech signals, we can rewrite (2.7) as

$$\mathbf{y}_{t,f} = \mathbf{h}_f^{(k)} S^{(k)}(t,f), \tag{2.60}$$

where k is an index of the source, i.e., $S^{(1)}(t,f)$ corresponds to noisy speech and $S^{(2)}(t,f)$ to noise, and $\mathbf{h}_f^{(k)}$ is a pseudo-steering vector associated with noisy speech ($k = 1$) or noise ($k = 2$). If the noisy speech and noise can be considered point sources, the pseudo-steering vectors correspond to actual physical steering vectors that represent the directional information of the sources. In general, this is not the case because of reverberation and the diffuseness of noise, but it is not a problem here because we do not estimate steering vectors but rather directly estimate spatial correlation matrices and allow them to be full rank as discussed below.

To derive a model of the microphone signals, let us assume that the source signals $S^{(k)}(t,f)$ follow a complex Gaussian distribution with a zero mean and a variance $\Phi_{t,f}^{(k)} = |S^{(k)}(t,f)|^2$,

$$p(S^{(k)}(t,f); \Phi_{t,f}^{(k)}) = \mathcal{N}_C(S^{(k)}(t,f); 0, \Phi_{t,f}^{(k)}). \tag{2.61}$$

From (2.60) and (2.61) we can express the distribution of the microphone signals as a complex Gaussian with a rank-1 covariance matrix given by $\Phi_{t,f}^{(k)} \mathbf{h}_f^{(k)} (\mathbf{h}_f^{(k)})^H$. However, to make the model more robust to fluctuations in the steering vectors due to speaker or microphone position changes, we allow the covariance matrix to be full rank. Consequently, the microphone signals are modeled as

$$p(\mathbf{y}_{t,f} | C_{t,f} = k; \theta_{t,f}^{(k)}) = \mathcal{N}_C(\mathbf{y}_{t,f}; 0, \Phi_{t,f}^{(k)} \mathbf{R}_f^{(k)}), \tag{2.62}$$

where $C_{t,f}$ is a random variable that indicates whether the time–frequency bin (t,f) corresponds to noisy speech ($k = 1$) or noise ($k = 2$), $\theta_{t,f}^{(k)} = \{\Phi_{t,f}^{(k)}, \mathbf{R}_f^{(k)}\}$, and $\mathbf{R}_f^{(k)}$ is the spatial correlation matrix of source k.

Given the above model of the sources, the probability distribution of the microphone signals can thus be expressed as

$$p(\mathbf{y}_{t,f}; \Theta) = \sum_k \alpha_f^{(k)} p(\mathbf{y}_{t,f} | C_{t,f} = k; \theta_{t,f}^{(k)}), \tag{2.63}$$

where $\alpha_f^{(k)}$ is the mixture weight or priors, and $\Theta = \{\alpha_f^{(k)}, \Phi_{t,f}^{(k)}, \mathbf{R}_f^{(k)}\}$ is a set of model parameters. We can obtain a time–frequency mask $\Omega^{(k)}(t,f)$ for noisy speech ($k = 1$) or noise ($k = 2$), as the following posterior probabilities:

$$\Omega^{(k)}(t,f; \Theta) = p(C_{t,f} = k | \mathbf{y}_{t,f}; \Theta) \tag{2.64}$$

$$= \frac{\alpha_f^{(k)} p(\mathbf{y}_{t,f} | C_{t,f} = k; \theta_{t,f}^{(k)})}{\sum_{k'} \alpha_f^{(k')} p(\mathbf{y}_{t,f} | C_{t,f} = k'; \theta_{t,f}^{(k')})}. \tag{2.65}$$

Expectation-Maximization-Based Parameter Estimation

We now review how to estimate the set of parameters Θ of the mixture model shown in (2.63). We can estimate Θ to maximize the log likelihood-function defined as

$$\mathcal{L}(\Theta) = \sum_t \sum_f \log(p(\mathbf{y}_{t,f}; \Theta)). \tag{2.66}$$

This optimization problem can be solved with the expectation maximization (EM) algorithm. Let us define the Q function as

$$Q(\Theta, \Theta') = \sum_t \sum_f \sum_k \Omega^{(k)}(t, f; \Theta') \log(\alpha_f^{(k)} p(\mathbf{y}_{t,f} | C_{t,f} = k; \theta_{t,f}^{(k)})), \quad (2.67)$$

where Θ' is a previous estimate of Θ. The Q function is maximized by iterating between an expectation step (E-step) and a maximization step (M-step) as follows.

E-step. Compute the posterior $\lambda^{(k)}(t, f; \Theta')$ using (2.65).

M-step. Update the parameters as follows:

$$\Phi_{t,f}^{(k)} = \frac{1}{M} \mathbf{y}_{t,f}^H (\mathbf{R}_f^{(k)})^{-1} \mathbf{y}_{t,f}, \quad (2.68)$$

$$\mathbf{R}_f^{(k)} = \frac{\sum_t \Omega^{(k)}(t, f; \Theta') \mathbf{y}_{t,f}^H \mathbf{y}_{t,f} / \Phi_{t,f}^{(k)}}{\sum_t \Omega^{(k)}(t, f; \Theta')}, \quad (2.69)$$

$$\alpha_f^{(k)} = \frac{1}{T} \sum_t \Omega^{(k)}(t, f; \Theta'). \quad (2.70)$$

The above EM equations are very similar to those obtained for estimating the parameters of a Gaussian mixture model (GMM) [6], except that the mean is zero and that the correlation matrix takes the form of $\Phi_{t,f}^{(k)} \mathbf{R}_f^{(k)}$. Moreover, since we are dealing with complex numbers, $\mathbf{R}_f^{(k)}$ can take complex values except for the terms on the diagonal.

Note that this algorithm can be used to separate multiple sources [4]. When considering only a single target speaker, we can assume uniform priors $\alpha_f^{(k)}$ to simplify computation.

CGMM-based mask estimation was first developed for utterance-based batch processing. However, it was recently extended to online processing [20].

Practical Considerations

Selection of Noise Mask

After the convergence of the EM algorithm, we obtain two masks, one associated with noisy speech and one associated with noise. However, we do not know a priori which mask is associated with noise. We can use both masks to compute the spatial correlation matrices of noisy speech and noise and select the noise correlation matrix by choosing the matrix that has the highest entropy among its eigenvalues [20]. Intuitively, for a free-field condition, the spatial correlation of speech would have rank 1, meaning that it would have one nonzero eigenvalue. In contrast, the noise spatial correlation matrix may present a more uniform distribution of its eigenvalues, on the assumption that noise comes from many directions.

Initialization

There are several approaches that can be used to initialize the spatial correlation matrices. For example, we can initialize the noise spatial correlation matrix to an identity matrix and the noisy speech correlation matrix to the correlation matrix of the observed microphone signals. If training data are available, it is also possible to use the spatial correlation matrices computed with noise-only and speech-only training data as initial values.

Convergence

The EM algorithm can be run for a fixed number of iterations. In practice, about 20 EM iterations appear to be sufficient if the spatial correlation matrices can be properly initialized [20].

2.4   Examples of Robust Front Ends

In this section, we present two examples of ASR systems that were developed for the recent REVERB and CHiME-3 challenges. Figure 2.4 shows a schematic diagram of an ASR system for distant speech recognition. It consists of a speech enhancement front end, which performs dereverberation with the WPE algorithm, followed by noise reduction with beamforming, and finally an ASR back end. The two systems that we describe in this section have the same overall structure, but the systems differ in the details of their implementation and the ASR back end that is used.

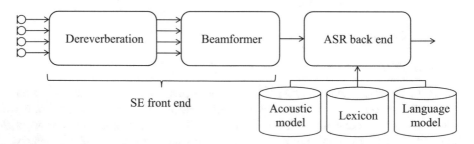

Fig. 2.4 Schematic diagram of a robust ASR system using dereverberation and beamforming

2.4.1   A Reverberation-Robust ASR System

Let us first introduce the system we used for the REVERB challenge. The REVERB challenge task is described in more detail in Chap. 15 and in [27]. This system demonstrates the power of multichannel dereverberation.

2.4.1.1   Experimental Settings

We used an ASR back end consisting of a DNN with seven hidden layers. The input features consist of 40 log mel filterbank coefficients with Δ and $\Delta\Delta$ coefficients appended with five left and right context frames. The features were processed with global mean and variance normalization and utterance-level mean normalization. Initial experimental results were obtained for an acoustic model trained on the REVERB challenge baseline training dataset, which consists of 17 h of multicondition training data generated artificially by convolving clean speech with measured room impulse responses and adding noise. The back end uses a trigram language model. The experimental settings and configuration are described in detail in [12].

The parameters of the speech enhancement front end are detailed in Table 2.2. For all the experiments, we used an 8-channel microphone array setting. Because the noise is relatively stationary in this task, the MVDR parameters were calculated using a noise spatial correlation matrix estimated from the first frames of each utterance.

2.4.1.2   Experimental Results

Figure 2.5 plots the word error rate (WER) as a function of different front-end configurations for the development set of the REVERB challenge task. These results were obtained without retraining the back-end acoustic model on the enhanced speech, as we confirmed that performance was better if we used an acoustic model trained on noisy and reverberant speech when there was a mismatch between the training and testing conditions.

Table 2.2 Settings for the speech enhancement front end

WPE
$\delta = 3, L = 7$
Window length: 32 ms, frame shift: 8 ms
Number of FFT points: 512 (number of frequency bands: 257)
MVDR
Window length: 32 ms, frame shift: 8 ms

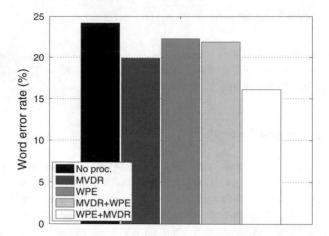

Fig. 2.5 Comparison of different speech enhancement front-end configurations for the development set of the REVERB challenge (RealData)

Table 2.3 Average WER for the RealData evaluation set of the REVERB challenge

Front end	WER (%)
No proc.	19.2
WPE (8 ch)	12.9
WPE (8 ch)+MVDR	9.3

Note that decoding headset and lapel microphone recordings using the same back end give WERs of 6.1% and 7.3%, respectively

The results in Fig. 2.5 demonstrate the effectiveness of multichannel dereverberation and beamforming. Moreover, from the results we clearly see that the performance is significantly better when performing dereverberation prior to beamforming than the other way around. Indeed, the WPE algorithm preserves spatial information but not the MVDR beamformer. Therefore, when dereverberation is performed after MVDR, only single-channel dereverberation is possible, which is far less effective. In addition, the steering vectors may be influenced by the effect of reverberation, making beamforming less effective, although this issue may not be severe in this case since the noise spatial correlation matrix was simply estimated from the first frames of the utterances.

Note that it is also possible to employ spectral subtraction to suppress late reverberation estimated using the WPE algorithm instead of the linear subtraction of (2.14). However, this approach performs significantly worse than linear filtering, possibly because nonlinear processing introduces distortions to the speech signal that are harmful to ASR [12].

We also tested our proposed front end with a more advanced back end, which uses extended training data and unsupervised environmental adaptation, and employed a recurrent-neural-network (RNN)-based language model [12]. Table 2.3 summarizes the results for the stronger back end and demonstrates the great performance improvement brought about by the SE front end. With the WPE algorithm and MVDR, the WER is close to that obtained when using the same back end to decode lapel microphone recordings (WER of 7.3%) or headset recordings (WER of 6.1%). The remaining difference of a couple of percentage points suggests room for further improvement.

2.4.2   Robust ASR System for Mobile Devices

Here we discuss the ASR system we proposed for the third CHiME challenge [48]. In this case, noise is more significant than reverberation. However, we still use dereverberation as the first component of the front end as it was shown to help in reverberant environments. To tackle nonstationary noise, we employ beamforming and estimate the filter parameters using spatial correlation matrices estimated with the time–frequency masking scheme described in Sect. 2.3.2.3.

2.4.2.1   Experimental Settings

For the SE front end, we used the WPE algorithm with the same configuration as in Sect. 2.4.1. Table 2.4 summarizes the settings of the beamformer. We tested different types of beamformers but estimated their statistics using the same time–frequency masking scheme described in Sect. 2.3.2.3. We also performed a comparison with the BeamformIt toolkit [2].

The ASR back end consists of a deep convolutional neural network (CNN) architecture, which consists of five convolutional layers followed by three fully connected layers before the output softmax layer. The features consist of 40 log mel filterbank coefficients with Δ and $\Delta\Delta$ coefficients appended with five left and right context frames. The features were processed with global mean and variance normalization and utterance-level mean normalization. We augmented the training data by using all the microphone signals of the training dataset separately. We used an RNN-based language model for decoding. The experimental settings and configuration are described in detail in [20, 48].

Table 2.4 Settings for the SE front end	Beamformer
	Window length: 25 ms
	Frame overlap: 75%

Table 2.5 Average WER for
the real data evaluation set of
the third CHiME challenge

Front end		WER (%)
No proc.		15.60
WPE		14.66
Masking		15.21
Beamforming	BeamformIt [2]	10.29
	Max-SNR	9.43
	MCWF	8.63
	MVDR	8.03
WPE + MVDR		7.60

The results are averaged over the four recording
conditions, i.e., bus, cafe, pedestrian, and street

2.4.2.2 Experimental Results

Table 2.5 shows the WER for different front-end configurations. We confirmed
that the WPE algorithm improves recognition in reverberant conditions (bus and
cafe) and does not harm the performance in open spaces (pedestrian and Street).
We compare performance using a time–frequency mask to enhance signals directly
(*Masking*) or to compute the spatial correlation matrix of noise and derive from
it different beamformers. The masking-based approach does not greatly improve
performance compared with our baseline. In contrast, beamforming is shown to be
very effective, with a relative WER reduction of up to 50%.

MVDR outperforms the other beamforming schemes, most probably because
of the distortionless constraint, which ensures that speech characteristics are not
degraded by the enhancement process. Nevertheless, for some conditions where
noise was particularly nonstationary, such as in cafes, the max-SNR beamformer
outperformed MVDR. This seems to indicate that the choice of the optimum
beamformer may depend on the target acoustic conditions. Note that with a stronger
back end, which includes speaker adaptation, we could further reduce the WER to
5.83% [48].

2.5 Concluding Remarks and Discussion

We have introduced several speech enhancement approaches that are effective
for improving distant speech recognition performance in noisy and reverberant
conditions. A key characteristic of the developed front ends is that they rely on linear
filtering and can thus cause few distortions in the processed speech. We have focused
our discussions on batch processing and assumed a fixed position for the sources.
Further research is required to tackle online processing and moving speakers.

In this chapter, we have not discussed another important class of front ends,
namely those that rely on neural networks to reduce acoustic interference as
discussed in Chaps. 4, 5, and 7. Such front ends have the advantage of enabling the

joint optimization of the front end and the ASR back end to achieve optimal ASR performance. However, this may also lead to overfitting to the acoustic conditions seen during training. In contrast, the approaches we have described in this chapter do not use any trained model, and may therefore be more appropriate when deployed in conditions unseen during training. The combination of both schemes constitutes an important future research direction.

References

1. Anguera, X.: BeamformIt. http://www.xavieranguera.com/beamformit/ (2014)
2. Anguera, X., Wooters, C., Hernando, J.: Acoustic beamforming for speaker diarization of meetings. IEEE Trans. Audio Speech Lang. Process. **15**(7), 2011–2023 (2007)
3. Araki, S., Sawada, H., Makino, S.: Blind speech separation in a meeting situation with maximum SNR beamformers. In: Proceedings of ICASSP'07, vol. 1, pp. I-41–I-44 (2007)
4. Araki, S., Okada, M., Higuchi, T., Ogawa, A., Nakatani, T.: Spatial correlation model based observation vector clustering and MVDR beamforming for meeting recognition. In: Proceedings of ICASSP'16, pp. 385–389 (2016)
5. Barker, J., Marxer, R., Vincent, E., Watanabe, S.: The third "CHiME" speech separation and recognition challenge: dataset, task and baselines. In: Proceedings of ASRU'15, pp. 504–511 (2015)
6. Bishop, C.M.: Pattern Recognition and Machine Learning. Information Science and Statistics. Springer, New York (2006)
7. Boll, S.: Suppression of acoustic noise in speech using spectral subtraction. IEEE Trans. Acoust. Speech Signal Process. **27**(2), 113–120 (1979)
8. Bradley, J.S., Sato, H., Picard, M.: On the importance of early reflections for speech in rooms. J. Acoust. Soc. Am. **113**(6), 3233–3244 (2003)
9. Brutti, A., Omologo, M., Svaizer, P.: Comparison between different sound source localization techniques based on a real data collection. In: Hands-Free Speech Communication and Microphone Arrays, 2008, HSCMA 2008, pp. 69–72 (2008)
10. Carletta, J., Ashby, S., Bourban, S., Flynn, M., Guillemot, M., Hain, T., Kadlec, J., Karaiskos, V., Kraaij, W., Kronenthal, M., et al.: The AMI Meeting Corpus: A Pre-announcement. Springer, Berlin (2005)
11. Chen, J., Benesty, J., Huang, Y.: Time delay estimation in room acoustic environments: an overview. EURASIP J. Adv. Signal Process. **2006**, 170–170 (2006). doi:10.1155/ASP/2006/26503. http://dx.doi.org/10.1155/ASP/2006/26503
12. Delcroix, M., Yoshioka, T., Ogawa, A., Kubo, Y., Fujimoto, M., Ito, N., Kinoshita, K., Espi, M., Araki, S., Hori, T., Nakatani, T.: Strategies for distant speech recognition in reverberant environments. EURASIP J. Adv. Signal Process. **2015**, 60 (2015). doi:10.1186/s13634-015-0245-7
13. Dennis, J., Dat, T.H.: Single and multi-channel approaches for distant speech recognition under noisy reverberant conditions: I2R'S system description for the ASpIRE challenge. In: Proceedings of ASRU'15, pp. 518–524 (2015)
14. Doclo, S., Moonen, M.: GSVD-based optimal filtering for single and multimicrophone speech enhancement. IEEE Trans. Signal Process. **50**(9), 2230–2244 (2002)
15. Erdogan, H., Hershey, J.R., Watanabe, S., Le Roux, J.: Phase-sensitive and recognition-boosted speech separation using deep recurrent neural networks. In: Proceedings of ICASSP'15, pp. 708–712 (2015)
16. Frost, O.L.: An algorithm for linearly constrained adaptive array processing. Proc. IEEE **60**(8), 926–935 (1972)
17. Harper, M.: The automatic speech recognition in reverberant environments (ASpIRE) challenge. In: Proceedings of ASRU'15, pp. 547–554 (2015)

18. Haykin, S.: Adaptive Filter Theory, 3rd edn. Prentice-Hall, Upper Saddle River, NJ (1996)
19. Heymann, J., Drude, L., Chinaev, A., Haeb-Umbach, R.: BLSTM supported GEV beamformer front-end for the 3RD CHiME challenge. In: Proceedings of ASRU'15, pp. 444–451. IEEE, New York (2015)
20. Higuchi, T., Ito, N., Yoshioka, T., Nakatani, T.: Robust MVDR beamforming using time-frequency masks for online/offline ASR in noise. In: Proceedings of ICASSP'16, pp. 5210–5214 (2016)
21. Hori, T., Araki, S., Yoshioka, T., Fujimoto, M., Watanabe, S., Oba, T., Ogawa, A., Otsuka, K., Mikami, D., Kinoshita, K., Nakatani, T., Nakamura, A., Yamato, J.: Low-latency real-time meeting recognition and understanding using distant microphones and omni-directional camera. IEEE Trans. Audio Speech Lang. Process. 20(2), 499–513 (2012)
22. Hori, T., Chen, Z., Erdogan, H., Hershey, J.R., Roux, J., Mitra, V., Watanabe, S.: The MERL/SRI system for the 3rd CHiME challenge using beamforming, robust feature extraction, and advanced speech recognition. In: Proceedings of ASRU'15, pp. 475–481 (2015)
23. Huang, X., Acero, A., Hon, H.W.: Spoken Language Processing: A Guide to Theory, Algorithm, and System Development, 1st edn. Prentice-Hall, Upper Saddle River, NJ (2001)
24. Jukic, A., Doclo, S.: Speech dereverberation using weighted prediction error with Laplacian model of the desired signal. In: Proceedings of ICASSP'14, pp. 5172–5176 (2014)
25. Kinoshita, K., Delcroix, M., Nakatani, T., Miyoshi, M.: Suppression of late reverberation effect on speech signal using long-term multiple-step linear prediction. IEEE Trans. Audio Speech Lang. Process. 17(4), 534–545 (2009)
26. Kinoshita, K., Delcroix, M., Yoshioka, T., Nakatani, T., Habets, E., Sehr, A., Kellermann, W., Gannot, S., Maas, R., Haeb-Umbach, R., Leutnant, V., Raj, B.: The REVERB challenge: a common evaluation framework for dereverberation and recognition of reverberant speech. In: Proceedings of WASPAA'13. New Paltz, NY (2013)
27. Kinoshita, K., Delcroix, M., Gannot, S., Habets, E., Haeb-Umbach, R., Kellermann, W., Leutnant, V., Maas, R., Nakatani, T., Raj, B., Sehr, A., Yoshioka, T.: A summary of the REVERB challenge: state-of-the-art and remaining challenges in reverberant speech processing research. EURASIP J. Adv. Signal Process. (2016). doi:10.1186/s13634-016-0306-6
28. Kuttruff, H.: Room Acoustics, 5th edn. Taylor & Francis, London (2009)
29. Lebart, K., Boucher, J.M., Denbigh, P.N.: A new method based on spectral subtraction for speech dereverberation. Acta Acustica 87(3), 359–366 (2001)
30. Nakatani, T., Yoshioka, T., Kinoshita, K., Miyoshi, M., Juang, B.H.: Blind speech dereverberation with multi-channel linear prediction based on short time Fourier transform representation. In: Proceedings of ICASSP'08, pp. 85–88 (2008)
31. Nakatani, T., Yoshioka, T., Kinoshita, K., Miyoshi, M., Juang, B.H.: Speech dereverberation based on variance-normalized delayed linear prediction. IEEE Trans. Audio Speech Lang. Process. 18(7), 1717–1731 (2010)
32. Narayanan, A., Wang, D.: Ideal ratio mask estimation using deep neural networks for robust speech recognition. In: Proceedings of ICASSP'13, pp. 7092–7096. IEEE, New York (2013)
33. Renals, S., Swietojanski, P.: Neural networks for distant speech recognition. In: 2014 4th Joint Workshop on Hands-free Speech Communication and Microphone Arrays (HSCMA), pp. 172–176 (2014)
34. Sivasankaran, S., Nugraha, A.A., Vincent, E., Morales-Cordovilla, J.A., Dalmia, S., Illina, I., Liutkus, A.: Robust ASR using neural network based speech enhancement and feature simulation. In: Proceedings of ASRU'15, pp. 482–489 (2015)
35. Souden, M., Araki, S., Kinoshita, K., Nakatani, T., Sawada, H.: A multichannel MMSE-based framework for speech source separation and noise reduction. IEEE Trans. Audio Speech Lang. Process. 21(9), 1913–1928 (2013)
36. Tachioka, Y., Narita, T., Weninger, F., Watanabe, S.: Dual system combination approach for various reverberant environments with dereverberation techniques. In: Proceedings of REVERB'14 (2014)
37. Van Trees, H.L.: Detection, Estimation, and Modulation Theory. Part IV, Optimum Array Processing. Wiley-Interscience, New York (2002)

38. Virtanen, T.: Monaural sound source separation by nonnegative matrix factorization with temporal continuity and sparseness criteria. IEEE Trans. Audio Speech Lang. Process. **15**(3), 1066–1074 (2007)
39. Warsitz, E., Haeb-Umbach, R.: Blind acoustic beamforming based on generalized eigenvalue decomposition. IEEE Trans. Audio Speech Lang. Process. **15**(5), 1529–1539 (2007)
40. Weninger, F., Watanabe, S., Roux, J.L., Hershey, J.R., Tachioka, Y., Geiger, J., Schuller, B., Rigoll, G.: The MERL/MELCO/TUM system for the REVERB challenge using deep recurrent neural network feature enhancement. In: Proceedings of REVERB'14 (2014)
41. Weninger, F., Erdogan, H., Watanabe, S., Vincent, E., Le Roux, J., Hershey, J.R., Schuller, B.: Speech enhancement with LSTM recurrent neural networks and its application to noise-robust ASR. In: Proceedings of Latent Variable Analysis and Signal Separation, pp. 91–99. Springer, Berlin (2015)
42. Xu, Y., Du, J., Dai, L.R., Lee, C.H.: A regression approach to speech enhancement based on deep neural networks. IEEE/ACM Trans. Audio Speech Lang. Process. **23**(1), 7–19 (2015)
43. Yoshioka, T., Nakatani, T.: Generalization of multi-channel linear prediction methods for blind MIMO impulse response shortening. IEEE Trans. Audio Speech Lang. Process. **20**(10), 2707–2720 (2012)
44. Yoshioka, T., Tachibana, H., Nakatani, T., Miyoshi, M.: Adaptive dereverberation of speech signals with speaker-position change detection. In: 2009 IEEE International Conference on Acoustics, Speech and Signal Processing, pp. 3733–3736 (2009)
45. Yoshioka, T., Nakatani, T., Miyoshi, M.: Integrated speech enhancement method using noise suppression and dereverberation. IEEE Trans. Audio Speech Lang. Process. **17**(2), 231–246 (2009)
46. Yoshioka, T., Sehr, A., Delcroix, M., Kinoshita, K., Maas, R., Nakatani, T., Kellermann, W.: Making machines understand us in reverberant rooms: robustness against reverberation for automatic speech recognition. IEEE Signal Process. Mag. **29**(6), 114–126 (2012)
47. Yoshioka, T., Chen, X., Gales, M.J.F.: Impact of single-microphone dereverberation on DNN-based meeting transcription systems. In: Proceedings of ICASSP'14 (2014)
48. Yoshioka, T., Ito, N., Delcroix, M., Ogawa, A., Kinoshita, K., Fujimoto, M., Yu, C., Fabian, W.J., Espi, M., Higuchi, T., Araki, S., Nakatani, T.: The NTT CHiME-3 system: advances in speech enhancement and recognition for mobile multi-microphone devices. In: Proceedings of ASRU'15, pp. 436–443 (2015)

Chapter 3
Multichannel Spatial Clustering Using Model-Based Source Separation

Michael I. Mandel and Jon P. Barker

Abstract Recent automatic speech recognition results are quite good when the training data is matched to the test data, but much worse when they differ in some important regard, like the number and arrangement of microphones or the reverberation and noise conditions. Because these configurations are difficult to predict a priori and difficult to exhaustively train over, the use of unsupervised spatial-clustering methods is attractive. Such methods separate sources using differences in spatial characteristics, but do not need to fully model the spatial configuration of the acoustic scene. This chapter will discuss several approaches to unsupervised spatial clustering, with a focus on model-based expectation maximization source separation and localization (MESSL). It will discuss the basic two-microphone version of this model, which clusters spectrogram points based on the relative differences in phase and level between pairs of microphones, its generalization to more than two microphones, and its use to drive minimum variance distortionless response (MVDR) beamforming. These systems are evaluated for speech enhancement as well as automatic speech recognition, for which they are able to reduce word error rates by between 9.9 and 17.1% relative over a standard delay-and-sum beamformer in mismatched train–test conditions.

3.1 Introduction

While automatic speech recognition (ASR) systems using deep neural networks (DNNs) as acoustic models have recently provided remarkable improvements in recognition performance [23], their discriminative nature makes them prone to overfitting the conditions used to train them. For example, in the recent REVERB challenge [27], far-field multichannel ASR systems consistently performed more accurately in the simulated conditions that matched their training than in the real recording conditions that did not. In order to address generalization, DNN acoustic

M.I. Mandel (✉)
Brooklyn College (CUNY), 2900 Bedford Ave, Brooklyn, NY 11210, USA
e-mail: mim@sci.brooklyn.cuny.edu

J.P. Barker
University of Sheffield, Regent Court, 211 Portobello, Sheffield S1 4DP, UK

© Springer International Publishing AG 2017   51
S. Watanabe et al. (eds.), *New Era for Robust Speech Recognition*,
DOI 10.1007/978-3-319-64680-0_3

models should be trained on data that reflect the conditions in which the model will
be operating. One common approach to such training is the use of multicondition
data [32], in which the recognizer is trained on speech mixed with many different
kinds of noise, in the hope that the noise at test time will resemble one of the training
noises. Multicondition training provides benefits for both Gaussion-mixture-model
(GMM)- and DNN-based acoustic models [39]. DNN enhancement systems can
similarly be trained explicitly to generalize across source positions for a fixed
microphone array [24], or even to generalize across microphone spacings in linear
arrays [47].

While explicit generalization to new spatial configurations of microphones,
sources, and rooms is expensive to include in discriminative training procedures,
it can be naturally factored out of the data through beamforming. Traditional
beamforming assumes a known array geometry, which hinders generalization to new
conditions, but unsupervised localization-based clustering avoids this assumption.
Successful systems of this type have been introduced for two-microphone separation
[37, 45, 61], and in larger ad hoc microphone arrays for localization [33], calibration [18], and construction of time–frequency (T–F) masks [4]. It can be applied to
distributed microphone arrays [22], but this chapter describes three similar systems
for performing unsupervised spatial clustering and beamforming with compact
microphone arrays [29, 37, 49].

These spatial-clustering approaches are based on the idea of time–frequency
masking, a technique for suppressing unwanted sound sources in a mixture by
applying different attenuations to different T–F points in a spectrogram [58]. The
time–frequency masking technique is also discussed in Chap. 2. Clustering T–F
points results in groups of points with similar spatial characteristics. Arranging the
weight of each T–F point's membership in each group results in a T–F mask that
can be used to isolate an individual source. This mask-based approach is in contrast
to traditional approaches to blind source separation (BSS), which aim to model all
sources at all time–frequency points. A good overview of BSS methods for audio
is presented in [57], including various types of additional information that can be
utilized to aid more in the source separation process.

3.2 Multichannel Speech Signals

Let a signal of interest in the time domain be denoted by $x_1[n]$. If it is recorded with
$I-1$ other signals, $x_i[n]$, at J microphones, with $y_j[n]$ the signal at the jth microphone,
then

$$y_j[n] = \sum_{i=1}^{I} \sum_{l=1}^{L} h_{ij}[l] x_i[n-l] + u_j[n] \qquad (3.1)$$

$$= \sum_{i=1}^{I} (h_{ij} * x_i)[n] + u_j[n] \qquad (3.2)$$

where $h_{ij}[n]$ is the impulse response between source i and microphone j and $u_j[n]$ are noise terms. In the time–frequency domain, assuming that impulse responses are shorter than the Fourier transform analysis window, this relation becomes

$$Y_j(t,f) = \sum_{i=1}^{I} H_{ij}(f)X_i(t,f) + U_j(t,f), \qquad (3.3)$$

where $Y_j(t,f)$, $H_{ij}(f)$, $X_i(t,f)$, and $U_j(t,f)$ are all complex scalar values.

The impulse responses $H_{ij}(f)$ capture the communication channel between source and microphone, which includes all of the paths that the sound from the source can take to get to the microphone. This includes the direct sound path, paths coming from a distinct direction that have experienced one or more specular reflections off walls, and paths that come from no distinct direction after having bounced or scattered off many walls, resulting in diffuse reverberation. In general, this channel is time-varying, but many models, including the spatial-clustering methods described below, make the assumption that it is time-invariant, i.e., that the sources, microphones, and reflectors are fixed in space.

3.2.1 Binaural Cues Used by Human Listeners

Human listeners are able to attend to and understand the speech of a talker of interest even when it co-occurs with speech from several competing talkers. They are able to do this using certain cues from individual impulse responses, but mainly by utilizing differences between impulse responses of the same source at the two ears [38]. By comparing the two observed signals to each other, it is easier to differentiate between the effects of the original sound source and the channel on the observations. Performing this same task on single-channel observations requires a strong prior model of the sound source, the channel, or both.

The difference between two microphone channels that human listeners utilize comes from the ratio of two complex spectrograms,

$$C_{jj'}(t,f) = \frac{Y_j(t,f)}{Y_{j'}(t,f)}. \qquad (3.4)$$

The log magnitude of this quantity is known as the interaural level difference (ILD),

$$\alpha_{jj'}(t,f) = 20\log_{10}|C_{jj'}(t,f)| = 20\log_{10}|Y_j(t,f)| - 20\log_{10}|Y_{j'}(t,f)|. \quad (3.5)$$

When $Y_j(t,f)$ and $Y_{j'}(t,f)$ are dominated by the contribution of a single source, $X_{i^*}(t,f)$, where i^* is the index of that dominant source at (t,f),

$$\alpha_{jj'}(t,f) \approx 20\log_{10}|H_{i^*j}(t,f)||X_{i^*}(t,f)| - 20\log_{10}|H_{i^*j'}(t,f)||X_{i^*}(t,f)| \quad (3.6)$$

$$= 20\log_{10}|H_{i^*j}(t,f)| - 20\log_{10}|H_{i^*j'}(t,f)|. \quad (3.7)$$

Note that this quantity is entirely independent of the source signals $X_{i^*}(t,f)$ because it is common to both channels. The property of a single source dominating each individual time–frequency point is known as W-disjoint orthogonality [61], and has been observed in binaural recordings of anechoic speech signals. In addition, in single-channel source separation systems, the log magnitude of a mixture of signals is commonly approximated as the log magnitude of the most energetic signal [46], known as the log-max approximation. This approximation holds for multiple channels as well, as long as the same source is the loudest in all channels. Both of these approximations support the idea of a single source dominating each time–frequency point, which will be used heavily in the remainder of this chapter. Note that different points can be dominated by different sources, so that each source has a set of points at which it dominates all other sources, including the noise.

The phase of $C_{jj'}(t,f)$ is known as the interaural phase difference (IPD). Again assuming W-disjoint orthogonality,

$$\phi_{jj'}(t,f) = \angle C_{jj'}(t,f) = \angle Y_j(t,f) - \angle Y_{j'}(t,f) + 2\ell\pi \quad (3.8)$$

$$\approx \angle H_{i^*j}(t,f)X_{i^*}(t,f) - \angle H_{i^*j'}(t,f)X_{i^*}(t,f) + 2\ell\pi \quad (3.9)$$

$$= \angle H_{i^*j}(t,f) - \angle H_{i^*j'}(t,f) + 2\ell'\pi, \quad (3.10)$$

where ℓ and ℓ' are integers that capture the 2π ambiguity in phase measurements. In the case that $h_{j'}[n]$ is related to $h_j[n]$ by a pure delay of $\Delta_{\tau_{jj'}}$ samples, then through basic Fourier transform properties

$$h_{ij'}[n] = h_{ij}[n] * \delta[n - \Delta_{\tau_{jj'}}], \quad (3.11)$$

$$\therefore \phi_{jj'}(t,f) = \angle\exp(\iota 2\pi f \Delta_{\tau_{jj'}}/f_s) = 2\pi f \Delta_{\tau_{jj'}}/f_s + 2\pi\ell, \quad (3.12)$$

where $\iota = \sqrt{-1}$ is the imaginary unit, ℓ is an unknown integer, and f_s is the sampling rate. This pure delay, $\Delta_{\tau_{jj'}}$, is known as the interaural time difference (ITD) and models the nonzero time it takes for a sound to physically traverse a microphone array. As can be seen in (3.12), this ITD corresponds to an IPD that increases linearly with frequency.

For a pure delay between two microphones, $\Delta_{\tau_{jj'}} = f_s d_{jj'}/c$, where c is the speed of sound in air, approximately 340 m/s, and $d_{jj'}$ is the distance between microphones j and j'. When $\ell = 0$, it is trivial to map from an observed IPD to an unobserved ITD using $\Delta_{\tau_{jj'}} = \phi_{jj'}f_s/2\pi f$. This is only possible when $f < c/2d_{jj'}$. For frequencies close to or above this critical value, it is less straightforward to map from IPD to ITD because ℓ must be estimated in some way, a problem known as spatial aliasing [16, 43]. This estimation becomes more difficult in the presence of noise and reverberation. For human listeners, ITD can be measured directly in an anechoic chamber to establish the critical frequency. For a set of 45 subjects, [2] found the average maximum ITD to be 646 μs, corresponding to a distance of 22.0 cm, for which spatial aliasing begins at approximately 800 Hz. Thus this problem is clearly relevant to the processes of source localization and separation in humans.

Figure 3.1 shows example interaural parameters for a recording from the CHiME-3 dataset [6], which collected speech interactions with a six-microphone tablet device in several noisy environments. The top row shows log magnitude spectrograms for the individual channels, 0–3. The microphone for channel 0 was located very close to the talker's mouth, so has a much higher signal-to-noise ratio than the other channels. This can be seen from the lower noise level, after the entire mixture has been attenuated to maintain a consistent speech level. The microphone for channel 2 was facing away from the talker on the back side of the tablet device, leading to a much lower signal-to-noise ratio than the other channels. This can be seen in the lower speech levels relative to channels 1 and 3.

These differences in the levels of the speech and noise signals in each channel lead to characteristic ILDs between pairs of channels. For example, between channels 1 and 2, there is a clear ILD for the speech that distinguishes it from the noise. The IPD, on the other hand, does not discriminate between them. For channels 1 and 3, differences in time of arrival cause the IPD to be much more useful in discrimination between target and noise than the ILD is. Spatial-clustering systems take advantage of these differences to identify time–frequency points from the same source and group them together.

3.2.2 Parameters for More than Two Channels

For recordings with more than two channels, the spatial parameters can be generalized in two ways. The first is a direct generalization of the interaural computation, arranging the $C_{jj'}(t,f)$ terms from (3.4) into a matrix, $\mathbf{C}(t,f)$. In this matrix, the phase term is the difference between the phases in channels j and j' at time–frequency point (t,f), and the log magnitude is the difference between the log magnitudes at that point.

The linearly constrained minimum variance (LCMV) beamformer [8] uses a slightly different matrix to characterize the relationship between the microphone channels. In particular, it uses quantities such as the spatial covariance matrices

$$\Phi_{UU}(f) = \mathbb{E}_t[\mathbf{U}(t,f)\mathbf{U}^\dagger(t,f)] \qquad \Phi_{YY}(f) = \mathbb{E}_t[\mathbf{Y}(t,f)\mathbf{Y}^\dagger(t,f)], \qquad (3.13)$$

where $\mathbf{U}(t,f) = [U_1(t,f), \ldots, U_J(t,f)]^\top$ and $\mathbf{Y}(t,f) = [Y_1(t,f), \ldots, Y_J(t,f)]^\top$. In these computations, the phase of element j, j' at time–frequency point (t,f) is again the difference in phases between $Y_j(t,f)$ and $Y_{j'}(t,f)$, but the log magnitude is the

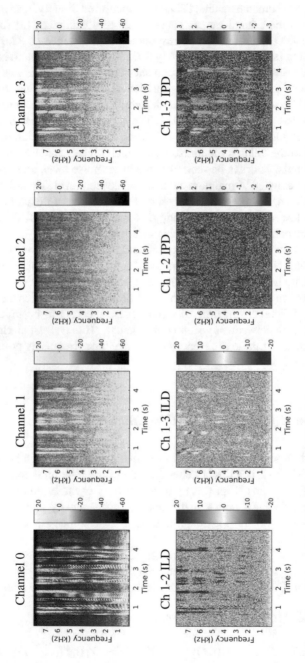

Fig. 3.1 Example multichannel recording from the CHiME-3 development dataset [6]. Real recording of talker F01 in pedestrian noise saying, "Excluding autos, sales increased zero point three percent." *Top row* shows individual channels. Channel 0 is the reference close-talking mic, channel 2 is rear-facing on the device, and channels 1 and 3 are front-facing. *Bottom row* shows interaural differences between channel 1 and channels 2 and 3

sum of the log magnitudes of $Y_j(t,f)$ and $Y_{j'}(t,f)$. If, however, it is assumed that there are no acoustic obstructions between any of the microphones in the array and the source is far away from the array (i.e., equidistant from all microphones), then

$$|Y_j(t,f)| = |Y_{j'}(t,f)| = 1, \tag{3.14}$$

$$\therefore \left| \frac{Y_j(t,f)}{Y_{j'}(t,f)} \right| = \left| Y_j(t,f) Y_{j'}^*(t,f) \right| = 1, \tag{3.15}$$

and the two sets of cues are equivalent. When the magnitudes are not unity, as in the CHiME-3 setup for channel 2, the perceptually motivated $C(t,f)$ matrix is not Hermitian symmetric because it contains the ratios of the channel observations, while the LCMV-related observation matrices and parameters are Hermitian symmetric.

3.3 Spatial-Clustering Approaches

Due to spatial aliasing, it is not possible to unambiguously map from noisy IPD estimates to the ITD at individual time–frequency points. The ambiguity can be resolved using spatial-clustering approaches. There are two main approaches: narrowband and wideband.

Narrowband spatial clustering (e.g., [49]) takes advantage of the fact that at almost all frequencies, sounds from two sources located at different positions will have different interaural parameters (phase and level differences). It typically does not make strong predictions about what those parameters will be, just that they will be different for different sources, thus permitting the separation of mixtures that include spatial aliasing. Once separation is performed in each individual frequency band, the sources identified in each band must be permuted to "match up" with one another in a second step.

In contrast, wideband models, such as [29] and [37], make stronger predictions about the connection between the interaural parameters at each frequency. In so doing, they are able to pool information across frequencies and avoid the potentially error-prone step of source alignment. The cost of this approach is that it must make certain assumptions about the form of the relationship across frequencies, and a failure of the observations to meet these assumptions could cause the failure of the separation process. In addition, care must be taken in developing this model so that it is robust to spatial aliasing.

All of these algorithms (i.e., [29, 37, 49]) have a similar structure. They first define each source by a set of frequency-dependent model parameters, $\Theta_i(f)$. They then alternate between two steps, an assignment of individual time–frequency points to source models using a soft or hard mask, $z_i(t,f)$, and an update of the source model parameters $\Theta_i(f)$ based on the observations at the points assigned to that source. This follows the expectation and maximization steps of the expectation maximization (EM) algorithm [15] in the case of soft masks or the two alternating steps of the k-means algorithm [34] in the case of hard masks.

3.3.1 Binwise Clustering and Alignment

The narrowband approach is exemplified by Sawada et al. [49]. Instead of clustering vectors of ILD and IPD measurements, the observed multichannel spectrogram signals are clustered directly. Building upon the notation of (3.3) to make a vectorial observation at each time–frequency point, let

$$\mathbf{H}_{i*}(f) = [H_{i*1}(f), \ldots, H_{i*J}(f)]^\top, \tag{3.16}$$

$$\mathbf{Y}(t,f) = [Y_1(t,f), \ldots, Y_J(t,f)]^\top \tag{3.17}$$

$$\approx [H_{i*1}(f)X_{i*}(t,f), \ldots, H_{i*J}(f)X_{i*}(t,f)]^\top = \mathbf{H}_{i*}(f)X_{i*}(t,f). \tag{3.18}$$

Independent processing is performed at each frequency, with no dependence on the frequency itself, so we will drop the f index from our notation in the remainder of this section. In order to separate the contribution of the target source from its spatial characteristics, the multichannel observations are magnitude-normalized at each time–frequency point:

$$\tilde{\mathbf{Y}}(t) = \frac{\mathbf{Y}(t)}{\|\mathbf{Y}(t)\|} = \frac{\mathbf{H}_{i*}}{\|\mathbf{H}_{i*}\|} \frac{X_{i*}(t)}{|X_{i*}(t)|}. \tag{3.19}$$

This normalization removes the magnitude of the source signal, but not its phase, $X_{i*}(t)/|X_{i*}(t)|$, which must be accounted for in the clustering procedure.

These observations are then clustered in a way that is similar to the line orientation separation technique (LOST) [40]. In this model, sources correspond to directions in a multidimensional complex space. These directions are represented by complex unit vectors, $\mathbf{a}_i$, and the distance between sources and observations is measured by projecting the observation onto the source direction. These distances are assumed to follow a circular complex Gaussian distribution with scalar variance σ_i^2,

$$p(\mathbf{Y}(t) \mid \mathbf{a}_i, \sigma_i) = \frac{1}{(\pi\sigma_i^2)^{J-1}} \exp\left(\frac{1}{\sigma_i^2} \|\tilde{\mathbf{Y}}(t) - (\mathbf{a}_i^H \tilde{\mathbf{Y}}(t))\mathbf{a}_i\|^2\right). \tag{3.20}$$

Note that, as required, this likelihood is invariant to a scalar phase applied to all channels of $\mathbf{Y}(t)$, because it is applied identically to $\tilde{\mathbf{Y}}(t)$ and $(\mathbf{a}_i^H \tilde{\mathbf{Y}}(t))\mathbf{a}_i$ and then removed by the magnitude computation. Thus this likelihood is invariant to the original phase of the source, $X_{i*}(t)/|X_{i*}(t)|$. For the same reason, it is also invariant to an additional scalar phase applied to the impulse response, so without loss of generality, we assume that $\angle H_{i*1} = 0$. Similarly, a scalar phase applied to $\mathbf{a}_i$ will cancel out, so without loss of generality, we assume that $\angle [\mathbf{a}_i]_1 = 0$.

If considered in relation to the interaural parameters described above, it can be seen that for two channels, $\angle H_{i*2} = \phi_{12}$, i.e., this parametrization is equivalent to the IPD. In addition,

$$\frac{[\mathbf{H}_{i*}]_1}{\|\mathbf{H}_{i*}\|} = \frac{H_{i*1}}{\sqrt{|H_{i*1}|^2 + |H_{i*2}|^2}} = \sqrt{\frac{|H_{i*1}|^2}{|H_{i*1}|^2 + |H_{i*2}|^2}} \tag{3.21}$$

$$= \sqrt{\frac{1}{1 + |H_{i*2}|^2/|H_{i*1}|^2}} = \sqrt{\frac{1}{1 + 10^{\alpha_{12}/10}}}, \tag{3.22}$$

showing that this parametrization is also equivalent to a pointwise transformation of the ILD. For every channel that is added beyond the second, this formulation adds an additional degree of freedom in phase and another in level for each source model. This linear growth is unlike the quadratic growth in degrees of freedom displayed by the spatial covariance matrix of each source in (3.13). This behavior implies that this parametrization can model point sources, but perhaps not diffuse sources, which require the full spatial covariance.

3.3.1.1 Cross-Frequency Source Alignment

In the narrowband clustering formulation, the frequency bands are processed independently and the source clusters can have a different arbitrary ordering in each band. Further processing is therefore required to assign clusters to sources in a consistent manner across frequency. Earlier techniques, e.g., [48], solved this same problem for frequency-domain independent component analysis (ICA) by correlating the extracted sources' magnitudes in adjacent frequency bands. For masking-based approaches, however, [49] found that performing the same sort of correlational alignment using the posterior probabilities from the masks, $z_i(t,f)$, yielded better alignments and thus better separation performance.

To perform this alignment exhaustively would take $O(J!F^2)$ time, where J is the number of sources and F the number of frequency bands. This is quite expensive, but [49] describes several heuristics for reducing the cost. The first is to perform a global exemplar-based clustering of the source posteriors across frequency. Instead of comparing all frequencies to each other, the posteriors at each frequency are compared to those of J exemplars, which reduces the cost to $O(J!FJ)$. While J and $J!$ are relatively small (typically $J < 5$), [49] suggests a greedy approach to the alignment calculation between a given pair of source sets, leading to an overall cost of $O(J^2FJ)$. This initial rough alignment is then refined using a fine-grained alignment based on comparing frequency bands that are either close to each other or harmonically related.

Overall, being a narrowband approach, this system is quite flexible in modeling impulse responses that vary a great deal across frequency. Such flexibility is not always required, however, and sacrifices some amount of noise robustness that comes from pooling information across frequencies. Instead, narrowband approaches tend to require longer temporal observations with stationary sources to achieve good separation performance. In addition, a good solution to the alignment problem requires careful tuning of the above heuristics. This can be difficult, for example, for wideband speech, where activity in frequencies up to 4 kHz containing sonorant phonemes is uncorrelated or even negatively correlated with activity in frequencies above 4 kHz containing obstruent phonemes, as can be seen in Fig. 3.1.

3.3.2   Fuzzy c-Means Clustering of Direction of Arrival

An example of a wideband approach is that of [29], which combines ideas from [10, 28, 53]. This approach performs clustering based solely on IPD converted to ITD using the Stepwise Phase dIfference REstoration (SPIRE) method [53], which resolves the spatial aliasing issue for certain kinds of arrays. SPIRE uses closely spaced pairs of microphones within a larger array to estimate the phase-wrapping terms in (3.12). Specifically, by sorting microphone pairs from the smallest separation to the largest, SPIRE identifies the unknown ℓ term in (3.12), expanded as

$$\phi_k = 2\pi f \Delta_{\tau_k}/f_s + 2\pi \ell_k = 2\pi d_k f/c + 2\pi \ell_k, \tag{3.23}$$

where k indexes the microphone pair and all terms indexed by k are specific to time–frequency point (t,f). For two different microphone pairs, most of these quantities are identical, allowing the correct ℓ_k to be identified recursively by

$$(\phi_{k-1} + 2\pi \ell_{k-1})\frac{d_k}{d_{k-1}} - \pi \le \phi_k + 2\pi \ell_k \le (\phi_{k-1} + 2\pi \ell_{k-1})\frac{d_k}{d_{k-1}} + \pi. \tag{3.24}$$

Once these IPD terms are identified for each time–frequency point, they can be directly converted to ITDs by

$$\Delta_{\tau_k}(t,f) = \frac{f_s}{2\pi f}(\phi_k(t,f) - 2\pi \ell_k(t,f)). \tag{3.25}$$

The scalar ITDs for the outermost microphone pair, $\Delta_{\tau_K}(t,f)$, are then clustered using an alternating approach similar to the GMM expectation maximization described above. Specifically, the parameters in this clustering are the direction for each source, denoted θ_i, and the soft cluster assignment for each time–frequency point, $z_i(t,f)$. These two quantities are updated using

$$z_i(t,f) = \frac{\|\Delta_{\tau_K}(t,f) - \theta_i\|^{2/(\gamma-1)}}{\sum_{i'} \|\Delta_{\tau_K}(t,f) - \theta_{i'}\|^{2/(\gamma-1)}}, \qquad \theta_i = \frac{\sum_{t,f} z_i^{\gamma}(t,f)\Delta_{\tau_K}(t,f)}{\sum_{t,f} z_i^{\gamma}(t,f)},$$

(3.26)

where $\gamma > 1$ is a user-defined parameter controlling the softness of the likelihoods. Aside from this γ parameter, which effectively scales the log-likelihood, these updates are equivalent to GMM EM with a spherical unit variance.

Because it is wideband, this approach is able to pool information across frequency and requires fewer temporal observations than narrowband approaches. The use of the microphone pair with the widest spacing for localization provides the most precise estimates. In order to do so, however, it makes the assumption that the ITD is a pure delay between microphones, which appears in the form of (3.25). This is generally not the case in reverberant environments when early specular reflections disrupt this relationship. It also implies that sounds come from point sources and have no diffuse component, which is also unlikely in reverberant environments.

3.3.3 Binaural Model-Based EM Source Separation and Localization (MESSL)

The binaural model-based EM source separation and localization (MESSL) algorithm [37] explicitly models IPD and ILD observations using a Gaussian mixture model. In order to avoid spatial aliasing, MESSL models the ITD as a discrete random variable, and the IPD as a mixture over these ITDs, computing the source assignment variables as $z_i(t,f) = \sum_{\tau} z_{i\tau}(t,f)$. Intuitively, while an IPD does not correspond to a unique ITD in the presence of spatial aliasing, every ITD does correspond to a unique IPD, and so, by comparing the likelihoods of a set of ITDs, the most likely explanation for a set of observed IPDs can be found. The probability distribution for the Gaussian observations for a source i and discrete delay Δ_{τ} is

$$p(\phi(t,f), \alpha(t,f) \mid i, \tau, \Theta)$$
$$= p(\phi(t,f) \mid \tau, \xi_{i\tau}(f), \sigma_{i\tau}(f)) \cdot p(\alpha(t,f) \mid \mu_i(f), \eta_i(f)).$$

(3.27)

The distributions of individual features are given by

$$p(\phi(t,f) \mid \tau, \xi_{i\tau}(f), \sigma_{i\tau}(f)) = \mathcal{N}\left(\hat{\phi}(t,f; \tau, \xi_{i\tau}(f)) \mid 0, \sigma^2(f)\right),$$

(3.28)

$$\text{where } \hat{\phi}(t,f; \tau, \xi_{i\tau}(f)) = \angle \exp\left(\iota(\phi(t,f) - 2\pi f \Delta_{\tau}/f_s - \xi(f))\right),$$

(3.29)

$$p(\alpha(t,f) \mid \mu_i(f), \eta_i(f)) = \mathcal{N}\left(\alpha(t,f) \mid \mu_i(f), \eta_i^2(f)\right).$$

(3.30)

The phase residual, $\hat{\phi}(t,f;\tau,\xi_{i\tau}(f))$, computes the distance between the observed phase difference, $\phi(t,f)$, and the phase difference that should be observed from source i at frequency f, namely $2\pi f \Delta_\tau / f_s + \xi_{i\tau}(f)$. The first term in this expression is the phase difference predicted at frequency f by the ITD model with delay Δ_τ, and the second term is a frequency-dependent phase offset parameter, which permits variations from the pure delay model caused by early echoes. Furthermore, this difference is constrained to lie in the interval $(-\pi, \pi]$. Note that Δ_τ comes from a discrete grid of delays on which the above expressions must be evaluated, so that their likelihoods may be compared to one another. This step is computationally expensive and could be avoided by using a more sophisticated optimization scheme to find the most likely ITD.

These likelihoods are then used in the expectation and maximization steps of the EM algorithm. In the expectation step, the assignment of time–frequency points to sources is computed by

$$z_{i\tau}(t,f) = \frac{p(\phi(t,f),\alpha(t,f) \mid i, \tau, \Theta)p(i,\tau)}{\sum_{i'\tau'} p(\phi(t,f),\alpha(t,f) \mid i', \tau'\Theta)p(i',\tau')}. \tag{3.31}$$

In the maximization step, the model parameters are all updated by taking weighted sums of sufficient statistics of the observations using the assignments as weights. For details, see [37].

As a wideband method, MESSL can pool localization information across frequency, requiring temporally shorter observations than narrowband methods. Its statistical formulation permits the incorporation of additional parameters, like the IPD means, $\xi_{i\tau}(f)$, that can model early echoes in addition to the direct-path pure delay. Because the model is so flexible, however, it requires careful initialization to avoid local minima that do effectively separate the sources. It also permits the use of a prior on the ILD means given the ITDs.

This flexibility has facilitated several extensions of MESSL. Weiss et al. [59] combined the spatial separation of MESSL with a probabilistic source model. Instead of estimating a single maximum likelihood setting of parameters, [14] used variational Bayesian inference to estimate posterior distributions over the MESSL parameters. Instead of a grid of ITDs, [54] used random sampling to extract the best IPD-ILD parameters for a multichannel configuration.

3.3.4 Multichannel MESSL

Multichannel MESSL [5] models every pair of microphones separately using the binaural model described in Sect. 3.3.3. These models are coordinated through a global T–F mask for each source. For a spatial-clustering system to be as flexible as possible, it should not require calibration information for the microphone array. This flexibility will allow it to be used in applications from ad hoc microphone arrays to databases of user-generated content that lack specifications of the hardware that produced the recordings: source separation that is blind to the microphone array geometry. Without calibration, model parameters are difficult to translate between microphone pairs, but T–F masks are much more consistent across pairs and can be used to coordinate sources and models. This is the strategy adopted by multichannel MESSL, which maximizes the following total log-likelihood for J microphones:

$$\mathcal{L}(\Theta) = \frac{2}{J} \sum_{j<j'=1}^{J} \mathcal{L}(\Theta_{jj'}) \tag{3.32}$$

$$= \frac{2}{J} \sum_{j<j'=1}^{J} \sum_{tf} \log \sum_{i\tau} \Big[p(z_{i\tau}(t,f) \mid \Theta_{jj'}) \cdot p(\phi_{jj'}(t,f), \alpha_{jj'}(t,f) \mid z_{i\tau}(t,f), \Theta_{jj'}) \Big].$$

Averaging over all pairs in this way assumes that all microphone pairs are independent of one another, whereas in reality only $J-1$ are. This false assumption leads to an overconfidence in the likelihoods that is compensated by the $2/J$ term. This factor has much the same effect as the γ coefficient in (3.26) for the fuzzy c-means clustering approach. Preliminary experiments showed that using all pairs of microphones with this correction factor led to higher-quality separations than designating a single microphone as a reference and using $J-1$ pairs. The E and M steps for the model then proceed almost as in the two-channel algorithm. In the E step, the likelihood of the observations for each microphone pair is calculated under each source model. These likelihoods are then multiplied across microphone pairs and normalized across sources to give the final global posterior masks. In the M step, these global masks are used to reestimate the parameters of each pairwise model.

Initializing the multichannel model requires initializing the pairwise models and coordinating the source models across microphone pairs. We explored two different initializations. The first used the PHAT-histogram approach [1] to find the dominant peaks in cross-correlations between pairs of channels, followed by several iterations of binaural MESSL to estimate a mask for each source. These masks were then used to align the sources across microphone pairs. This approach has the advantage of being self-contained. The second initialization used a T–F mask derived from level differences between a beamformer output and a reference microphone. In the experiments below on CHiME-3 data, this was between the output of BeamformIt [3] and channel 2, the microphone facing away from the talker. The initial mask is then constructed from the 30% of points where the beamformer output is maximally greater in energy than the reference. This initialization has the advantage of automatically aligning the source models across microphone pairs, but can fail if the baseline beamformer fails in localization or separation.

Multichannel MESSL modeling all pairs of microphones has enough parameters to model both point sources and diffuse sources. The models can be arranged into a $J \times J$ matrix of sorts to reflect the observations $\mathbf{C}(t,f)$, where each pair of microphones corresponds to an entry in the matrix. This parametrization comes at the cost of a running time that is quadratic in the number of microphones. Preliminary experiments to reduce this computational complexity showed that subsampling the microphone pairs could trade off separation performance for complexity. For the six microphone recordings in CHiME-3, this was not necessary, so we will leave this investigation for future work.

3.4 Mask-Smoothing Approaches

One widely recognized problem that arises in mask-based separation is musical noise due to isolated false positive T–F points in the mask. Several approaches have attempted to alleviate this problem by applying a separate smoothing process after mask estimation [13, 19, 35, 56]. This section discusses the incorporation of these smoothing procedures into the spatial clustering process itself.

3.4.1 Fuzzy Clustering with Context Information

[29] introduced a mask-smoothing approach based on a heuristic modification of the source assignments $z_i(t,f)$ after each expectation step, following an approach first applied in image segmentation [12]. In particular, they defined

$$\bar{z}_i(t,f) = \frac{1}{|N(t,f)|} \sum_{t',f' \in N(t,f)} z_i(t',f'), \qquad (3.33)$$

where $N(t,f)$ is a set of time–frequency indices for points that neighbor point (t,f). In [30], N is a rectangular neighborhood of 15 frequency bands and 9 time frames centered on the target point, equivalent to a rectangle of size 118 Hz by 90 ms. This averaged mask is applied in the expression for the update of the source memberships,

$$\tilde{z}_i(t,f) = \frac{z_i^\gamma(t,f)\bar{z}_i^\beta(t,f)}{\sum_{i'} z_{i'}^\gamma(t,f)\bar{z}_{i'}^\beta(t,f)}, \qquad (3.34)$$

where β is a parameter controlling the relative contribution of the smoothed masks. [29] ran an initial iteration of the separation process with $\beta = 0$ to provide an unbiased estimate of θ_i, followed by five iterations with $\beta = 10$ to provide robustness to noise and reverberation. After iteration, a median filter was run over the masks to further reduce spurious classifications and musical noise.

3.4.2 MESSL in a Markov Random Field

With the same motivation, [36] proposed embedding the MESSL algorithm into a grid-shaped pairwise Markov random field (MRF) to simultaneously estimate model parameters and smooth T–F masks. This MRF penalizes the assignment of neighboring T–F points to different sources, smoothing the masks and reducing musical noise. The combined model is referred to as MESSL-MRF. In image segmentation applications, these models have been shown to be effective in combining evidence across neighboring pixels, e.g., [7]. While exact inference in these models is intractable, a number of approximation methods have been shown to be effective, including graph-cuts and loopy belief propagation (LBP) [52]. In addition, learning the parameters of an MRF model is also typically intractable, but it has been shown that approximate learning using expectation maximization can provide a reasonable approximation in practice for segmenting noisy images [20, 62]. MRFs have been used in several speech separation systems recently for both single- [25, 31] and multichannel approaches [26].

3.4.2.1 Pairwise Markov Random Fields

An MRF is an undirected graphical model, representing the joint probability of several random variables as a product of potential functions over subsets of those variables [7]. Depending on the structure of the graph, certain quantities can be estimated much more efficiently because of this factorization. This section focuses on pairwise MRFs, in which only pairwise interactions between variables are nonzero and thus only pairwise potential functions are necessary. In such models, the joint distribution of random variables $z_1, z_2, \ldots, z_N$ can be written as

$$p(z_1, z_2, \ldots, z_N) = \frac{1}{Z} \prod_{kk'} \psi_{kk'}(z_k, z_{k'}) \prod_k \psi_k(z_k), \qquad (3.35)$$

where $\psi_k(z_k)$ is the potential function of variable z_k by itself, perhaps induced by a corresponding observation, and $\psi_{kk'}(z_k, z_{k'})$ is the pairwise potential function between z_k and $z_{k'}$, representing compatibilities between their various configurations. Using the sum–product variant of the belief propagation algorithm [41], it is possible to estimate the distribution of each individual variable when all of the others are marginalized away. In the case of tree-structured graphs, belief propagation can compute these quantities exactly. In the case of graphs with loops, it can only approximate these quantities, but it has been shown that such approximations perform well in practice [60].

3.4.2.2   MESSL-MRF

We propose smoothing MESSL masks by using the MESSL likelihood as the local potential in a grid-shaped pairwise MRF. In the context of such a model, z_k is the random variable representing the source number responsible for the majority of the energy at time–frequency point k.[1] If there are I sound sources, then z_k is a discrete I-dimensional multinomial random variable. In the experiments below, I was 2. The grid-shaped MRF then has potentials between every T–F point and its four direct neighbors in time and frequency. Thus the potential function $\psi_{kk'}(z_k, z_{k'})$ represents the compatibility between source z_k dominating T–F point k and source $z_{k'}$ dominating T–F point k'. We set the compatibility potentials, $\psi_{kk'}(z_k, z_{k'})$, to

$$\psi_{kk'}(z_k, z_{k'}) = \exp(-\beta \delta(z_k, z_{k'})), \tag{3.36}$$

where $\delta(z_k, z_{k'})$ is the discrete Dirac delta function, which is 1 when $z_k = z_{k'}$ and 0 otherwise, and β is a parameter that we tuned on a separate validation dataset. While simple, this potential is standard in MRF approaches to image segmentation.

More sophisticated compatibility potentials are possible and can be learned from training data. In particular, at low frequencies, ground truth masks tend to be more correlated across time because of the presence of strong lower harmonics. At high frequencies, they are more correlated across frequency because of wideband bursts and frication noise. Thus a frequency-dependent compatibility potential could be useful, but we leave this approach for future work.

In MESSL-MRF, the local potential is defined as

$$\psi_{tf}(z_{tf}) = \sum_{\tau} z_{i\tau}(t, f), \tag{3.37}$$

where we have changed the notation back from indexing hidden variables by k to t, f, and $z_{i\tau}(t, f)$ is defined in (3.31). We find the maximum likelihood parameters Θ from the test data using the EM algorithm [15, 20, 62]. Although learning in this MRF is intractable, it can be approximated by inserting the MRF belief propagation step between the E and M steps of a standard EM algorithm. In MESSL, it thus becomes a mask-smoothing step. MESSL's E step computes $z_{i\tau}(t, f)$, which defines the local potential $\psi_{tf}(z_{tf})$ in (3.37). From these, LBP is run until convergence to compute the soft masks, $b_{tf}(z_{tf})$, which are used to compute updated posteriors

$$\bar{z}_{i\tau}(t, f) = z_{i\tau}(t, f) \frac{b_{tf}(z_{tf})}{\sum_{\tau'} z_{i\tau'}(t, f)}. \tag{3.38}$$

And these are used in the standard MESSL M-step updates.

---

[1]For the purposes of the MESSL-MRF discussion, the indices k and k' are a shorthand for the T–F coordinates (t_k, f_k) and $(t_{k'}, f_{k'})$.

This approach has a similar effect to the context incorporation described in Sect. 3.4.1, namely encouraging neighboring points to belong to the same source. The probabilistic formulation of MESSL-MRF allows it to easily incorporate prior information about the relationships between neighboring points. It permits the substitution of the solution algorithm from loopy belief propagation [52] if desired. It also makes clear the approximations being made and their effect on the solution. The cost of the approach, however, is that to maintain these desirable properties, it must not utilize too large a neighborhood in its smoothing. Large neighborhoods in grid-shaped graphical models reduce the benefits of factorizing the joint distribution and lead to longer convergence times, if convergence is achieved at all.

3.5 Driving Beamforming from Spatial Clustering

Beamforming is the process of combining signals recorded from a microphone array into a single estimate of a target signal. This estimate is typically driven by an optimality criterion. One popular criterion for fixed (nonadaptive) filter-and-sum beamforming is that of minimum variance distortionless response (MVDR) [8], which aims to minimize the output power of the beamformer while preserving signals from a target "look" direction. For signals recorded as in (3.3), a filter-and-sum beamformer can be represented as a frequency-dependent vector, $\mathbf{w}(f)$, and the signal estimated by the beamformer is

$$\hat{X}_1(t,f) = \mathbf{w}^H(f)\mathbf{Y}(t,f). \tag{3.39}$$

For a steering vector $\mathbf{d}(f)$, which should have unity gain, the MVDR beamformer is

$$\mathbf{w}^*(f) = \min_{\mathbf{w}} E\left\{|\mathbf{w}^H \mathbf{X}(t,f)|^2\right\} \text{ s.t.} \mathbf{w}^H \mathbf{d}(f) = 1. \tag{3.40}$$

Recently, [50] showed that this can be solved without the use of an explicit steering vector by

$$\mathbf{w}^*(f) = \frac{\Phi_{UU}^{-1}(f)\Phi_{HH}(f)e_{\text{ref}}}{\text{tr}\left(\Phi_{UU}^{-1}(f)\Phi_{HH}(f)\right)} = \frac{(\Phi_{UU}^{-1}(f)\Phi_{YY}(f) - I)e_{\text{ref}}}{\text{tr}\left(\Phi_{UU}^{-1}(f)\Phi_{YY}(f)\right) - J}, \tag{3.41}$$

where I is the $J \times J$ identity matrix, and e_{ref} is a vector of zeros with a single one selecting a reference microphone. This method allows the MVDR beamformer to be estimated without the use of an explicit steering vector, but still requires the estimation of $\Phi_{UU}(f)$, the noise spatial covariance, and either $\Phi_{YY}(f)$, the mixture spatial covariance, or $\Phi_{XX}(f) \propto \Phi_{HH}(f)$, the target source spatial covariance (note that the constant of proportionality divides out in (3.41)). In our experiments, the denominator of these expressions was sometimes close to zero or even negative for

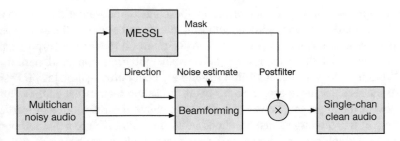

Fig. 3.2 Three ways that spatial-clustering outputs can drive minimum variance distortionless response beamforming: IPD parameters for look direction and masks for noise estimation and/or nonlinear postfiltering

a small set of frequencies, causing a large gain in the output at those frequencies and poor overall sound quality. We overcame this issue by enforcing that it be at least 1.

In the experiments discussed below, we explore the use of spatial clustering, and specifically MESSL, in driving MVDR beamforming in several ways, as illustrated in Fig. 3.2. Masks from spatial clustering can be used to estimate the noise spatial covariance $\Phi_{UU}(f)$, model parameters from spatial clustering can be used to compute a steering vector $\mathbf{d}(f)$, and the masks can also be used as a nonlinear postfilter applied to the output of the beamformer. This use of spatial clustering to drive MVDR beamforming was suggested by Cermak et al. [10], [11], and Kühne et al. [29].

The complement of the mask for a single source, $z_i(t,f)$, can be used as a frequency-dependent noise activity detector to estimate $\Phi_{UU}(f)$ as

$$\Phi_{UU}(f) \approx \frac{\sum_{t=1}^{T} (1 - z_i(t,f))\, \mathbf{X}(t,f)\mathbf{X}^H(t,f)}{\sum_{t=1}^{T} (1 - z_i(t,f))}. \tag{3.42}$$

Alternatively, [10] models and separates $I - 1$ noise sources individually, and computes $\Phi_{UU}(f)$ from the sum of these noise sources. To avoid speech damage, observations can be excluded from this sum from frames in which more than 40% of frequencies are predicted to be speech. To ensure that $\Phi_{UU}(f)$ is invertible, a certain number of frames from the beginning and end of the signal can be included in estimating it. We have found that the first M frames and the last $2M$ frames of an utterance work well for this empirically.

The steering vector can also be computed from the output of spatial clustering. From the estimated ITDs, assuming a pure delay,

$$\mathbf{d}^{(i)}(f) = [1, \exp(-\imath 2\pi f \Delta_{\tau_{12}^{(i)}}(f)/f_s), \ldots, \exp(-\imath 2\pi f \Delta_{\tau_{1J}^{(i)}}(f)/f_s)]. \tag{3.43}$$

Another possibility [11] is to find the $\mathbf{d}^{(i)}(f)$ that produces the best resynthesis of the observation from an estimate of the target signal, captured by the cost function

$$\mathcal{L}(\mathbf{d}) = \mathbb{E}_t \left[(\mathbf{x}(t,f) - \mathbf{d}(f)z_i(t,f)y_1(t,f))^2 \right], \tag{3.44}$$

which is solved by

$$\mathbf{d}^{(i)}(f) = \frac{\sum_t \mathbf{x}(t,f)z_i(t,f)y_1^*(t,f)}{\sum_t |z_i(t,f)y_1(t,f)|^2}. \tag{3.45}$$

And, finally, it is possible to use the IPD estimates from multichannel MESSL to directly compute a full-rank Φ_{HH} for use in (3.41). While ILD is not useful for beamforming, as shown in (3.15), it is close to 1 for arrays without acoustic obstructions between microphones. Using just the IPD,

$$\Phi_{H_j H_{j'}}^{(i)}(f) = \frac{\phi_{ijj'f}}{|\phi_{ijj'f}|} \text{ for } \phi_{ijj'f} = \mathbb{E}_\tau \left[\exp(-\iota 2\pi f(\Delta_\tau + \xi_{jj'\tau}^{(i)}(f))/f_s) \right], \tag{3.46}$$

where $\Delta_\tau + \xi_{jj'\tau}^{(i)}(f)$ is the fine-grained mean of the IPD Gaussian between microphones j and j' for source i at the ITD indexed by τ. This approach takes advantage of MESSL's frequency-varying IPD estimates and does not assume a pure delay between microphones, as the first steering-vector formulation does.

Finally, masks estimated through spatial filtering can be used as nonlinear postfilters for the output of the MVDR beamformer. Suppressing points where $z_i(t,f) = 0$ to silence leads to musical noise, which can be avoided by suppressing them by some maximum amount. We found that using a maximum suppression of $-9\,\mathrm{dB} = 0.355$ gave good noise suppression without causing noticeable musical noise.

3.6 Automatic Speech Recognition Experiments

This section describes experiments that examine the performance of MVDR beamforming driven by spatial clustering as a means of adapting far-field DNN-based automatic speech recognition to mismatched conditions. In particular, it uses the baseline recognizer from the AMI Meeting Corpus [9, 44], and tests it on the CHiME-3 corpus. These two conditions are mismatched in many ways, including signal-to-noise ratio, amount of reverberation, the distance to the microphone array, and the number and arrangement of the microphones. These experiments show that spatial clustering can provide significant recognition performance gains towards overcoming mismatched far-field conditions.

The recognizer was trained on the AMI Meeting Corpus, which contains speech recorded on an 8-microphone circular array of diameter 10 cm. We used the multiple-distant-mic (MDM) condition processed by the BeamformIt tool [3], which performs delay-and-sum beamforming using time-varying source localization. We used the AMI Full-ASR partition training set (about 78 h of speech) proposed in [51] and the corresponding Kaldi recipe with the provided automatic segmentations (version 1.6.1). The final acoustic model was a fully connected DNN that takes as input 40-dimensional log mel filterbank features with first and second time derivatives [55]. This DNN was trained on labels aligned by a GMM–hidden-Markov-model (HMM) model trained on mel-frequency cepstral coefficient (MFCC) features followed by linear discriminant analysis [21] and semi-tied covariance transforms [17], and discriminatively trained using the boosted maximum mutual information [42] criterion. The number of tied states was roughly 4000.

This recognizer was tested on the live-data portion of the CHiME-3 [6] dataset, which records speech input to a simulated tablet device in noisy environments. It used a 6-microphone rectangular array built around the edge of the tablet, to which a talker whose mouth was 30–50 cm away read sentences from the Wall Street Journal corpus (WSJ0). The recordings were made in four different noisy environments with an estimated signal-to-noise ratio averaging around 0 dB. The acoustic model described above was used with the default CHiME-3 language model. Thus the training and test sets differed significantly in the number of microphones, array geometry, amount of reverberation, microphone array distance, amount and type of noise, speaking style, and vocabulary. MESSL was used only on the development and test sets, not in training. The variant of multichannel MESSL used in the experiments had fully frequency-dependent parameters and smoothed its masks using MESSL-MRF.

For estimating the noise spatial covariance matrices $\Phi_{UU}(f)$, we compared using MESSL's masks to using the 400–800 ms of audio preceding the speech of each utterance, assumed to be noise only, which is the approach taken by the baseline CHiME-3 system (see [6]). For estimating the steering vector, we compared an estimate of $\Phi_{HH}(f)$ based on MESSL's IPD parameters to a derivation from (3.41) using Φ_{YY}. For a nonlinear postfilter, we compared the use of MESSL's masks to apply a gain to each T–F point of the beamformed signal to the use of the unmodified output of the beamformer.

3.6.1   Results

Table 3.1 shows the results of these experiments. The best system on the development set is shown in row 15 and used the MESSL noise estimate, the MESSL postfilter, cross-correlation initialization for MESSL, and the mixture spatial covari-

Table 3.1 Word error rates for recognizer trained on AMI data and tested on enhanced CHiME-3 real recordings

	Noise est	Postfilt	MESSL init	Look dir	WER (%) Dev	Test
1	Prev	None	–	Mix	29.2	48.6
2	Prev	None	BeamformIt	MESSL	26.1	39.7
3	Prev	None	Xcorr	MESSL	24.6	40.2
4	Prev	MESSL	BeamformIt	MESSL	22.8	35.4
5	Prev	MESSL	BeamformIt	Mix	23.2	39.5
6	Prev	MESSL	Xcorr	MESSL	20.8	35.6
7	Prev	MESSL	Xcorr	Mix	22.5	40.1
8	MESSL	None	BeamformIt	MESSL	26.7	43.9
9	MESSL	None	BeamformIt	Mix	22.4	32.4
10	MESSL	None	Xcorr	MESSL	23.1	41.3
11	MESSL	None	Xcorr	Mix	22.1	34.8
12	MESSL	MESSL	BeamformIt	MESSL	23.9	39.5
13	MESSL	MESSL	BeamformIt	Mix	20.8	**30.0**
14	MESSL	MESSL	Xcorr	MESSL	20.4	36.1
15	MESSL	MESSL	Xcorr	Mix	**19.7**	32.6
16	–	None	–	–	22.7	36.2
17	–	MESSL	–	–	20.6	31.0

Key: Noise estimates from the previous 400–800 ms (Prev) or MESSL mask. Postfilter not used (None) or MESSL mask. MESSL initialized from BeamformIt or cross-correlation (Xcorr). Look direction from mixture (Mix) or from MESSL IPD. Bottom: BeamformIt baselines. Rows that are discussed in the text are shaded with $N = 27,119$, system 15 is significantly better than system 14 on the dev set ($p < 0.05$) and system 13 is significantly better than system 17 on the test set ($p < 0.01$) according to a one-sided binomial test

ance for (3.41). The columns of the table are ordered by the increase in word error rate (WER) on the development set caused by changing one of these parameters from this best setting. The rows of the table are ordered by the settings in each column. The parameter with the largest effect on this system is the noise estimate. Using the preceding 800 ms instead of the MESSL mask to estimate the noise results in a 2.75% absolute (14.0% relative) increase in the development set WER (row 7 vs. 15). The second largest effect comes from the postfilter. Removing the postfilter results in a 2.4% absolute (12.2% relative) increase in WER (row 11 vs. 15). The last two parameters have smaller effects on the development set. Initializing MESSL from BeamformIt instead of using cross-correlations results in a 1.1% absolute (5.4% relative) increase in WER (row 13 vs. 15). Using the look direction from the MESSL IPD instead of the mixture results in a 0.7% absolute (3.7% relative) increase in WER (row 14 vs. 15).

Baseline systems using BeamformIt are shown in the bottom two rows. The MESSL postfilter decreases WER by 2.1% absolute (9.3% relative) (row 16 vs. 17). Without a postfilter, two MESSL-MVDR systems (rows 9 and 11) achieve lower development and test WERs than the corresponding baseline (row 16), showing that

MESSL can be used to effectively drive beamforming. With the postfilter, the same two systems (rows 13 and 15) perform comparably to the baseline (row 17). The MESSL-MVDR system that performs best on the development set (row 15) reduces the WER on the test set by 3.6% absolute (9.9% relative) compared to the plain BeamformIt baseline. Consistent differences in performance have been seen on test and development sets for CHiME-3 [6], which might suggest looking directly for the best system on the test set, in which case, the best MESSL-MVDR system (row 13) reduces the WER by 6.2% absolute (17.1% relative).

3.6.2 Example Separations

Figure 3.3 shows example outputs of several of the systems described above for the input mixture shown in Fig. 3.1. The leftmost column shows a noisy input channel (channel 1) and the close microphone recording for reference. The remaining plots show system outputs. The top row of the figure shows the effect of the postfilter mask, with system 11 using no postfilter, system 15 using a postfilter with 9 dB maximum suppression, and the unnumbered system in the rightmost plot showing 40 dB maximum suppression. System 15 gave the best performance on the development set, and it can be seen that using too little postfilter suppression leaves too much noise in the output, while using too much suppression leads to artifacts such as musical noise. These artifacts, including the lack of noise suppression at the lowest frequencies, are due to the postfilter being purely based on spatial characteristics of the recordings. The incorporation of a speech-aware model into the mask estimation procedure could mitigate these artifacts, permitting a greater maximum suppression to be used with the postfilter.

The bottom row of plots in Fig. 3.3 shows the output of systems that differ in a single component from the best system (number 15), paralleling the discussion in Sect. 3.6.1. System 14 is the same as system 15, except that it uses MESSL's estimate of the look direction, based on its IPD model. It can be seen that in this separation, MESSL's look direction estimate leads to a residual noise that is more uniform across frequency in regions where the speech is inactive, although with slightly more noise at frequencies between 500 and 1000 Hz. System 13 uses BeamformIt to initialize MESSL instead of cross-correlations between channels. Its performance looks quite similar to that of system 15 for this separation. System 7 uses the 400–800 ms of noise preceding the speech to estimate the noise parameters. Its output on this example contains slightly more residual noise than system 15s, for example around 1200 Hz.

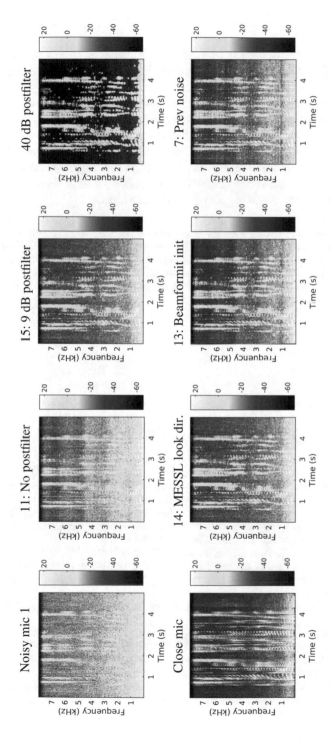

Fig. 3.3 Enhanced versions of the recording shown in Fig. 3.1 from several systems described in Table 3.1. The numbers above each plot identify the row in Table 3.1 corresponding to each system

3.7 Conclusion

This chapter has described the use of multichannel spatial clustering to drive minimum variance distortionless response beamforming. By clustering time–frequency points based on their spatial characteristics, these systems are able to generalize to quite different recording conditions. Experiments recognizing data from CHiME-3 with a recognizer trained on AMI show that there are several ways of utilizing the outputs of spatial clustering with MVDR beamforming, including incorporating its mask into the noise spatial covariance estimate, using the mask as a post–filter, or using estimated interaural phase differences to form the target spatial covariance matrix. In the future, generalizing the speech models of [59] from binaural to multichannel recordings could improve performance further.

Acknowledgements The work reported here was carried out during the 2015 Jelinek Memorial Summer Workshop on Speech and Language Technologies at the University of Washington, Seattle, and was supported by Johns Hopkins University via NSF Grant No. IIS 1005411, and gifts from Google, Microsoft Research, Amazon, Mitsubishi Electric, and MERL. It is also based upon work supported by the NSF under Grant No. IIS 1409431. Any opinions, findings, and conclusions or recommendations expressed in this material are those of the author(s) and do not necessarily reflect the views of the National Science Foundation.

References

1. Aarabi, P.: Self-localizing dynamic microphone arrays. IEEE Trans. Syst. Man Cybern. C **32**(4), 474–484 (2002)
2. Algazi, V.R., Duda, R.O., Thompson, D.M., Avendano, C.: The CIPIC HRTF database. In: Proceedings of WASPAA, pp. 99–102 (2001)
3. Anguera, X., Wooters, C., Hernando, J.: Acoustic beamforming for speaker diarization of meetings. IEEE Trans. Audio Speech Language Process. **15**(7), 2011–2022 (2007)
4. Araki, S., Sawada, H., Mukai, R., Makino, S.: Underdetermined blind sparse source separation for arbitrarily arranged multiple sensors. Signal Process. **87**, 1833–1847 (2007)
5. Bagchi, D., Mandel, M.I., Wang, Z., He, Y., Plummer, A., Fosler-Lussier, E.: Combining spectral feature mapping and multi-channel model-based source separation for noise-robust automatic speech recognition. In: Proceedings of ASRU (2015)
6. Barker, J., Marxer, R., Vincent, E., Watanabe, S.: The third "CHiME" speech separation and recognition challenge: dataset, task and baselines. In: Proceedings of ASRU (2015)
7. Besag, J.: On the statistical analysis of dirty pictures (with discussion). J. R. Stat. Soc. B **48**(3), 259–302 (1986)
8. Capon, J.: High-resolution frequency-wavenumber spectrum analysis. Proc. IEEE **57**(8), 1408–1418 (1969)
9. Carletta, J.: Unleashing the killer corpus: experiences in creating the multi-everything AMI meeting corpus. Lang. Resour. Eval. **41**(2), 181–190 (2007)
10. Cermak, J., Araki, S., Sawada, H., Makino, S.: Blind speech separation by combining beamformers and a time frequency binary mask. In: Proceedings of IWAENC, Paris (2006)
11. Cermak, J., Araki, S., Sawada, H., Makino, S.: Blind source separation based on a beamformer array and time frequency binary masking. In: Proceedings of ICASSP, vol. 1, pp. 145–148. IEEE, New York (2007)

12. Chuang, K.S., Tzeng, H.L., Chen, S., Wu, J., Chen, T.J.: Fuzzy c-means clustering with spatial information for image segmentation. Comput. Med. Imaging Graph. **30**(1), 9–15 (2006)
13. Cobos, M., Lopez, J.: Maximum a posteriori binary mask estimation for underdetermined source separation using smoothed posteriors. IEEE Trans. Audio Speech Language Process. **20**(7), 2059–2064 (2012)
14. Deleforge, A., Forbes, F., Horaud, R.: Variational EM for binaural sound-source separation and localization. In: Proceedings of ICASSP, pp. 76–79 (2013)
15. Dempster, A., Laird, N., Rubin, D.: Maximum likelihood from incomplete data via the EM algorithm. J. R. Stat. Soc. B **39**, 1–38 (1977)
16. Dmochowski, J., Benesty, J., Affes, S.: On spatial aliasing in microphone arrays. IEEE Trans. Signal Process. **57**(4), 1383–1395 (2009)
17. Gales, M.: Semi-tied covariance matrices for hidden Markov models. IEEE Trans. Audio Speech Language Process. **7**(3), 272–281 (1999)
18. Gaubitch, N.D., Kleijn, W.B., Heusdens, R.: Auto-localization in ad-hoc microphone arrays. In: Proceedings of ICASSP, pp. 106–110. IEEE, New York (2013)
19. Grais, E., Erdogan, H.: Spectro-temporal post-smoothing in NMF based single-channel source separation. In: Proceedings of EUSIPCO, pp. 584–588 (2012)
20. Gu, D.B., Sun, J.: EM image segmentation algorithm based on an inhomogeneous hidden MRF model. IEEE Vis. Image Signal Process. **152**(2), 184–190 (2004)
21. Haeb-Umbach, R., Ney, H.: Linear discriminant analysis for improved large vocabulary continuous speech recognition. In: Proceedings of ICASSP, pp. 13–16 (1992)
22. Himawan, I., McCowan, I., Sridharan, S.: Clustered blind beamforming from ad-hoc microphone arrays. IEEE Trans. Audio Speech Language Process. **19**(4), 661–676 (2011)
23. Hinton, G., Deng, L., Yu, D., Dahl, G.E., Mohamed, A., Jaitly, N., Senior, A., Vanhoucke, V., Nguyen, P., Sainath, T.N., Kingsbury, B.: Deep neural networks for acoustic modeling in speech recognition: the shared views of four research groups. IEEE Signal Process. Mag. **29**(6), 82–97 (2012)
24. Jiang, Y., Wang, D., Liu, R.: Binaural deep neural network classification for reverberant speech segregation. In: Proceedings of Interspeech, pp. 2400–2403 (2014)
25. Kim, M., Smaragdis, P.: Single channel source separation using smooth nonnegative matrix factorization with Markov random fields. In: Proceedings of MLSP, pp. 1–6 (2013)
26. Kim, M., Smaragdis, P., Ko, G.G., Rutenbar, R.A.: Stereophonic spectrogram segmentation using Markov random fields. In: Proceedings of MLSP, pp. 1–6 (2012)
27. Kinoshita, K., Delcroix, M., Yoshioka, T., Nakatani, T., Habets, E., Sehr, A., Kellermann, W., Gannot, S., Maas, R., Haeb-Umbach, R., Leutnant, V., Raj, B.: The REVERB challenge: a common evaluation framework for dereverberation and recognition of reverberant speech. In: Proceedings of WASPAA, New Paltz, NY (2013)
28. Kühne, M., Togneri, R., Nordholm, S.: Smooth soft mel-spectrographic masks based on blind sparse source separation. In: Proceedings of Interspeech (2007)
29. Kühne, M., Togneri, R., Nordholm, S.: Adaptive beamforming and soft missing data decoding for robust speech recognition in reverberant environments. In: Proceedings of Interspeech, pp. 976–979 (2008)
30. Kühne, M., Togneri, R., Nordholm, S.: A novel fuzzy clustering algorithm using observation weighting and context information for reverberant blind speech separation. Signal Process. **90**(2), 653–669 (2010)
31. Liang, S., Liu, W., Jiang, W.: Integrating binary mask estimation with MRF priors of cochleagram for speech separation. IEEE Signal Process. Lett. **19**(10), 627–630 (2012)
32. Lippmann, R., Martin, E., Paul, D.: Multi-style training for robust isolated-word speech recognition. In: Proceedings of ICASSP, vol. 12, pp. 705–708 (1987)
33. Liu, Z., Zhang, Z., He, L.W., Chou, P.: Energy-based sound source localization and gain normalization for ad hoc microphone arrays. In: Proceedings of ICASSP, vol. 2, pp. 761–764. IEEE, New York (2007)

34. Lloyd, S.P.: Least squares quantization in PCM. IEEE Trans. Inf. Theory **28**(2), 129–137 (1982)
35. Madhu, N., Breithaupt, C., Martin, R.: Temporal smoothing of spectral masks in the cepstral domain for speech separation. In: Proceedings of ICASSP, pp. 45–48 (2008)
36. Mandel, M.I., Roman, N.: Enforcing consistency in spectral masks using Markov random fields. In: Proceedings of EUSIPCO (2015)
37. Mandel, M.I., Weiss, R.J., Ellis, D.P.W.: Model-based expectation maximization source separation and localization. IEEE Trans. Audio Speech Language Process. **18**(2), 382–394 (2010)
38. Middlebrooks, J.C., Green, D.M.: Sound localization by human listeners. Annu. Rev. Psychol. **42**, 135–159 (1991)
39. Narayanan, A., Wang, D.: Investigation of speech separation as a front-end for noise robust speech recognition. IEEE Trans. Audio Speech Language Process. **22**(4), 826–835 (2014)
40. O'Grady, P.D., Pearlmutter, B.A.: Soft-LOST: EM on a mixture of oriented lines. In: Independent Component Analysis and Blind Signal Separation, vol. 3195, 1270 pp. Springer, Berlin (2004)
41. Pearl, J.: Probabilistic Reasoning in Intelligent Systems: Networks of Plausible Inference. Morgan Kaufmann, San Francisco, CA (1988)
42. Povey, D., Kanevsky, D., Kingsbury, B., Ramabhadran, B., Saon, G., Visweswariah, K.: Boosted MMI for model and feature-space discriminative training. In: Proceedings of ICASSP, pp. 4057–4060 (2008)
43. Rafaely, B., Weiss, B., Bachmat, E.: Spatial aliasing in spherical microphone arrays. IEEE Trans. Signal Process. **55**(3), 1003–1010 (2007)
44. Renals, S., Hain, T., Bourlard, H.: Recognition and understanding of meetings: the AMI and AMIDA projects. In: Proceedings of ASRU, Kyoto (2007)
45. Roman, N., Wang, D.L., Brown, G.J.: Speech segregation based on sound localization. J. Acoust. Soc. Am. **114**, 2236–2252 (2003)
46. Roweis, S.: Factorial models and refiltering for speech separation and denoising. In: Proceedings of Eurospeech, Geneva, pp. 1009–1012 (2003)
47. Sainath, T.N., Weiss, R.J., Senior, A., Wilson, K.W., Vinyals, O.: Learning the speech front-end with raw waveform CLDNNS. In: Proceedings of Interspeech (2015)
48. Sawada, H., Mukai, R., Araki, S., Makino, S.: A robust and precise method for solving the permutation problem of frequency-domain blind source separation. IEEE Trans. Speech Audio Process. **12**(5), 530–538 (2004)
49. Sawada, H., Araki, S., Makino, S.: Underdetermined convolutive blind source separation via frequency bin-wise clustering and permutation alignment. IEEE Trans. Audio Speech Language Process. **19**(3), 516–527 (2011)
50. Souden, M., Benesty, J., Affes, S.: On optimal frequency-domain multichannel linear filtering for noise reduction. IEEE Trans. Audio Speech Language Process. **18**(2), 260–276 (2010)
51. Swietojanski, P., Ghoshal, A., Renals, S.: Hybrid acoustic models for distant and multichannel large vocabulary speech recognition. In: Proceedings of ASRU (2013)
52. Szeliski, R., Zabih, R., Scharstein, D., Veksler, O., Kolmogorov, V., Agarwala, A., Tappen, M., Rother, C.: A comparative study of energy minimization methods for Markov random fields with smoothness-based priors. IEEE Trans. Pattern Anal. Mach. Intell. **30**(6), 1068–1080 (2008)
53. Togami, M., Sumiyoshi, T., Amano, A.: Stepwise phase difference restoration method for sound source localization using multiple microphone pairs. In: Proceedings of ICASSP (2007)
54. Traa, J., Kim, M., Smaragdis, P.: Phase and level difference fusion for robust multichannel source separation. In: Proceedings of ICASSP, pp. 6687–6690 (2014)
55. Veselý, K., Ghoshal, A., Burget, L., Povey, D.: Sequence-discriminative training of deep neural networks. In: Proceedings of Interspeech, pp. 2345–2349 (2013)
56. Vincent, E.: An experimental evaluation of Wiener filter smoothing techniques applied to under-determined audio source separation. In: International Conference on Latent Variable Analysis and Signal Separation, pp. 157–164. Springer, Berlin, Heidelberg (2010)

57. Vincent, E., Bertin, N., Gribonval, R., Bimbot, F.: From blind to guided audio source separation: how models and side information can improve the separation of sound. IEEE Signal Process. Mag. **31**(3), 107–115 (2014)
58. Wang, D.: On ideal binary mask as the computational goal of auditory scene analysis. In: Divenyi, P. (ed.) Speech Separation by Humans and Machines, pp. 181–197. Springer US, Boston, MA (2005)
59. Weiss, R., Mandel, M.I., Ellis, D.W.P.: Combining localization cues and source model constraints for binaural source separation. Speech Commun. **53**(5), 606–621 (2011)
60. Yedidia, J., Freeman, W., Weiss, Y.: Generalized belief propagation. In: Advances in Neural Information Processing Systems, pp. 689–695. MIT, Cambridge (2000)
61. Yilmaz, O., Rickard, S.: Blind separation of speech mixtures via time–frequency masking. IEEE Trans. Audio Speech Language Process. **52**(7), 1830–1847 (2004)
62. Zhang, Y., Brady, M., Smith, S.: Segmentation of brain MR images through a hidden Markov random field model and the expectation-maximization algorithm. IEEE Trans. Med. Imaging **20**(1), 45–57 (2001)

Chapter 4
Discriminative Beamforming with Phase-Aware Neural Networks for Speech Enhancement and Recognition

Xiong Xiao, Shinji Watanabe, Hakan Erdogan, Michael Mandel, Liang Lu, John R. Hershey, Michael L. Seltzer, Guoguo Chen, Yu Zhang, and Dong Yu

Abstract Speech-processing systems such as automatic speech recognition (ASR) usually consist of a large number of steps to accomplish their tasks. Due to the long processing pipeline, the processing steps are usually designed to optimize cost functions that are not directly related to the task, leading to suboptimal performance. In this chapter, we introduce a beamforming (BF) network to perform spatial filtering that is optimal for the ASR task. The BF network takes in array signals and predicts the optimal beamforming parameters in the frequency domain, assuming that the array geometry does not change. The network consists of both deterministic

X. Xiao (✉)
Temasek Laboratories, Nanyang Technological University, 9th Storey, BorderX Block, Research Techno Plaza, 50 Nanyang Drive, Singapore 637553, Singapore
e-mail: xiaoxiong@ntu.edu.sg

S. Watanabe
Mitsubishi Electric Research Laboratories (MERL), Cambridge, MA, USA

H. Erdogan
Microsoft Research, City center square, Bellevue, WA 98004, USA

M. Mandel
Brooklyn College (CUNY), 2900 Bedford Ave, Brooklyn, NY 11210, USA

L. Lu
Toyota Technological Institute at Chicago, 6045 S. Kenwood Ave., Chicago, IL 60637, USA

J.R. Hershey
Mitsubishi Electric Research Laboratories (MERL), Cambridge, MA, USA

M.L. Seltzer
Microsoft Research, City Center Square, Bellevue, WA 98004, USA

G. Chen
Johns Hopkins University, Baltimore, MD, USA

Y. Zhang
Massachusetts Institute of Technology, Cambridge, MA, USA

D. Yu
Tencent AI Lab, 10900 NE 8th Street, Bellevue, WA 98004, USA

© Springer International Publishing AG 2017
S. Watanabe et al. (eds.), *New Era for Robust Speech Recognition*,
DOI 10.1007/978-3-319-64680-0_4

79

processing steps and trainable steps realized by neural networks and trained to minimize the cross-entropy cost function of ASR. In our experiments, the BF network is trained with both artificially generated and real microphone array signals. On the AMI meeting transcription, we found that the trained BF network produces competitive ASR results compared to traditional delay-and-sum beamforming on unseen array signals.

4.1 Introduction

Beamforming algorithms combine multiple microphone signals recorded in slightly different locations in such a way that they emphasize signals of interest, while attenuating all other signals. The spatial diversity of the channels allows this selection to be performed based on the spatial characteristics of the sources in addition to their characteristics in time and frequency. The development of beamforming algorithms has generally followed a trajectory from methods based on geometrical considerations of the microphone array and the spatial positions of the signals to those based on data-driven considerations of the microphones, sources, and their spatial characteristics. The current chapter describes a new approach that adjusts the beamforming filters to directly maximize automatic speech recognition (ASR) performance, and allows the ASR acoustic model to simultaneously be adjusted to accommodate the output of the beamformer.

In the rest of this chapter, we will first review multichannel speech-processing techniques for ASR in Sect. 4.2. The review includes both classic geometric and statistical beamforming methods, as well as learning-based methods developed in recent years. Then, we will describe a new learning-based approach, called the beamforming network, in Sect. 4.3. The network predicts beamforming weights and can be trained by using the ASR's cost function, hence is optimizable for the ASR task. The beamforming network is experimentally studied in Sect. 4.4, where we analyze its behavior on unseen array data and also present the ASR results. Finally, we summarize the findings and discuss future research directions in Sect. 4.5.

4.2 Beamforming for ASR

In this section, we review beamforming and related methods with ASR as the intended application. We classify the beamforming methods into three categories. The first category is geometric beamforming, which mainly relies on the array geometry and spatial location of the source to determine the parameters of the beamforming. One example of this category is delay-and-sum (DS) beamforming [29], which does not consider the spectral characteristics of the target signal or noise. The second category is statistical methods that rely on the characteristics of the target signal and noise in addition to the geometric information. Example methods include the linearly constrained minimum variance (LCMV) beamformer [9] and minimum variance distortionless response (MVDR) beamformer [6, 11]. These methods critically rely on the estimation of the target steering vectors/spatial covariance matrices and the noise spatial covariance matrix to perform well. The third category is called the learning-based approach, which has been developed in recent years for speech recognition. The major characteristic of this category is that besides using the previously mentioned information sources, a model is trained on a large amount of single/multichannel signals to capture the prior knowledge about the signal of interest, which is human speech in this chapter. Generally speaking, from the geometric to the learning-based approaches, more and more information is used for optimally determining the beamforming parameters for the ASR task. In the following text, these three categories of methods will be described in detail.

4.2.1 Geometric Beamforming

A sound recorded by a microphone array from a single direction will arrive at each microphone at a slightly different time. If the original signal of interest in the time domain is $x_1[n]$, and it is recorded with $I - 1$ other signals at J microphones, then the recorded signals $y_j[n]$ at the jth microphone are

$$y_j[n] = \sum_{i-1}^{I} \sum_{l-1}^{L} h_{ij}[l] x_i[n - l] + u_j[n] \tag{4.1}$$

$$= \sum_{i=1}^{I} h_{ij} * x_i + u_j[n]. \tag{4.2}$$

$u_j[n]$ is the jth noise term and L is the length of the impulse responses. In the anechoic case, the impulse responses $h_{ij}[l]$ are pure delays, which can be counteracted by appropriate delays in the opposite direction. The estimated signal is then

$$\hat{x}_1[n] = \sum_{j=1}^{J} \sum_{l=1}^{L} w_j[l] y_j[n - l], \tag{4.3}$$

where $w_j[l] = \delta_{l,\hat{\tau}_j}$ equals 1 when $l = \hat{\tau}_j$ and 0 otherwise. $\hat{\tau}_j$ is the estimated delay of channel j. Equation (4.3) should constructively reinforce the signal of interest. Signals coming from most other directions will reinforce less, leading to a relative amplification of the target. This method is called delay-and-sum beamforming. With a known array geometry, the necessary delays can be computed analytically as a function of direction of arrival. This is particularly straightforward for certain array geometries, like uniformly spaced linear, planar, circular, and spherical arrays.

Other geometries can be accommodated, however, through calibration procedures. Calibration is helpful in all geometric approaches, however, because of variabilities in the manufacturing process affecting microphone placement and sensitivity.

Delay-and-sum beamforming is optimal in the sense of maximizing the output signal-to-noise ratio (SNR) in the case of a target signal arriving from a single known direction in the presence of uncorrelated noise [12]. In practice, however, the direction of arrival is not perfectly known and the noise is not uncorrelated. Furthermore, multipath propagation causes echoes of the target signal to arrive from multiple directions. It is, however, useful to think about finding the optimal beamformer given an optimality criterion, a microphone array configuration, and a configuration of sound sources. A more general optimization problem is to consider filter-and-sum beamformers, which have the same form as (4.3) but allow $w_i[l]$ to be arbitrary filters instead of pure delays. Equation (4.3) can also be written in the frequency domain, for filters $w_i[l]$ that are shorter than the analysis frame length, as

$$\hat{X}_1(t,f) = \sum_{j=1}^{J} W_j(f) Y_j(t,f). \tag{4.4}$$

In the case of the delay-and-sum beamformer,

$$W_j(f) = \exp\left(\frac{2\pi \mathrm{j} f \tau_j}{f_s}\right), \tag{4.5}$$

where f_s is the sampling frequency and τ_j is the compensatory delay in the sample at microphone j, and $\mathrm{j} = \sqrt{-1}$.

The superdirective beamformer [5] is a simple geometric filter-and-sum beamformer that improves upon the delay-and-sum beamformer by achieving maximum output SNR for a sound arriving from a single known direction in the presence of purely diffuse noise. It also has better spatial selectivity at low frequencies than the delay-and-sum beamformer and a more uniform spatial selectivity across frequency. If we denote the frequency-dependent coefficients of the delay-and-sum beamformer as the vector $d(f)$, then the filter-and-sum beamformer is

$$w(f) = \frac{\Phi^{-1}(f) d(f)}{d^H(f) \Phi^{-1}(f) d(f)}, \tag{4.6}$$

where Φ is the spatial correlation matrix between microphones for diffuse noise, i.e.,

$$\phi_{ij}(f) = \mathrm{sinc}\left(\frac{2\pi f d_{ij}}{c}\right), \tag{4.7}$$

where c is the speed of sound and d_{ij} is the distance between microphones i and j. Diffuse noise is noise that comes equally from all directions, in either two or three dimensions (also known as cylindrically or spherically isotropic noise) [8]. It is a good model for noise that does not come from a point source, or has been reflected and diffracted off many surfaces. At high frequencies, it is uncorrelated between microphones, but at low frequencies (where the wavelength is larger than the microphone spacing) it is correlated across microphones. Purely uncorrelated noise, as assumed by the delay-and-sum beamformer, is a more accurate model of noise originating from within sensors and recording equipment, but not from the actual acoustical measurements.

4.2.2 Statistical Methods

Typically, however, interfering signals do not consist of purely diffuse sources, so data-driven methods that estimate the spatial statistics of the noise from a recording can provide better performance than those based purely on geometric considerations. A filter-and-sum beamformer, being a sum of various input terms without feedback, acts as a spatial finite impulse response filter, and its transfer function consists only of (spatial) zeros. While these zeros can be used to shape a beam pattern with a reasonable gain in the direction of the target signal, they are more effective at cancelling out unwanted signals. This cancellation is much easier to achieve when using measured noise statistics as opposed to theoretically derived noise statistics. Measured noise statistics contain information about both point sources of noise and diffuse noise and allow both of them to be canceled effectively.

A popular statistics-based beamformer is the LCMV beamformer [9], which minimizes the output power subject to one or more linear constraints. The MVDR beamformer [11] is a particular LCMV beamformer where the constraint is to have unit gain in the direction of a target signal. The solution to the MVDR beamformer is actually the same as (4.6), where Φ comes from an estimate of the spatial correlation of the noise,

$$\phi_{ij}(f) = \mathbb{E}\left[U_i^*(t,f)U_j(t,f)\right], \tag{4.8}$$

where $U_i(t,f)$ is the Fourier coefficients of the noise received by microphone i at frame t and frequency bin f. Other LCMV beamformers can include constraints on multiple target and noise directions. With the constraint that the target signal is unaffected by the beamformer, the minimization of the output power corresponds to minimizing the amount of nontarget energy in the output.

In order to derive the beamformers discussed so far, a target direction must be specified. This direction usually comes from a localization algorithm, which can sometimes be unreliable. In order to avoid a separate localization step, [27] showed that this information can be extracted from an estimate of the noise spatial statistics and the mixture spatial statistics.

Another important statistical beamformer is the multichannel Wiener filter (MWF) [7]. The MWF uses a different optimality criterion than the MVDR beamformer, namely the minimum mean squared error (MMSE) criterion. Doing so leads to a very similar formulation of the filter weights, requiring estimates of the same spatial statistics. The slight differences between the two formulations, however, show that the two optimality criteria prioritize slightly different qualities. The MVDR criterion sacrifices some noise suppression performance to provide maximal preservation of the target signal. The MWF sacrifices some preservation of the target signal to provide maximal noise suppression. Intermediate trade-offs are possible by varying a μ parameter as defined in Eq. (30) of [7]. Manipulating the definition of the MWF shows that it can be factored into the MVDR beamformer followed by a single-channel Wiener filter postfilter.

4.2.3 Learning-Based Methods

Because traditional ASR systems are greatly hindered by signal-processing artifacts, the distortionless response in the speech direction of arrival is useful in maintaining good ASR performance. If speech recognition performance is the desired goal of a multimicrophone system, however, it should be better to select the beamforming filter weights that directly optimize this metric. While in the past this has been difficult for computational and systems design reasons, recent tools like the Computational Network Toolkit (CNTK) [33] make the creation of a combined beamforming and ASR acoustic model feasible to construct from and train on data. Thus there has recently been an increase of interest in other optimality criteria for beamforming.

In this section, we review the learning-based methods for beamforming. By learning, we mean the system first learns some knowledge from a large amount of single-channel/multichannel speech signals, then applies this knowledge when performing beamforming at run time. Another characteristic of a learning-based system is that the beamforming module and the acoustic-modeling module for ASR can often be integrated into a single computational network and jointly optimized on training data.

4.2.3.1 Maximum Likelihood Approach

One of the first learning-based approaches for beamforming is the likelihood-maximizing beamforming (LIMABEAM) proposed for the hidden-Markov-model (HMM)/Gaussian-mixture-model (GMM) acoustic model [26]. The basic concept of LIMABEAM is to tune the beamforming weights to generate a sequence of feature vectors that fit the acoustic model for speech recognition. At training stage, the acoustic model is trained to capture the feature distribution of (clean) speech signals. At run time, a filter-and-sum beamformer is used in the time or frequency domain to combine multichannel inputs into a single enhanced channel, which is used for feature extraction. As no closed-form solution exists due to the nonlinearity involved in the feature extraction process, back-propagation is used to estimate the beamforming weights. It was reported in [26] that LIMABEAM outperforms the classic DS beamformer on speech recognition tasks.

LIMABEAM can be seen as a nonlinear feature adaptation method and shares similarity with its single-channel and linear counterparts, such as feature space maximum likelihood linear regression (fMLLR) [10]. Given multichannel training data, it is possible to achieve joint training of LIMABEAM filters and the acoustic model, just like speaker-adaptive training (SAT) with fMLLR. The model consists of a canonical acoustic model and multiple sets of beamforming filter weights, one set for each training utterance. The joint training will alternate the estimation of the canonical acoustic model and the estimation of beamforming weights. It is also possible to perform maximum likelihood linear regression (MLLR) acoustic-model adaptation to deal with speaker variation simultaneously [26].

4.2.3.2   Neural Network Approaches with Multichannel Inputs

In the past few years, there have been studies on using multichannel inputs directly for neural-network-based acoustic modeling without explicitly using beamforming. Early work was reported in [18], where the features of all channels we are simply concatenated and used in a deep-neural-network (DNN) acoustic model. This method outperforms the DNN acoustic model with a single channel. It also performs equally to or slightly better than applying beamforming before the feature extraction acoustic model, especially when the number of channels is small (two to four channels). In other early work [28], a convolutional neural network (CNN) was used together with a DNN as the acoustic model. It was found that using cross-channel maxpooling before frequency band maxpooling on filterbank features helped to improve performance by selecting informative channels for speech recognition. Although better than single-microphone performance can be obtained by the methods proposed in [18] and [28], these improvements are not coming from beamforming as no phase information has been used by the networks.

Recently, there have been several studies from Google (see Chap. 5) on multichannel acoustic modeling using raw waveforms as inputs [14, 24, 25]. In [14], a temporal convolution layer is first applied to the multichannel (two channels) waveforms to learn spatial and spectral filterbanks. The output of the convolutional filters is maxpooled with a 25 ms window and 10 ms shift, then rectified and compressed by the natural logarithm function to mimic the filterbank energy features. Then the features are used as the input of a feedforward DNN acoustic model. The temporal convolutional filters and the DNN acoustic model are trained jointly on the ASR cost function. It was reported that the learned convolutional filters exhibit both spectral selectivity and spatial selectivity, i.e., each filter has one or more fixed looking directions and extracts speech energies in a certain frequency band. The work has been improved by moving to a convolutional long short-term memory deep neural network (CLDNN) architecture which has long short-term memory (LSTM) layers that can better deal with temporal variations [24].

A limitation of the work in [14, 24] is that the temporal convolutional filters are both spatially and spectrally selective; hence it requires a large number of such filters to cover the combination of spatial directions and frequency bands. To alleviate this issue, a factored model was proposed in [25]. In the factored model, two temporal convolution layers are used in sequence. The first temporal convolution uses a small number (up to 10 in [25]) of multichannel filters with 5 ms length (80 taps for a 16 kHz sampling rate). The filters are expected to do processing like filter-and-sum beamforming and convert the multichannel waveforms into single-channel without pooling and nonlinearity. The second temporal convolution layer uses 128 single-channel filters (each of which is much longer than the first layer's multichannel filters) to extract filterbank-like features for speech recognition from the output of the first convolution layer. These two convolution layers are jointly trained with the acoustic model. It is reported that the factored model in [25] outperforms the unfactored model in [24] in terms of word error rate (WER) on a voice search database with two microphones. However, the learned filters in the first convolution layer still exhibit both spectral and spatial selectivity, which suggests that the factorization of spatial filtering and spectral filtering is not complete.

4.2.3.3 Neural Networks for Better Spatial-Statistics Estimation

The performance of traditional beamforming often depends on how accurately the statistics of speech and noise are estimated. The statistics are usually the target source steering vector, and speech and noise spatial covariance matrices at different frequencies. Recent work [13, 15] uses a neural network for more accurate estimation about the statistics. In this work, a bidirectional LSTM is used to predict whether a time–frequency (TF) bin is dominated by speech or noise. The speech-dominant TF bins are used to estimate the speech spatial covariance matrix, while the noise-dominant TF bins are used to estimate the noise covariance matrix. These two types of covariance matrices are then used to construct statistical beamformers, such as MVDR or generalized eigenvalue (GEV) beamformers. Good performance has been reported on the CHiME-3 task [4]. The mask-predicting network can also be jointly optimized for the ASR task, as shown for the single-channel case in [19] and the multichannel case in [31]. In [31], the mask-predicting LSTM (based on a single channel) is connected to the acoustic model to form a global computational network. Both the mask-predicting LSTM and the acoustic-model DNN are jointly refined to reduce the cross-entropy of phone classification. Results on CHiME-4 confirmed the advantage of the joint training approach over the separate training approach.

4.3   Beamforming Networks

4.3.1   Motivation

The major motivation of the learning-based approach is to optimize the beamforming directly for the final ASR task by using a lot of training examples. Google's approach [14, 24, 25] can be considered as a "black box" approach that lets the network determine all the processing steps, including spatial filtering, feature learning, and acoustic modeling with little human intervention. With the introduction of factored temporal convolution layers, the approach becomes more "transparent" in the sense that different layers of the network have more clear functions. With more and more domain knowledge incorporated into the network design, we expect this approach to become more and more "transparent." On the other hand, the mask estimation approach [13, 15] described in Sect. 4.2.3.3 only uses the neural networks to predict speech masks for spatial-covariance estimation, and still uses traditional GEV or MVDR rules to determine the beamforming parameters. Hence, this approach can be considered as a more "transparent" approach.

In this chapter, we present a new learning-based beamforming approach. We believe that both array-processing domain knowledge and the learning capability of neural networks are required to perform optimal multichannel ASR. Therefore, instead of starting with a "black box" approach and then making it more and more "transparent" by incorporating domain knowledge, we take an opposite approach. Specifically, we start with traditional beamforming methods and gradually replace suitable processing steps with neural networks. As the research continues, we will learn about which ingredients in the traditional array signal processing are important to achieve good performance and which can be replaced by neural networks.

The work presented in this chapter is the first step towards our goal. We keep most of the modules in the traditional beamforming and acoustic-modeling pipeline and only realize the beamforming-weight determination module with a neural network, which is called the beamforming network. The beamforming network and the acoustic-model network together with deterministic processing modules form a computational network that converts the multichannel waveforms to senone posterior probabilities used for speech recognition. Gradients of the ASR cost function can flow back to the beamforming network and optimize it for the ASR task. In the remainder of this section, we will describe the proposed approach in detail.

4.3.2   System Overview

The joint training of beamforming and acoustic model networks is illustrated in Fig. 4.1. The input of the system is multichannel speech signals in the time domain. In the left branch of the network, a DNN (or other network type such as an LSTM) is

Fig. 4.1 System diagram of
the joint training of a
beamforming network and
acoustic-model network. The
two shaded boxes denote
neural networks that are
trained together, while other
blocks denote deterministic
processing

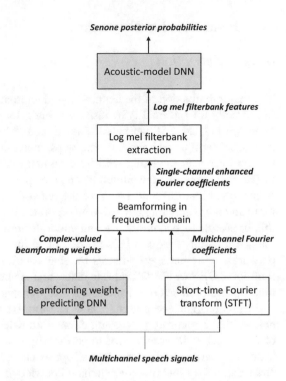

used to predict the complex-valued beamforming weights in the frequency domain.
In the right branch of the network, the time domain signal is converted to the
frequency domain by using a short-time Fourier transformation (STFT). Then, the
predicted beamforming weights are applied to the multichannel Fourier coefficients
in the same way as conventional frequency domain beamformers. The enhanced
single-channel Fourier coefficients are then fed to the feature extraction block to
generate log mel filterbanks for acoustic modeling. The acoustic-model network
maps filterbanks to senone posterior probabilities. In the rest of this section, we will
explain each block in detail.

Compared to conventional approaches, the major difference between the proposed joint beamforming–acoustic-modeling approach as shown in Fig. 4.1 is that
the estimation of beamforming weights is now implemented by a *neural network*,
which has trainable parameters. As a result, the system is able to learn how to perform beamforming automatically through training its parameters on multichannel
speech signals. Furthermore, the weight prediction DNN can be trained together
with the acoustic-model DNN using an ASR cost function such as the cross-entropy.
This allows the interaction between the two DNNs and theoretically can achieve
beamforming that is more optimal for ASR than conventional beamforming.

Fig. 4.2 Illustration of the array geometry used in this chapter

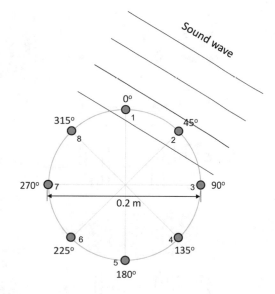

Although the joint beamforming–acoustic-modeling approach studied in this chapter can be applied to any microphone array geometry in theory, we need to choose one geometry for illustration and experiments. We choose to use a circular array with eight omnidirectional microphones as shown in Fig. 4.2. The diameter of the array is 0.2 m. The array is intended to be used in far-field scenarios, such as meeting rooms. Such an array geometry is also used in several robust ASR corpora, including the REVERB challenge [16] and the AMI meeting corpora [22]. In this chapter, we will use these corpora to train and evaluate the proposed approach.

4.3.3 Predicting Beamforming Weights by DNN

Given multichannel speech signals, a feedforward DNN is used to determine the beamforming weights in the frequency domain (filter-and-sum beamforming). We assume that there is only one target speech source, and the beamformer will retrieve the speech from the target direction while attenuating interference from other directions.

The process of predicting beamforming weights is illustrated in Fig. 4.3. From the multichannel speech signals, we first extract generalized cross-correlation (GCC) features for each frame, resulting in 588-dimensional feature vectors that encode the phase delay information about the channels. Each of the GCC feature vectors is mapped to a beamforming weight vector of 4112 dimensions by the DNN. Finally, the average beamforming weights are obtained by taking the mean of the beamforming weight vectors over an utterance. In the following two sections, we will describe the details of the GCC feature extraction and the beamforming weight vector.

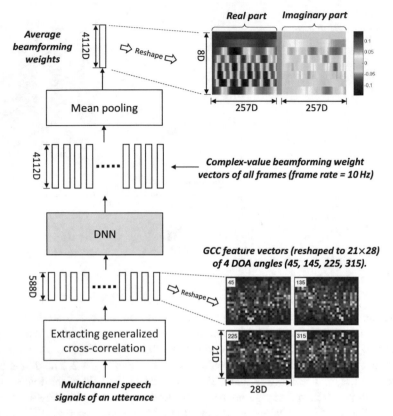

Fig. 4.3 Block diagram of beamforming weight-predicting network

4.3.3.1 Extraction of GCC Features

Beamforming requires information about the target direction. For far-field scenario, this leads to the problem of determining the time difference of arrival (TDOA) of the channels in the time domain, or the phase difference in the frequency domain. Similarly to traditional beamforming methods, the DNN also requires information about the TDOA or phase difference to predict beamforming weights.

In theory, the DNN should be able to learn the phase information directly from raw waveforms. We can also make use of existing methods that have been proven to work well, such as the generalized cross-correlation phase transform (GCC-PHAT) method [17]. For signals recorded by two microphone channels $y_i[n]$ and $y_j[n]$, the cross-correlation in the frequency domain can be computed using the GCC-PHAT method by

$$G_{i,j}(f) = \frac{Y_i(f)Y_j^*(f)}{|Y_i(f)Y_j^*(f)|}, \tag{4.9}$$

where $Y_i(f)$ and $Y_j(f)$ are the Fourier transforms of $y_i[n]$ and $y_j[n]$, respectively, at frequency bin f. $G_{i,j}(f)$ measures the phase difference between the two channels. The cross-correlation in the time domain can be obtained by

$$R_{i,j}(\tau) = \text{IFT}(G_{i,j}(f)), \tag{4.10}$$

where IFT() denotes the inverse Fourier transform. In classic methods, we can estimate the TDOA between the microphones i and j by finding the peak of the cross-correlation function:

$$\hat{\tau}_{i,j} = \arg\max_{\tau} R_{i,j}(\tau). \tag{4.11}$$

However, it is not suitable to use the estimated TDOA as features to predict beamforming weights due to two reasons. First, if an error occurs in estimating the TDOA, the error will propagate to the beamforming network and cannot be corrected. Second, a single TDOA estimate contains much less information than the whole correlation function $R_{i,j}(\tau)$. Due to these reasons, we use the whole correlation function as the input features.

In practice, it is not necessary to use the whole correlation function as input, since most of its elements are not informative for either TDOA estimation or predicting beamforming weights for normal microphone array geometries. Let's use an example for illustration. Assume the speech signals are sampled at 16,000 Hz, and we use a window of 0.2 s (i.e., 3200 samples) to estimate the correlation function. The resulting correlation function will have a length of 3200 elements. Suppose the maximum distance between two microphones in the array is 0.2 m; the maximum time delay between microphones is (0.2 m/340 m/s) × 16,000 sample/s ≈ 9.4 samples. Hence, it will suffice to keep only the correlation function up to ±10 samples of time delays, i.e., we extract a feature vector of 21 elements for each microphone pair.

To improve the robustness of beamforming weight prediction, it is necessary to use the correlation function of all microphone pairs even for known array geometries. For example, if there are eight microphones, there will be $C(8, 2) = 28$ combinations of microphones. Our preliminary study shows that using the correlation of all microphone pairs outperforms using only the correlation between a reference microphone and other microphones. This could be because there is complementary information when using all microphone pairs that could help the prediction. As speakers may move, the GCC features are extracted for every 0.2 s long window with a shift of 0.1 s.

In summary, we use the GCC function as the features for beamforming weight prediction. For a circular array with eight channels and a diameter of 0.2 m, we will have 28 correlation vectors, each containing 21 elements. If we put the 28 vectors as the columns of a matrix, we will find different DOA angles corresponding to different patterns, as shown in the bottom of Fig. 4.3. This shows that the

GCC features are informative for determining the DOA of the source, and also determining the beamforming weights. More details on using GCC features for DOA estimation with neural networks can be found in [30].

4.3.3.2 Beamforming Weight Vector

For each GCC feature vector, a set of frequency domain beamforming weights are predicted by the beamforming DNN as shown in Fig. 4.3. If the FFT length is 512, there are 257 frequency bins to cover 0 to 8000 Hz. There are eight complex-valued weights for each frequency bin, one for each channel. As a standard DNN is not able to handle complex numbers directly, the beamforming DNN predicts the real and imaginary parts of the weights independently. Therefore, the weight vector contains $257 \times 8 \times 2 = 4112$ real-valued elements.

If the speakers are known to be stationary over an utterance, we can achieve more stable weight prediction by averaging the weight vectors over an utterance. As speakers in the AMI corpus are stationary most of the time, we used mean pooling in the experiments. If the speakers move significantly within one utterance, we can use smoothing instead of mean pooling to allow tracking of slow speaker moves. Another choice is to use recurrent neural networks such as LSTMs that have temporal memory.

An example of beamforming weights predicted by the DNN is shown at the top of Fig. 4.3. The real part and imaginary part are shown separately, each reshaped to an 8×257 matrix. From the figure, the predicted weights are quite stable, although they are predicted independently. In the experiments section, it will be shown that the predicted weights have consistent-looking directions at different frequencies most of the time. Note that as we choose the first microphone as the reference microphone, it always has zero imaginary value.

4.3.4 Extraction of Log Mel Filterbanks

After the beamforming block, a series of steps are used to implement the typical speech recognition feature extraction. In this chapter, we choose to use log mel filterbank energies as features for ASR. We will introduce all the feature extraction blocks in detail below. The feature extraction blocks are also illustrated in Fig. 4.4

- **Power spectrum.** Given the enhanced complex-valued spectrum $X(t,f)$, compute the power spectrum as $\|X(t,f)\|^2 = X(t,f)X^*(t,f)$.
- **Mel filterbank.** Group the power spectrum into mel filterbanks that have equal bandwidths on the Mel frequency scale [20]. This step is implemented by using

Fig. 4.4 Block diagram of feature extraction from complex-valued spectrum (MVN, mean and variance normalization; CMN, cepstral mean normalization)

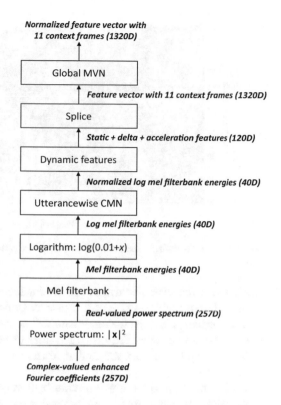

a linear transform $\mathbf{x}_{\text{mel}}(t) = \mathbf{M}\mathbf{x}(t)$, where M is a matrix of size 40×257, with 40 and 257 being the number of filterbanks and number of frequency bins, respectively. $\mathbf{x}(t) = [||X(t, 1)||^2, \ldots, ||X(t, K)||^2]^T$ is the vector of power spectrum coefficients for frame t, and $\mathbf{x}_{\text{mel}}(t)$ is its transformed version on the mel frequency scale. The parameters of the transform M are precomputed (shown in Fig. 4.5) and not updated during network training.

- **Logarithm.** Apply the natural logarithm to the mel filterbank energies individually to compress their dynamic range, i.e., $\mathbf{x}_{\text{lm}}(t) = \log(\mathbf{x}_{\text{mel}}(t))$. As the gradient of the log function $\partial\log(x)/\partial x = 1/x$ is not numerically stable when x approaches 0, we add a small constant to the function, $\log(\mathbf{x}_{\text{mel}}(t) + \text{const})$, where const was set to 0.01 in our experiments. To make sure that the speech filterbank energies in most of the time–frequency bins are larger than const and not masked, we multiply the input waveforms by a big number like 10^4.

- **Utterancewise mean normalization.** Mean normalization is applied to each utterance independently to reduce the channel effect. For each utterance, the normalized feature vectors are obtained as $\tilde{\mathbf{x}}_{\text{lm}}(t) = \mathbf{x}_{\text{lm}}(t) - \overline{\mathbf{x}}_{\text{lm}}$, where $\overline{\mathbf{x}}_{\text{lm}}$ is the mean vector of the current sentence.

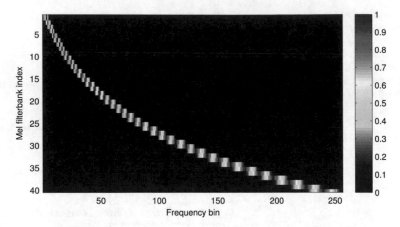

Fig. 4.5 Linear transform that converts power spectrum to mel filterbank power

- **Dynamic features.** Delta and acceleration features are computed using Eq. (5.16) of [32] with a window size of 2. These dynamic features are appended to the 40D log filterbanks. Hence, the final feature dimension for acoustic modeling is 120D.
- **Concatenate context frames.** 11 frames of contextual frames (5 from left and 5 from right) are concatenated to form the final input to the acoustic-model network.
- **Global mean and variance normalization (MVN).** A global MVN is applied to make sure that every feature dimension has zero mean and unit variance on the training corpus. The MVN is implemented as a diagonal affine transform which is fixed during the joint network training. The same transform is also used in testing.

4.3.5 Training Procedure

The joint training of the beamforming and acoustic model as shown in Fig. 4.1 optimizes both networks for ASR simultaneously. However, training the networks from random initialization may be difficult or slow to converge. In practice, we can initialize the beamforming weight prediction network and the acoustic-model network separately, and then put them together for joint training. We followed the following training procedure in the experiments:

1. Initialize the beamforming network in Fig. 4.3 by learning the behavior of DS beamforming. This step is carried out on simulated array data, where we know the true DOA and hence the DS beamforming weights. The network parameters are trained to minimize the mean squared error (MSE) between the predicted weights and the DS beamforming weights. Mean pooling is removed for this step.

2. (Optional) Refine the beamforming network by minimizing the MSE between the enhanced log spectrum and the clean log spectrum. This step can be carried out on simulated data, where the clean speech signal is available, or on real data, where a close-talk microphone signal can be used as the clean speech.
3. Initialize the acoustic model network by using beamformed log mel filterbank features from step 1 or 2. We can also use an existing acoustic model trained on a single-channel corpus if we want to optimize the beamforming network to be used with the existing acoustic model.
4. Further refine both the beamforming and the acoustic model network simultaneously to minimize ASR cost functions such as cross-entropy. It is also possible to use multitask learning, e.g., to optimize both the ASR cost function and the speech enhancement cost function in step 2. Note that for steps 2 and 4, sentence-based training is required due to the use of mean pooling and computing of dynamic features, etc. This means that an utterancewise minibatch is used instead of a framewise minibatch for training the networks.

4.4 Experiments

4.4.1 Settings

4.4.1.1 Corpus

We used both simulated and real array signals for the training of the beamforming and acoustic-model networks. The array geometry is shown in Fig. 4.2, and the sampling rate was 16 kHz. The simulated array signals were generated by convolving single-channel clean speech utterances with artificial room impulse responses (RIRs). The clean utterances came from the training set of the WSJCAM0 training set [23], which contains 7861 sentences. The RIRs were generated by using the image method [2] with various room sizes and T60 reverberation times. Three room sizes were used, including small, medium, and large rooms. The T60 reverberation time was randomly sampled from 0.1 to 1.0 s. After the reverberant array signal was simulated, additive noise samples from the REVERB Challenge corpus [16] were added at SNR levels randomly chosen from 0 to 30 dB. In total, 90 h of simulated array data were generated for steps 1 and 2 as listed in Sect. 4.3.5.

The real array signals were from the multiple distant microphone (MDM) scenario of the AMI meeting corpus [22]. The training set contained about 75 h of data, while the eval set contained about 8 h of data. The training set was used to jointly refine the beamforming and acoustic-model networks as described in steps 3 and 4 in Sect. 4.3.5. Besides the array signals, the AMI corpus also contains close-talk microphone data that was recorded in parallel with the array signals. The close-talk microphone data was used to train and test another acoustic model to show the upper bound of beamforming and other speech enhancement techniques.

The joint training of the beamforming and the acoustic model-networks was implemented in MATLAB using frame-based cross-entropy cost function. No special processing for the silent frames was applied during training. Once the networks were trained, they were used to generate enhanced speech as either waveform or filterbank features. The features were then used to train the DNN acoustic model from scratch using the Kaldi speech recognition toolkit [21]. The LSTM acoustic models were trained using CNTK [1]. Both the DNN and the LSTM acoustic model were first trained with a cross-entropy cost function, then a sequential cost function. For ASR decoding, a trigram language model trained with the word label of the 75 h training data was used.

4.4.1.2 Network Configurations

Although the beamforming network could be implemented by more advanced network types, we used a simple feedforward DNN in this study. There were two hidden layers in the DNN, each with 1024 sigmoid hidden nodes. As described previously, the input and output dimensions of the network were 588 and 4112, respectively. A linear activation function was used for the output layer.

Two types of acoustic model network were used. For joint cross-entropy (CE) training of the beamforming and acoustic-model networks, we used a feedforward DNN as the acoustic model which contained six hidden layers, each with 2048 sigmoid hidden nodes. The input and output dimensions were 1320 and 3968, respectively. To achieve better ASR performance, we also trained an LSTM-based acoustic model using the features processed by the beamforming network jointly trained with the DNN acoustic model. The reason for using a feedforward DNN as the acoustic model was mainly due to our implementation, not because of any limitation of the proposed beamforming network. We will investigate the use of LSTMs in both the beamforming and the acoustic-model network in the future.

4.4.2 Beam Patterns

To understand the behavior of the beamforming networks, we look into the beamforming weights they predicted and analyze the beam patterns. In Fig. 4.6, we show the beam patterns at four different frequencies for a simulated sentence that was not seen during training. Four beam patterns are compared, i.e., the beam pattern of the DS beamformer given the true DOA as marked by the red square boxes, and the beam patterns generated by the networks trained by steps 1, 2, and 4 listed in Sect. 4.3.5. From the figure, it can be observed that the beam patterns all have the maximum gain near the true DOA direction, except for step 4 (joint training) at frequencies of 1250 and 2500 Hz. There is spatial aliasing at the high frequencies, such as 3750 and 5000 Hz. The beam patterns of step 1 are very close to that of the DS beamformer. This is reasonable, as the DS beamformer is used to

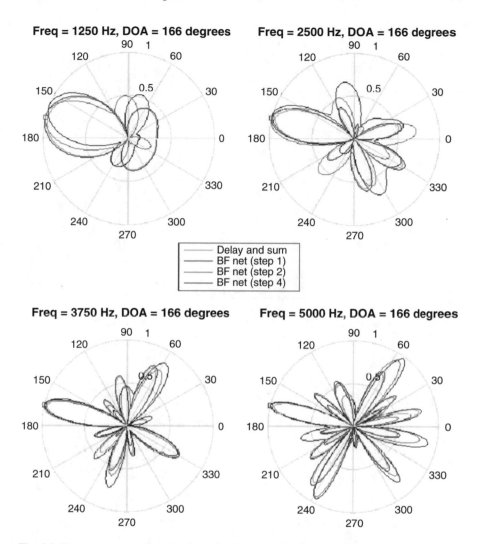

Fig. 4.6 Beam patterns predicted by beamforming networks for a simulated sentence. The *small square boxes on the left side* of each pattern denote the true DOA direction

teach the network in training step 1. The beam patterns of steps 2 and 4 deviate from the beam pattern of the DS beamformer as they are refined with speech enhancement and ASR cost functions, respectively.

In Fig. 4.7, we show beam patterns for a real sentence from the eval set of the AMI corpus. As we don't know the ground truth DOA for the AMI data, we only show the beam patterns generated by the beamforming networks obtained in training steps 1, 2, and 4. It can be observed that the beam patterns are similar to each other. At all frequencies, there is a beam pointing in a direction of about 120°, which is probably the direction of the speech source. Note that the beamforming networks

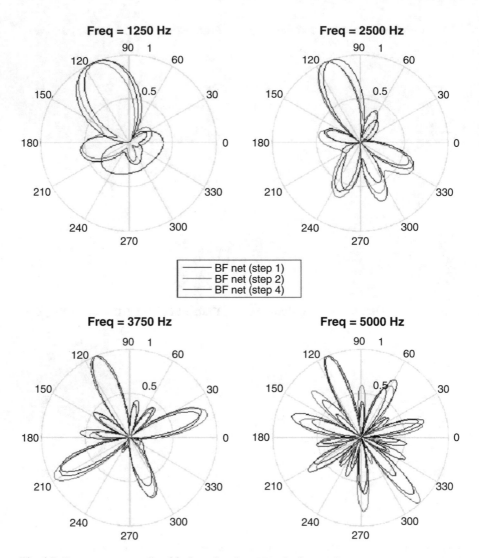

Fig. 4.7 Beam patterns predicted by beamforming networks for a real sentence

of steps 1 and 2 were only trained on simulated data, and they can still produce reasonable beam patterns for the AMI data.

From the beam patterns of both the simulated and the real array data, we can conclude that the beamforming networks are able to predict reasonable weights for unseen data as long as the array geometry does not change. The visual difference between the beam patterns of the 3 training steps is not significant. We will use ASR results to evaluate the performance of them in the following sections.

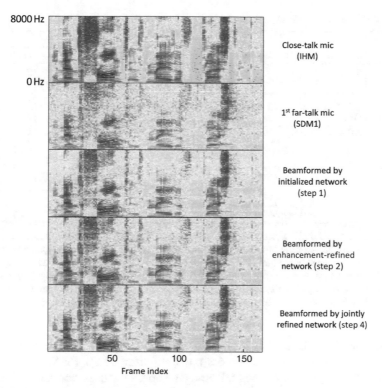

Fig. 4.8 Enhanced spectrograms produced by beamforming networks at various stages of training. The sentence is taken from the AMI corpus

4.4.3 Speech Enhancement Results

We also investigate the enhanced spectrograms produced by the beamforming networks, as shown in Fig. 4.8. Compared to the far-talk microphone's spectrogram (SDM1), the reverberation and noise are significantly reduced by the beamforming networks, e.g., in the frames after the 140th frame. However, it is not obvious to see the difference between the spectrograms enhanced by the networks of the three training steps. By an informal listening test, we noticed significantly better quality of the enhanced speech for steps 1, 2, and 4.

4.4.4 Speech Recognition Results

WER results for ASR are shown in Table 4.1. The results for several other systems are shown for comparison purposes, including the close-talk microphone (individual headset microphone, or IHM), a single distant microphone (SDM1, the first microphone of the circular array), and speech enhanced by applying the DS

Table 4.1 ASR results on eval set of AMI corpus

			Feature settings	Acoustic models		
Row no.	Method	Training steps	Use waveform	Feature type	DNN	LSTM
1	IHM	–	–	MFCC (CMNspk+ LDA+ MLLT+fMLLR)	25.5	–
2	SDM1	–	–		53.8	–
3	DSB	–	Yes		47.9	–
4	BF networks	Step 1	Yes		47.2	–
5		Step 2	Yes		45.7	–
6			Yes	fbank (CMNspk)	46.1	–
7			Yes	fbank (CMNutt)	45.3	–
8			No	fbank (CMNutt)	45.7	–
9		Step 4	No	fbank (CMNutt)	44.7	42.2
10	DSB	–	Yes	fbank (CMNutt)	–	44.8

For each row, the acoustic model was trained from scratch on the corresponding features. "Use waveform" specifies whether we resynthesized waveforms from the enhanced spectrum. "fbank" refers to log mel filterbank features. "CMNspk" and "CMNutt" are speaker-dependent and utterance-dependent CMN, respectively. "LDA," "MLLT," and "fMLLR" are feature projection/transforms usually used with met frequency cepstral coefficient (MFCC) features

beamformer (implemented by the BeamformIt toolkit [3]) to the eight-channel array signals. The DS beamformer reduces the WER from 53.8% for SDM1 to 47.9%, showing the effectiveness of beamforming in improving the performance of far-field ASR.

If we use the beamforming network trained in step 1 to process the array signals (row 4), we obtain similar performance to the DS beamformer. This is reasonable, as the beamforming network is trained to approximate the DS beamformer in the first step of training. It is worth noting that the BF network is applied to each segment (as defined by the Kaldi recipe of the AMI corpus, a few seconds long on average) independently, while DS beamforming is applied to entire audio files with the weights updated every few hundred milliseconds. So there is a minor difference between the two systems.

If the beamforming network is trained up to the second step (row 5), i.e., it optimizes the beamforming network for speech enhancement, the WER is further reduced to 45.7%. This is a significant improvement compared to training step 1 (row 4) and the DS beamformer (row 3). Until now, the BF network has not used the AMI corpus for training. This shows that the BF network is able to generalize well to unseen room types and speakers if the array geometry of the test data is the same as that of the simulated training data.

Until now, we extracted features from waveforms resynthesized from enhanced spectrum. In the joint beamforming and acoustic model training, the enhanced spectrum is used to generate log mel filterbank features directly without resynthesizing the waveforms. In the next three rows of the table (rows 6–8), we gradually move to using filterbank features generated from enhanced spectrum so the results will be comparable with joint training. In row 6, we use filterbank features generated from resynthesized waveform. By comparing to row 5, we observe that without the feature projections/transforms LDA+MLLT+fMLLR, the WER increases from 45.7% to 46.1%. We then replace speaker-dependent mean normalization (CMNspk) with utterance dependent mean normalization, and observe a 0.8% absolute WER reduction. Finally, we extract filterbanks directly from the enhanced spectrum and observe a 0.4% WER increase. The slightly better performance for resynthesizing the waveform (row 7) could be because the overlap and sum (OLS) operation used in waveform resynthesis may have a smoothing effect that reduced processing variations.

The joint training of the beamforming and acoustic-model networks using AMI data is shown in row 9. A WER reduction of 1.0% absolute is obtained, which proves the benefit of jointly refining the beamforming networks with the acoustic model. Finally, we replace the DNN acoustic model with the LSTM. The results in rows 9 and 10 show that the LSTM improves the performance over the DNN acoustic model, and the jointly trained beamforming network outperforms the DS beamformer by 2.6% WER absolute.

4.5 Summary and Future Directions

In this chapter, we have reviewed beamforming approaches for ASR, with a focus on learning-based approaches. We also described our recent work in detail, which uses a DNN to predict beamforming weights and jointly train the DNN with the acoustic model for optimal performance. From the experimental results, we have two major observations:

1. A feedforward DNN, or more generally a neural network, is able to learn the mapping from GCC features to beamforming weights in the frequency domain. Such a mapping is reasonably stable, as observed in the enhancement and ASR results where the DNN trained with simulated data works fine with unseen real recordings. The only requirement for such a mapping to be useful is that the array geometry in training and testing should be the same.
2. The beamforming network can be optimized for the ASR task if we jointly train it with the acoustic model using the ASR cost function. The experimental results in Table 4.1, row 8 and 9, prove this hypothesis. In the current version of joint training, we take only the filterbank features and then retrain the acoustic model from scratch. This shows that the improvement is solely coming from the improved beamforming network, not the acoustic model.

The work presented in this chapter can be improved further in several ways. We list some of them in the following:

- A Better network structure for beamforming weight prediction. For example, an LSTM could be used to predict beamforming weights. As an LSTM is able to capture long-term temporal dynamics of inputs, it may be able to produce better prediction, especially when the speech source is moving.
- More suitable inputs other than GCC. GCC is a classic way to extract phase information that is required for beamforming. The phase information can also be extracted by a network. It is possible to map other types of input to beamforming weights, e.g., raw waveforms or a multichannel complex-valued spectrum.
- Better target weights. In this chapter, we used a DS beamformer to teach the network. The DS beamformer uses only geometry information, but not noise information. It is possible to use other types of beamformer, such as MVDR, to teach the network.
- Multitask learning. We can train the beamforming and acoustic model networks to minimize multiple cost functions simultaneously. For example, one cost function could be ASR related (such as cross-entropy), and another could be speech enhancement related (MSE between enhanced speech and clean speech in suitable domains). Different cost functions provide training signals from different perspectives and may improve the robustness of the network.
- Beamforming for speech separation. In this work, we focused on a single-source scenario where the source with the highest energy is assumed to be the target. In practical applications, it is common to have multiple speakers speaking simultaneously. It is a challenging and promising direction to use a learning-based beamforming approach to solve the speech separation task.
- Array-geometry-independent beamforming. Current work assumes that the array geometry is fixed. Although this assumption is satisfied in many practical applications, it is interesting to investigate whether a single network is able to predict beamforming weights for multiple array geometries.

References

1. Agarwal, A., Akchurin, E., Basoglu, C., Chen, G., Cyphers, S., Droppo, J., Eversole, A., Guenter, B., Hillebrand, M., Hoens, R., et al.: An introduction to computational networks and the computational network toolkit. Microsoft Technical Report, MSR-TR-2014-112 (2014)
2. Allen, J., Berkley, D.: Image method for efficiently simulating small-room acoustics. J. Acoust. Soc. Am. **65**(4), 943–950 (1979)
3. Anguera, X., Wooters, C., Hernando, J.: Acoustic beamforming for speaker diarization of meetings. IEEE Trans. Audio Speech Lang. Process. **15**(7), 2011–2022 (2007)
4. Barker, J., Marxer, R., Vincent, E., Watanabe, S.: The third "CHiME" speech separation and recognition challenge: dataset, task and baselines. In: 2015 IEEE Automatic Speech Recognition and Understanding Workshop (ASRU 2015) (2015)

5. Bitzer, J., Simmer, K.U.: Superdirective microphone arrays. In: Brandstein, M.S., Ward, D. (eds.) Microphone Arrays: Signal Processing Techniques and Applications, Chap. 2, pp. 19–38. Springer, Berlin (2001)
6. Capon, J.: High-resolution frequency-wavenumber spectrum analysis. Proc. IEEE **57**(8), 1408–1418 (1969)
7. Doclo, S., Moonen, M.: GSVD-based optimal filtering for single and multimicrophone speech enhancement. IEEE Trans. Signal Process. **50**(9), 2230–2244 (2002)
8. Elko, G.W.: Spatial coherence functions for differential microphones in isotropic noise fields. In: Brandstein, M.S., Ward, D. (eds.) Microphone Arrays: Signal Processing Techniques and Applications, Chap. 4, pp. 61–85. Springer, Berlin (2001)
9. Er, M., Cantoni, A.: Solar wind monitor satellite: derivative constraints for broad-band element space antenna array processors. IEEE Trans. Audio Speech Lang. Process. **31**(6), 1378–1393 (1983)
10. Gales, M.J.: Maximum likelihood linear transformations for HMM-based speech recognition. Comput. Speech Lang. **12**(2), 75–98 (1998)
11. Griffiths, L.J., Jim, C.W.: An alternative approach to linearly constrained adaptive beamforming. IEEE Trans. Antennas Propag. **30**(1), 27–34 (1982)
12. Haeb-Umbach, R., Warsitz, E.: Adaptive filter-and-sum beamforming in spatially correlated noise. In: International Workshop on Acoustic Echo and Noise Control (IWAENC 2005) (2005)
13. Heymann, J., Drude, L., Chinaev, A., Haeb-Umbach, R.: BLSTM supported GEV beamformer front-end for the 3rd CHiME challenge. In: 2015 IEEE Workshop on Automatic Speech Recognition and Understanding (ASRU), pp. 444–451. IEEE, New York (2015)
14. Hoshen, Y., Weiss, R.J., Wilson, K.W.: Speech acoustic modeling from raw multichannel waveforms. In: IEEE International Conference on Acoustics, Speech and Signal Processing, pp. 4624 4628. IEEE, New York (2015)
15. Jahn Heymann, L.D., Haeb-Umbach, R.: Neural network based spectral mask estimation for acoustic beamforming. In: IEEE International Conference on Acoustics, Speech and Signal Processing. IEEE, New York (2016)
16. Kinoshita, K., Delcroix, M., Yoshioka, T., Nakatani, T., Sehr, A., Kellermann, W., Maas, R.: The REVERB challenge: a common evaluation framework for dereverberation and recognition of reverberant speech. In: IEEE Workshop on Applications of Signal Processing to Audio and Acoustics (WASPAA), pp. 1–4. IEEE, New York (2013)
17. Knapp, C.H., Carter, G.C.: The generalized correlation method for estimation of time delay. IEEE Trans. Acoust. Speech Signal Process. **24**(4), 320–327 (1976)
18. Liu, Y., Zhang, P., Hain, T.: Using neural network front-ends on far field multiple microphones based speech recognition. In: 2014 IEEE International Conference on Acoustics, Speech and Signal Processing (ICASSP), pp. 5542–5546. IEEE, New York (2014)
19. Narayanan, A., Wang, D.: Joint noise adaptive training for robust automatic speech recognition. In: 2014 IEEE International Conference on Acoustics, Speech and Signal Processing (ICASSP), pp. 2504–2508. IEEE, New York (2014)
20. Picone, J.W.: Signal modeling techniques in speech recognition. Proc. IEEE **81**(9), 1215–1247 (1993)
21. Povey, D., Ghoshal, A., Boulianne, G., Burget, L., Glembek, O., Goel, N., Hannemann, M., Motlicek, P., Qian, Y., Schwarz, P., Silovsky, J., Stemmer, G., Vesely, K.: The Kaldi speech recognition toolkit. In: IEEE 2011 Workshop on Automatic Speech Recognition and Understanding. IEEE Signal Processing Society (2011). IEEE Catalog No.: CFP11SRW-USB
22. Renals, S., Hain, T., Bourlard, H.: Recognition and understanding of meetings: the AMI and AMIDA projects. In: IEEE Workshop on Automatic Speech Recognition and Understanding, ASRU, Kyoto (2007). IDIAP-RR 07-46
23. Robinson, T., Fransen, J., Pye, D., Foote, J., Renals, S.: WSJCAM0: a British English speech corpus for large vocabulary continuous speech recognition. In: IEEE International Conference on Acoustics, Speech and Signal Processing (ICASSP), pp. 81–84 (1995)

24. Sainath, T.N., Weiss, R.J., Wilson, K.W., Narayanan, A., Bacchiani, M., Senior, A.: Speaker location and microphone spacing invariant acoustic modeling from raw multichannel waveforms. In: IEEE Workshop on Automatic Speech Recognition and Understanding (ARSU), pp. 30–36 (2015)

25. Sainath, T.N., Weiss, R.J., Wilson, K.W., Narayanan, A., Bacchiani, M.: Factored spatial and spectral multichannel raw waveform CLDNNs. In: IEEE International Conference on Acoustics, Speech and Signal Processing (2016)

26. Seltzer, M.L., Raj, B., Stern, R.M.: Likelihood-maximizing beamforming for robust hands-free speech recognition. IEEE Trans. Speech Audio Process. **12**(5), 489–498 (2004)

27. Souden, M., Benesty, J., Affes, S.: On optimal frequency-domain multichannel linear filtering for noise reduction. IEEE Trans. Audio Speech Lang. Process. **18**(2), 260–276 (2010)

28. Swietojanski, P., Ghoshal, A., Renals, S.: Convolutional neural networks for distant speech recognition. IEEE Signal Process. Lett. **21**(9), 1120–1124 (2014)

29. Van Veen, B.D., Buckley, K.M.: Beamforming: a versatile approach to spatial filtering. IEEE ASSP Mag. **5**(2), 4–24 (1988)

30. Xiao, X., Zhao, S., Zhong, X., Jones, D.L., Chng, E.S., Li, H.: A learning-based approach to direction of arrival estimation in noisy and reverberant environments. In: IEEE International Conference on Acoustics, Speech and Signal Processing (ICASSP), pp. 2814–2818. IEEE, New York (2015)

31. Xiao, X., Xu, C., Zhang, Z., Zhao, S., Sun, S., Watanabe, S., Wang, L., Xie, L., Jones, D.L., Chng, E.S., Li, H.: Investigation of neural networks based beamforming approaches for speech recognition: the NTU systems for CHiME-4 evaluation. In: CHiME 4 Workshop (2016)

32. Young, S., Evermann, G., Gales, M., Hain, T., Kershaw, D., Liu, X., Moore, G., Odell, J., Ollason, D., Povey, D., et al.: The HTK Book, 3.4 edn. Cambridge University Engineering Department, Cambridge (2006)

33. Yu, D., Eversole, A., Seltzer, M., Yao, K., Huang, Z., Guenter, B., Kuchaiev, O., Zhang, Y., Seide, F., Wang, H., et al.: An introduction to computational networks and the computational network toolkit. Tech. Rep. MSR, Microsoft Research (2014)

Chapter 5
Raw Multichannel Processing Using Deep Neural Networks

**Tara N. Sainath, Ron J. Weiss, Kevin W. Wilson, Arun Narayanan,
Michiel Bacchiani, Bo Li, Ehsan Variani, Izhak Shafran, Andrew Senior,
Kean Chin, Ananya Misra, and Chanwoo Kim**

Abstract Multichannel automatic speech recognition (ASR) systems commonly separate speech enhancement, including localization, beamforming, and postfiltering, from acoustic modeling. In this chapter, we perform multichannel enhancement jointly with acoustic modeling in a deep-neural-network framework. Inspired by beamforming, which leverages differences in the fine time structure of the signal at different microphones to filter energy arriving from different directions, we explore modeling the raw time-domain waveform directly. We introduce a neural network architecture which performs multichannel filtering in the first layer of the network and show that this network learns to be robust to varying target speaker direction of arrival, performing as well as a model that is given oracle knowledge of the true target speaker direction. Next, we show how performance can be improved by *factoring* the first layer to separate the multichannel spatial filtering operation from a single-channel filterbank which computes a frequency decomposition. We also introduce an adaptive variant, which updates the spatial filter coefficients at each time frame based on the previous inputs. Finally, we demonstrate that these approaches can be implemented more efficiently in the frequency domain. Overall, we find that such multichannel neural networks give a relative word error rate improvement of more than 5% compared to a traditional beamforming-based multichannel ASR system and more than 10% compared to a single-channel waveform model.

T.N. Sainath (✉) • R.J. Weiss • K.W. Wilson • A. Senior • M. Bacchiani
Google Inc., 76 9th Avenue, New York, NY 10011, USA
e-mail: tsainath@google.com

A. Narayanan • B. Li • E. Variani • I. Shafran • K. Chin • A. Misra • C. Kim
Google Inc., 1900 Charleston Road, Mountain View, CA 94043, USA

© Springer International Publishing AG 2017
S. Watanabe et al. (eds.), *New Era for Robust Speech Recognition*,
DOI 10.1007/978-3-319-64680-0_5

5.1 Introduction

While state-of-the-art automatic speech recognition (ASR) systems perform reasonably well in close-talking microphone conditions, performance degrades in conditions when the microphone is far from the user. In such far-field cases, the speech signal is degraded by reverberation and additive noise. To improve recognition in such cases, ASR systems often use signals from multiple microphones to enhance the speech signal and reduce the impact of reverberation and noise [2, 6, 10].

Multichannel ASR systems often use separate modules to perform recognition. First, microphone array speech enhancement is applied, typically broken into localization, beamforming, and postfiltering stages. The resulting single-channel enhanced signal is passed to a conventional acoustic model [15, 35]. A commonly used enhancement technique is filter-and-sum beamforming [2], which begins by aligning signals from different microphones in time (via localization) to adjust for the propagation delay from the target speaker to each microphone. The time-aligned signals are then passed through a filter for each microphone and summed to enhance the signal from the target direction and to attenuate noise coming from other directions. Commonly used filter design criteria are based on minimum variance distortionless response (MVDR) [10, 39] or multichannel Wiener filtering (MWF) [6].

When the end goal is to improve ASR performance, tuning the enhancement model independently from the acoustic model might not be optimal. To address this issue, [34] proposed likelihood-maximizing beamforming (LIMABEAM), which optimizes the beamformer parameters jointly with those of the acoustic model. This technique was shown to outperform conventional techniques such as delay-and-sum beamforming (i.e., filter-and-sum where the filters consist of impulses). Like most enhancement techniques, LIMABEAM is a model-based scheme and requires an iterative algorithm that alternates between acoustic-model inference and enhancement model parameter optimization. Contemporary acoustic models are generally based on neural networks, optimized using a gradient learning algorithm. Combining model-based enhancement with an acoustic model that uses gradient learning can lead to considerable complexity, e.g., [17].

In this chapter we extend the idea of performing beamforming jointly with acoustic modeling from [34], but do this within the context of a deep-neural-network (DNN) framework by training an acoustic model directly on the raw signal. DNNs are attractive because they have been shown to be able to perform feature extraction jointly with classification [23]. Previous work has demonstrated the possibility of training deep networks directly on raw, single-channel, time-domain waveform samples [11, 18, 19, 24, 30, 37]. The goal of this chapter is to explore a variety of different joint enhancement/acoustic-modeling DNN architectures that operate on multichannel signals. We will show that jointly optimizing both stages is more effective than techniques which cascade independently tuned enhancement algorithms with acoustic models.

Since beamforming takes advantage of the fine time structure of the signal at different microphones, we begin by modeling the raw time-domain waveform directly. In this model, introduced in [18, 29], the first layer consists of multiple time convolution filters, which map the multiple microphone signals down to a single time–frequency representation. As we will show, this layer learns bandpass filters which are spatially selective, often learning several filters with nearly identical frequency response, but with nulls steered toward different directions of arrival. The output of this spectral filtering layer is passed to an acoustic model, such as a convolutional long short-term memory deep-neural-network (CLDNN) acoustic model [28]. All layers of the network are trained jointly.

As described above, it is common for multichannel speech recognition systems to perform spatial filtering independently from single-channel feature extraction. With this in mind, we next investigate explicitly factorizing these two operations to be separate neural network layers. The first layer in this "factored" raw-waveform model consists of short-duration multichannel time convolution filters which map multichannel inputs down to a single channel, with the idea that the network might learn to perform broadband spatial filtering in this layer. By learning several filters in this "spatial filtering layer," we hypothesize that the network can learn filters for multiple different spatial look directions. The single-channel waveform output of each spatial filter is passed to a longer-duration time convolution "spectral filtering layer" intended to perform finer-frequency-resolution spectral decomposition analogous to a time-domain auditory filterbank as in [30]. The output of this spectral filtering layer is also passed to an acoustic model.

One issue with the two architectures above is that once weights are learned during training, they remain fixed for each test utterance. In contrast, some beamforming techniques, such as the generalized sidelobe canceller [14], update weights adaptively within each utterance. We explore an adaptive neural net architecture, where a long short-term memory (LSTM) is used to predict spatial filter coefficients that are updated at each frame. These filters are used to filter and sum the multichannel input, replacing the "spatial filtering layer" of the factored model described above, before passing the enhanced single-channel output to a waveform acoustic model.

Finally, since convolution between two time domain signals is equivalent to the elementwise product of their frequency-domain counterparts, we investigate speeding up the raw-waveform neural network architectures described above by consuming the complex-valued fast Fourier transform of the raw input and implementing filters in the frequency domain.

5.2 Experimental Details

5.2.1 Data

We conducted experiments on a dataset comprising about 2000 h of noisy training data consisting of three million English utterances. This dataset was created by artificially corrupting clean utterances using a room simulator to add varying degrees of noise and reverberation. The clean utterances were anonymized and hand-transcribed voice search queries, and are representative of Google's voice search traffic, as described in Chap. 18. Noise signals, which included music and ambient noise sampled from YouTube and recordings of "daily life" environments, were added to the clean utterances at signal-to-noise ratios (SNRs) ranging from 0 to 20 dB, with an average of about 12 dB. Reverberation was simulated using the image method [1]—room dimensions and microphone array positions were randomly sampled from 100 possible room configurations with T_{60}s ranging from 400 to 900 ms, with an average of about 600 ms. The simulation used an eight-channel uniform linear microphone array, with an inter-microphone spacing of 2 cm. Both the noise source location and the target speaker locations changed between utterances; the distance between the sound source and the microphone array varied between 1 and 4 m.

The primary evaluation set consisted of a separate set of about 30,000 utterances (over 20 h), and was created by simulating similar SNR and reverberation settings to the training set. Care was taken to ensure that the room configurations, SNR values, T_{60} times, and target speaker and noise positions in the evaluation set differed from those in the training set. The microphone array geometry in the training and simulated test sets was identical.

We obtained a second "rerecorded" test set by playing the evaluation set and the noises separately using a mouth simulator and a speaker, respectively, in a living room setting. The signals were recorded using a seven-channel circular microphone array with a radius of 3.75 cm. Assuming an x-axis that passed through two diagonally opposite mics on the circumference of the array, the angle of arrival of the target speaker ranged from 0° to 180°. Noise originated from locations different from the target speaker. The distance of the sources to the target ranged from 1 to 6 m. To create noisy rerecorded eval sets, the rerecorded speech and noise signals were mixed artificially after scaling the noise to obtain SNRs ranging from 0 to 20 dB. The distribution of the SNR matched the distribution used to generate the simulated evaluation set. We created four versions of the rerecorded sets to measure the generalization performance of our models. The first two had rerecorded speech without any added noise. The mic array was placed at the center of the room and closer to the wall, respectively, to capture reasonably different reverberation characteristics. The remaining two subsets corresponded to the noisy versions of these sets.

5.2.2   Baseline Acoustic Model

We compare the models proposed in this chapter to a baseline CLDNN acoustic model trained using log mel features [28] computed with a 25 ms window and a 10 ms hop. Single-channel models were trained using signals from channel 1, $C = $ two-channel models used channels 1 and 8 (14 cm spacing), and $C = $ four-channel models used channels 1, 3, 6, and 8 (14 cm array span, with a microphone spacing of 4 cm–6 cm–4 cm).

The baseline CLDNN architecture is shown in the CLDNN module in Fig. 5.1. First, the fConv layer performs convolution across the frequency dimension of the

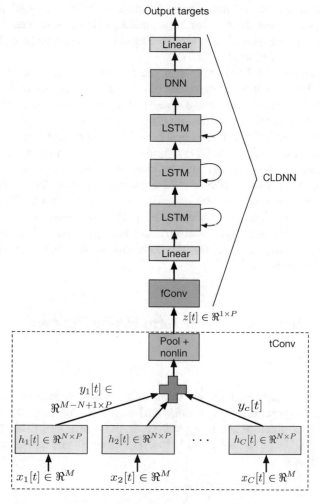

Fig. 5.1 Multichannel raw-waveform CLDNN architecture

input log mel time–frequency feature to gain some invariance to pitch and vocal tract length. The architecture used for this convolutional layer is similar to that proposed in [25]. Specifically, a single convolutional layer with 256 filters of size 1×8 in time–frequency is used. Our pooling strategy is to use nonoverlapping maxpooling along the frequency axis, with a pooling size of 3. The pooled output is given to a 256-dimensional linear low-rank layer.

The output of frequency convolution is passed to a stack of LSTM layers, which model the signal across long time scales. We used three LSTM layers, each comprising 832 cells, and a 512 unit projection layer for dimensionality reduction following [33]. Finally, we pass the final LSTM output to a single fully connected DNN layer comprising 1024 hidden units. Due to the high dimensionality of the 13,522 context-dependent state output targets used by the language model, a 512-dimensional linear-output low-rank projection layer is used prior to the softmax layer to reduce the number of parameters in the overall model [27]. Some experiments in the chapter did not use the frequency convolution layer, and we will refer to such acoustic models as LDNNs.

During training, the CLDNN is unrolled for 20 time steps and trained using truncated back-propagation through time (BPTT). In addition, the output state label is delayed by five frames, as we have observed that information about future frames helps to better predict the current frame [28].

Unless otherwise indicated, all neural networks were trained using asynchronous stochastic gradient descent (ASGD) optimization [9] to optimize a cross-entropy (CE) criterion. Additional sequence-training experiments also used distributed ASGD [16]. All networks had 13,522 context-dependent (CD) output targets. The weights for all convolutional neural network (CNN) and DNN layers were initialized using the Glorot–Bengio strategy [13], while those of all LSTM layers were randomly initialized using a uniform distribution between -0.02 and 0.02. We used an exponentially decaying learning rate, initialized to 0.004 and decaying by 0.1 over 15 billion frames.

5.3   Multichannel Raw-Waveform Neural Network

5.3.1   Motivation

The proposed multichannel raw-waveform CLDNN is related to filter-and-sum beamforming, a generalization of delay-and-sum beamforming which filters the signal from each microphone using a finite impulse response (FIR) filter before summing them. Using similar notation to [34], filter-and-sum enhancement can be written as follows:

$$y[t] = \sum_{c=0}^{C-1} \sum_{n=0}^{N-1} h_c[n] x_c[t - n - \tau_c], \qquad (5.1)$$

where $h_c[n]$ is the nth tap of the filter associated with microphone c, $x_c[t]$, is the signal received by microphone c at time t, τ_c is the steering time difference of arrival induced in the signal received by microphone c used to align it with the other array channels, and $y[t]$ is the output signal. C is the number of microphones in the array, and N is the number of FIR filter taps.

5.3.2 Multichannel Filtering in the Time Domain

Enhancement algorithms implementing (5.1) generally depend on an estimate of the steering delay τ_c obtained using a separate localization model, and they compute filter parameters $h_c[n]$ by optimizing an objective such as MVDR. In contrast, our aim is to allow the network to jointly estimate steering delays and filter parameters by optimizing an acoustic-modeling classification objective. The model captures different steering delays using a bank of P multichannel filters. The output of filter $p \in \{0, \ldots, P-1\}$ can be written as follows:

$$y^P[t] = \sum_{c=0}^{C-1} \sum_{n=0}^{N-1} h_c^p[n] x_c[t-n] = \sum_{c=0}^{C-1} x_c[t] * h_c^p, \tag{5.2}$$

where the steering delays are implicitly absorbed into the filter parameters $h_c^p[n]$. In this equation, "$*$" denotes the convolution operation

The first layer in our raw-waveform architecture implements (5.2) as a multichannel convolution (in time) with a FIR spatial filterbank $h_c = \{h_c^1, h_c^2, \ldots, h_c^P\}$, where $h_c \in \Re^{N \times P}$ for $c \in \{0, \ldots, C-1\}$. Each filter h_c^p is convolved with the corresponding input channel x_c, and the overall output for filter p is computed by summing the result of this convolution across all channels $c \in \{0, \ldots, C-1\}$. The operation within each filter is equivalent to an FIR filter-and-sum beamformer, except that it does not explicitly shift the signal in each channel by an estimated time difference of arrival. As we will show, the network learns the steering delay and filter parameters implicitly.

The output signal remains at the same sampling rate as the input signal, which contains more information than is typically relevant for acoustic modeling. In order to produce an output that is invariant to perceptually and semantically identical sounds appearing at different time shifts, we pool the outputs in time after filtering [18, 30], in an operation that has an effect similar to discarding the phase in the short-time Fourier transform. Specifically, the output of the filterbank is maxpooled across time to give a degree of short-term shift invariance, and then passed through a compressive nonlinearity.

As shown in [18, 30], single-channel time convolution layers similar to the one described above implement a conventional time-domain filterbank. Such layers are capable of implementing, for example, a standard gammatone filterbank, which consists of a bank of time-domain filters followed by rectification and averaging over a small window. Given a sufficiently large P, the corresponding multichannel layer can (and, as we will show, does in fact) similarly implement a frequency decomposition in addition to spatial filtering. We will therefore subsequently refer to the output of this layer as a "time–frequency" feature representation.

A schematic of the multichannel time convolution layer is shown in the tConv block Fig. 5.1. First, we take a small window of the raw waveform of length M samples for each channel C, denoted as $\{x_0[t], x_1[t], \ldots, x_{C-1}[t]\}$ for $t \in 1, \ldots, M$. The signal from each channel x_c is convolved with a bank of P filters with N taps $h_c = \{h_c^1, h_c^2, \ldots, h_c^P\}$. When the convolution is stridden by 1 in time across M samples, the output from the convolution in each channel is $y_c[t] \in \Re^{(M-N+1)\times P}$. After summing $y_c[t]$ across channels c, we maxpool the filterbank output in time (thereby discarding short-term phase information), over the entire time length of the output signal $M - N + 1$, to produce $y[t] \in \Re^{1\times P}$. Finally, we apply a rectified nonlinearity, followed by a stabilized logarithm compression,[1] to produce $z[l]$, a P-dimensional frame-level feature vector at frame l. We then shift the window around the waveform by 10 ms and repeat this time convolution, producing a sequence of feature frames at 10 ms intervals.

To match the timescale of the log mel features, the raw-waveform features are computed with an identical filter size of 25 ms, or $N = 400$ at a sampling rate of 16 kHz. The input window size is 35 ms ($M = 560$), giving a 10 ms fully overlapping pooling window. Our experiments explored varying the number of time-convolutional filters P.

As shown in the CLDNN block in Fig. 5.1, the output of the time-convolutional layer (tConv) produces a frame-level feature, denoted as $z[l] \in \Re^{1\times P}$. This feature is then passed to a CLDNN model [28] described in Sect. 5.2, which predicts context-dependent state output targets.

5.3.3 Filterbank Spatial Diversity

Figure 5.2 plots example multichannel filter coefficients and their corresponding spatial responses, or beam patterns, after training for tConv. The beam patterns show the magnitude response in dB as a function of frequency and direction of arrival, i.e., each horizontal slice of the beam pattern corresponds to the filter's magnitude response for a signal coming from a particular direction, and each vertical slice corresponds to the filter's response across all spatial directions in a particular frequency band. Lighter shades indicate regions of the frequency–direction space which are passed through the filter, while darker shades indicate regions which are filtered out. Within a given beam pattern, we refer to the frequency band containing the maximum overall response as the filter's *center frequency* (since the filters are

---

[1]We use a small additive offset to truncate the output range and avoid numerical instability with very small inputs: $\log(\cdot + 0.01)$.

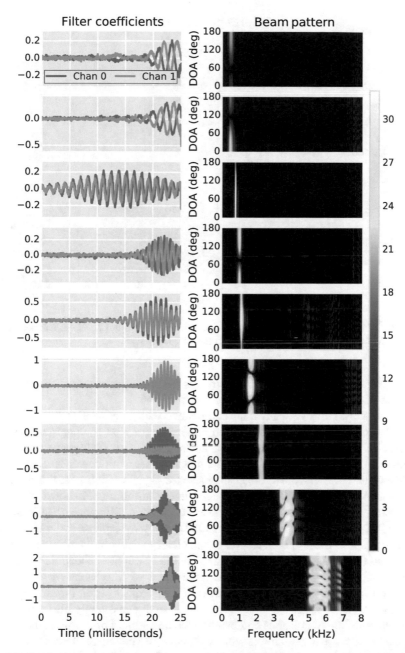

Fig. 5.2 Example filter coefficients and corresponding spatial-response beam patterns learned in a network with 128 tConv filters trained on two-channel inputs. Some filters learned by this network have nearly-identical center frequencies but different spatial responses. For example, the top two example filters both have center frequencies of about 440 Hz, but the first filter has a null at a direction of arrival of about 60°, while the second has a null at about 120°. The corresponding phase difference between the two channels of each filter is visible in the time-domain filter coefficients plotted on the left

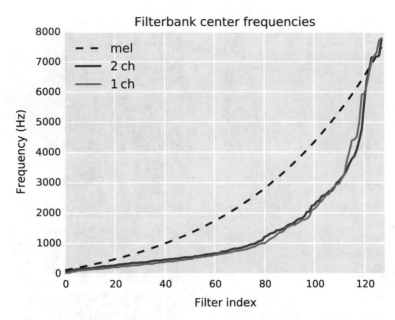

Fig. 5.3 Comparison of the peak response frequencies of waveform CLDNN filterbanks trained on one- and two-channel inputs to the standard mel frequency scale

primarily bandpass in frequency), and the direction corresponding to the minimum response in that frequency as the filter's *null direction*.

The network tends to learn filter coefficients with very similar shapes in each channel except they are slightly shifted relative to each other, consistent with the notion of a steering delay τ_c described in Sect. 5.3. Most filters have a bandpass response in frequency, with bandwidths that increase with center frequency, and many are steered to have a stronger response for signals arriving from a particular direction. Approximately two-thirds of the filters in the model shown in Fig. 5.2 demonstrate a significant spatial response, i.e., show a difference of at least 6 dB between the directions with the minimum and maximum response at the filter center frequency. Such strong spatial responses are clearly visible in the null near 120° in the second filter, and a similar null near 60° in the fourth filter shown in Fig. 5.2.

Figure 5.3 plots the peak response frequencies of filterbanks from networks trained on one- and two-channel networks of the form shown in Fig. 5.2. The two networks converge to similar frequency scales, both consistently allocating many more filters to low frequencies compared to the mel scale. For example, the learned filterbanks have roughly 80 filters with peak responses below 1000 Hz, while a 128-band mel scale has only 40 bands with center frequencies below 1000 Hz. The network also learns subsets of filters with the same overall shape and frequency response but tuned to have nulls in different directions, as illustrated by the top two example filters in Fig. 5.2. Such diversity in spatial response gives upstream layers

Table 5.1 WER for
raw-waveform multichannel
CLDNNs with different
numbers of input channels

Filters	2 ch (14 cm)	4 ch (4–6–4 cm)	8 ch (2 cm)
128	21.8	21.3	21.1
256	21.7	20.8	20.6
512	–	20.8	20.6

The intermicrophone spacing is given in parentheses

Table 5.2 WER for log mel
multichannel CLDNNs

Filters	2 ch (14 cm)	4 ch (4–6–4 cm)	8 ch (2 cm)
128	22.0	21.7	22.0
256	21.8	21.6	21.7

information that can be used to discriminate between signals arriving from different directions.

Because each filter has a fixed directional response, the ability of the network to exploit directional cues is constrained by the number of filters it uses. By increasing the number of filters, we can potentially improve the spatial diversity of the learned filters and therefore allow the network to better exploit directional cues. Table 5.1 demonstrates the effect of increasing the number of filters on the overall word error rate (WER). Improvements saturate at 128 filters for networks trained on two channel inputs, while four- and eight-channel networks continue to improve with 256 filters. With additional input channels, the tConv filters are able to learn more complex spatial responses (even though the total array span is unchanged), enabling the network to make use of additional filterbank capacity to improve performance.

5.3.4   Comparison to Log Mel

We trained baseline multichannel log mel CLDNNs by computing log mel features for each channel, and treating these as separate feature maps into the CLDNN. Since the raw-waveform model improves as we increase the number of filters, we performed the same experiment for log mel. It should be noted that concatenating magnitude-based features (i.e., log mel) from different channels into a neural network has been shown to give improvements over a single channel [22, 36].

Table 5.2 shows that for log mel, neither increasing the number of filters (frequency bands) nor increasing the number of microphone channels has a strong effect on word error rate. Since log mel features are computed from the fast Fourier transform (FFT) magnitude, the fine time structure (stored in the phase), and therefore information about inter-microphone delays, is discarded. Log mel models can therefore only make use of the weaker intermicrophone level difference cues. However, the multichannel time-domain filterbanks in the raw-waveform models utilize the fine time structure and show larger improvements as the number of filters increases.

Comparing Tables 5.1 and 5.2, we can see that raw-waveform models consistently outperform log mel, particularly for a larger number of channels where more spatial diversity is possible.

5.3.5   Comparison to Oracle Knowledge of Speech TDOA

Note that the models presented in the previous subsection do not explicitly estimate the time difference of arrival of the target source at different microphones, which is commonly done in beamforming [2]. Time difference of arrival (TDOA) estimation is useful because time-aligning and combining signals steers the array such that the target speech signal is enhanced relative to noise sources coming from other directions.

In this section, we analyze the behavior of raw-waveform CLDNNs when the signals are time-aligned using the true TDOA calculated using the room geometry. In the delay-and-sum (D+S) approach, we shift the signals in each channel by the corresponding TDOA, average them together, and pass the result into a one-channel raw-waveform CLDNN. In the time-aligned multichannel (TAM) approach, we align the signals in time and pass them as separate channel inputs to a multichannel raw-waveform CLDNN. Thus the difference between the multichannel raw waveform CLDNNs described in Sect. 5.2 and TAM is solely in how the data is presented to the network (whether or not they are first explicitly aligned to "steer" toward the target speaker direction); the network architectures are identical.

Table 5.3 compares the WER of D+S, TAM, and raw-waveform models when we do not shift the signals by the TDOA. First, notice that as we increase the number of channels, D+S continues to improve, since finer spatial sampling reduces the sidelobes of the spatial response, leading to increased suppression of noise and reverberation energy arriving from other directions. Second, notice that TAM always has better performance than D+S, as TAM is more general than D+S because it allows individual channels to be filtered before being combined. But notice that the raw-waveform CLDNN, without any explicit time alignment or localization (TDOA estimation), performs as well as TAM with time alignment. This shows us that the trained unaligned network is implicitly robust to varying TDOA.

Table 5.3   WER with oracle knowledge of the true target TDOA

Feature	1 ch	2 ch (14 cm)	4 ch (4–6–4 cm)	8 ch (2 cm)
Oracle D+S	23.5	22.8	22.5	22.4
Oracle TAM	23.5	21.7	21.3	21.3
Raw, no TDOA	23.5	21.8	21.3	21.1

All models use 128 filters

Table 5.4 Raw-waveform model WER after sequence training

Model	WER–CE	WER–Seq
Raw, 1 ch	23.5	19.3
D+S, 8 ch, oracle	22.4	18.8
MVDR, 8 ch, oracle	22.5	18.7
Raw, unfactored, 2 ch	21.8	18.2
Raw, unfactored, 4 ch	20.8	17.2
Raw, unfactored, 8 ch	20.6	17.2

All models use 128 filters

5.3.6   Summary

To conclude this section, we show the results after sequence training in Table 5.4. We also include results for eight-channel oracle D+S, where the true target speech TDOA is known, as well as oracle MVDR [39], where the true speech and noise estimates are known in addition to the target TDOA. Table 5.4 shows that the raw unfactored model, even using only two-channel inputs and no oracle information, outperforms the single-channel and oracle signal-processing methods. Using four-channel inputs, the raw-waveform unfactored model achieves between an 8–10% relative improvement over a single channel, D+S, and MVDR.

5.4   Factoring Spatial and Spectral Selectivity

5.4.1   Architecture

In multichannel speech recognition systems, multichannel spatial filtering is often performed separately from single-channel feature extraction. However, in the unfactored raw-waveform model, spatial and spectral filtering are done in one layer of the network. In this section, we factor out spatial and spectral filtering into separate layers, as shown in Fig. 5.4.

The motivation for this architecture is to design the first layer to be spatially selective, while implementing a frequency decomposition shared across all spatial filters in the second layer. Thus the combined output of the second layer will be the Cartesian product of all spatial and spectral filters.

The first layer, denoted by tConv1 in the figure, again models (5.2) and performs a multichannel time convolution with am FIR spatial filterbank. The operation of each filter $p \in \{0, \ldots, P - 1\}$, which we will refer to as a spatial look direction in the factored model, can again be interpreted as a filter-and-sum beamformer, except that any overall time shift is implicit in the filter coefficients rather than being explicitly represented as in (5.1). The main differences between the unfactored and factored approaches are as follows. First, both the filter size N and the number of filters P are much smaller in order to encourage the network

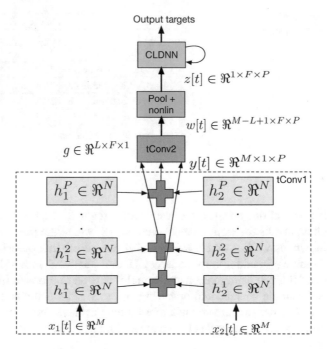

Fig. 5.4 Factored multichannel raw-waveform CLDNN architecture for P look directions. The figure shows two channels for simplicity

to learn filters with a broadband response in frequency that span a small number of spatial look directions needed to cover all possible target speaker locations. The shorter filters in this layer will have worse frequency resolution than those in the unfactored model, but that will be dealt with in the next layer. We hope that this poor frequency resolution will encourage the network to use this first layer to focus on spatial filtering, with a limited spectral response. To make the combination of the first two layers of the factored model conceptually similar to the first layer of the unfactored model (i.e., a bank of bandpassed beamformers), the multichannel (first) filter layer is not followed by any nonlinear compression (i.e., ReLU, log), and we do not perform any pooling between the first and second layers.

The second time-convolution layer, denoted by tConv2 in the figure, consists of longer-duration single-channel filters. It therefore can learn a decomposition with better frequency resolution than the first layer but is incapable of doing any spatial filtering. Given the P feature maps from the first layer, we perform a time convolution on each of these signals, very similar to the single-channel time-convolution layer described in [30], except that the time convolution is shared across all P feature maps or "look directions.". We denote this layer's filters as $g \in \Re^{L \times F \times 1}$, where 1 indicates sharing across the P input feature maps. The "valid" convolution produces an output $w[t] \in \Re^{(M-L+1) \times F \times P}$. The output of the spectral convolution layer, for each look direction p and each filter f, is given by (5.3):

$$w_f^p[t] = y^p[t] * g_f. \tag{5.3}$$

Next, we pool the filterbank output in time thereby discarding short-time (i.e. phase) information, over the entire time length of the output signal, to produce an output of dimension $1 \times F \times P$. Finally, we apply a rectified nonlinearity, followed by a stabilized logarithm compression, to produce a frame-level feature vector at frame l, i.e., $z_l \in \Re^{1 \times F \times P}$, which is then passed to a CLDNN model. We then shift the window of the raw waveform by a small (10 ms) hop and repeat this time convolution to produce a set of time–frequency–direction frames at 10 ms intervals.

5.4.2 Number of Spatial Filters

We first explore the behavior of the proposed factored multichannel architecture as the number of spatial filters P varies. Table 5.5 shows that we get good improvements up to ten spatial filters. We did not explore above ten filters due to the computational complexities of passing ten feature maps to the tConv2 layer. The factored network, with 10 spatial filters, achieves a WER of 20.4%, a 6% relative improvement over the two-channel unfactored multichannel raw-waveform CLDNN. It is important to note that since the tConv2 layer is shared across all look directions P, the total number of parameters is actually less than in the unfactored model.

5.4.3 Filter Analysis

To better understand what the tConv1 layer learns, Fig. 5.5 plots two-channel filter coefficients and the corresponding spatial responses, or beam patterns, after training.

Despite the intuition described in Sect. 5.4.1, the first-layer filters appear to perform both spatial and spectral filtering. However, the beam patterns can nevertheless

Table 5.5 WER when varying the size of the spatial filters in tConv1

# spatial filters P	WER
Baseline 2 ch, raw [29]	21.8
1	23.6
3	21.6
5	20.7
10	20.4

All models use 128 filters for tConv2, and results are presented for two channels

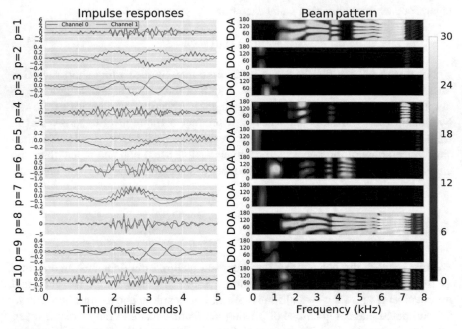

Fig. 5.5 Trained filters and spatial responses for ten spatial directions

Table 5.6 WER for training
vs. fixing the tConv1 layer,
two channels

# spatial filters P	tConv1 layer	WER
5	Fixed	21.9
5	Trained	20.9

be categorized into a few broad classes. For example, filters 2, 3, 5, 7, and 9
in Fig. 5.5 only pass through some low-frequency subbands below about 1.5 kHz,
where most vowel energy occurs, but are steered to have nulls in different directions.
Very little spatial filtering is done in high-frequency regions, where many fricatives
and stops occur. The low frequencies are most useful for localization because they
are not subject to spatial aliasing and because they contain much of the energy in
the speech signal; perhaps that is why the network exhibits this structure.

To further understand the benefit of the spatial and spectral filtering in tConv1,
we forced this layer to perform only spatial filtering by initializing the filters to
be an impulse centered at a delay of zero for channel 0, and offset from zero
in channel 1 by different delays for each filter. By not training this layer, this
amounts to performing delay-and-sum filtering across a set of fixed look directions.
Table 5.6 compares performance when fixing vs. training the tConv1 layer. The
results demonstrate that learning the filter parameters, and therefore performing
some spectral decomposition, improves performance over keeping this layer fixed.

Table 5.7 Factored-model
WER after sequence training,
simulated

Method	WER–CE	WER–Seq
Raw, unfactored, 2 ch	21.8	18.2
Raw, factored, 2 ch	20.4	17.2
Raw, unfactored, 4 ch	20.8	17.2
Raw, factored, 4 ch	19.6	16.3

5.4.4   Results Summary

To conclude this section, we show the results after sequence training in Table 5.7, comparing the factored and unfactored models. Notice that the two-channel factored model provides 6% relative improvement over the unfactored model, while the four-channel model provides 5% relative improvement. We do not go above four channels, as the results from Table 5.4 in Sect. 5.3.6 show that there is no difference between 4 and 8 channels.

5.5   Adaptive Beamforming

While the unfactored model improves over the factored model, the model also suffers from a few drawbacks. First, the learned filters in this model are fixed during decoding, which potentially limits the ability to adapt to previously unseen or changing conditions. In addition, since the factored model must perform spectral filtering for every look direction, this comes with a large computational complexity.

5.5.1   NAB Model

To address the limited adaptability and reduce the computational complexity of the models from [29, 32], we propose a neural network adaptive beamforming (NAB) model [21] which reestimates a set of spatial filter coefficients at each input frame. The NAB model is depicted in Fig. 5.6. At each time frame l, it takes in a small window of M waveform samples for each channel c from the C channel inputs, denoted as $x_0(l)[t], x_1(l)[t], \ldots, x_{C-1}(l)[t]$ for $t \in \{0, \ldots, M-1\}$. Additionally to previous notation, the frame index l is explicitly used in this section to emphasize the frame-dependent filtering coefficients. For simplicity, the figure shows an NAB model with $C = 2$ channels. We will describe the different NAB modules in subsequent subsections.

Fig. 5.6 Neural network adaptive beamforming (NAB) model architecture. It consists of filter prediction (FP), filter-and-sum (FS) beamforming, acoustic modeling (AM), and multitask learning (MTL). The figure shows two channels for simplicity

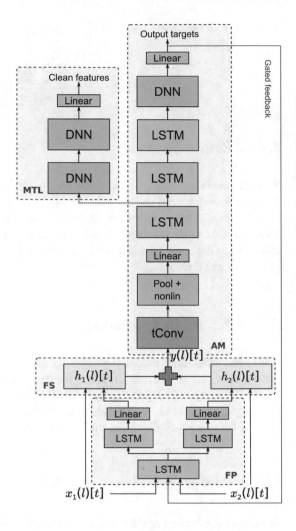

5.5.1.1 Adaptive Filters

The adaptive filtering layer is given by (5.4), where $h_c(l)[n]$ is the estimated filter for channel c at time frame l. This model is very similar to the FS model from (5.1), except now the steering delay τ_c is implicitly absorbed into the estimated filter parameters:

$$y(l)[t] = \sum_{c=0}^{C-1} \sum_{n=0}^{N-1} h_c(l)[n] x_c(l)[t-n]. \tag{5.4}$$

In order to estimate $h_c(l)[t]$, we train a filter prediction (FP) module with one shared LSTM layer, one layer of channel-dependent LSTMs, and linear output projection layers to predict N filter coefficients for each channel. The input to the *FP module* is a concatenation of frames of raw input samples $x_c(l)[t]$ from all the channels, and can also include features typically computed for localization such as cross-correlation features [20, 40, 41]. The estimation of FP module parameters is jointly done with the acoustic modeling (AM) parameters by directly minimizing a cross-entropy or sequence loss function. Following (5.4), the estimated filter coefficients $h_c(l)[t]$ are convolved with input samples $x_c(l)[t]$ for each channel. The outputs of the convolution are summed across channels to produce a single-channel signal $y(l)[t]$.

After adaptive FS, the single-channel enhanced signal $y(l)[t]$ is passed to an *AM module* (Fig. 5.6). We adopt the single-channel raw-waveform CLDNN model [30] for acoustic modeling, except that we now skip the frequency convolution layer as it has recently been shown in [26] to not help for noisier tasks. During training, the AM and FP (Fig. 5.6) are trained jointly.

5.5.1.2 Gated Feedback

Augmenting the network input at each frame with the prediction from the previous frame has been shown to improve performance [4]. To investigate the benefit of feedback in the NAB model, we pass the AM prediction at frame $l - 1$ back to the FP model at time frame l (line labeled "Gated feedback" in Fig. 5.6). Since the softmax prediction is very high-dimensional, we feed back the low-rank activations preceding the softmax to the FP module to limit the increase of model parameters [42].

This feedback connection gives the FP module high-level information about the phonemic content of the signal to aid in estimating beamforming filter coefficients. This feedback comprises model *predictions* which may contain errors, particularly early in training, and therefore might lead to poor model training [4]. A gating mechanism [8] is hence introduced into the connection to modulate the degree of feedback. Unlike conventional LSTM gates, which control each dimension independently, we use a global scalar gate to moderate the feedback. The gate $g^{\text{fb}}(l)$ at time frame l is computed from the input waveform samples $x(l)$, the state of the first FP LSTM layer $s(l - 1)$, and the feedback vector $v(l - 1)$, as follows:

$$g^{\text{fb}}(l) = \sigma(w_x^T \cdot x(l) + w_s^T \cdot s(l - 1) + w_v^T \cdot v(l - 1)), \tag{5.5}$$

where w_x, w_s, and w_v are the corresponding weight vectors and σ is an elementwise nonlinearity. We use a logistic function for σ which outputs values in the range $[0, 1]$, where 0 cuts off the feedback connection and 1 directly passes the feedback through. The effective FP input is hence $\left[x(l), \quad g^{\text{fb}}(l)v(l - 1) \right]$.

5.5.1.3   Regularization with MTL

Multitask learning has been shown to yield improved robustness [7, 12, 32]. We adopt an *MTL module* similar to [32] during training by configuring the network to have two outputs, one recognition output which predicts CD states and a second denoising output which reconstructs 128 log mel features derived from the underlying clean signal. The denoising output is only used in training to regularize the model parameters; the associated layers are discarded during inference. In the NAB model the MTL module branches off from the first LSTM layer of the AM module, as shown in Fig. 5.6. The MTL module is composed of two fully connected DNN layers followed by a linear output layer which predicts clean features. During training the gradients back-propagated from the two outputs are weighted by α and $1 - \alpha$ for the recognition and denoising outputs. respectively.

5.5.2   NAB Filter Analysis

The best NAB model found in [21] has the following configurations:

1. The FP module has one shared 512-cell LSTM layer across channels, one layer of channel-dependent 256-cell LSTMs, and one layer of channel-dependent 25-dimensional linear projection layer.
2. The FP module takes in the concatenation of raw waveform samples from each channel.
3. The FP module outputs a 1.5 ms filter (25-dimensional vector) for each channel.
4. The AM module is a single-channel raw-waveform LDNN model [30] with 256 tConv filters and without the frequency convolution layer [26], which is also similar to other multichannel models discussed in this chapter.
5. 128-dimensional clean log mel features are used as the secondary reconstruction objectives with a weight of 0.1 for MTL.
6. A per-frame gated feedback connection from the bottleneck layer right before the AM module's softmax layer is appended to the FP module's input.

Figure 5.7 illustrates the frequency responses of the predicted beamforming filters at the target speech and interfering noise directions. The SNR for this utterance is 12 dB. The responses in the target speech direction have relatively more speech-dependent variations than those in the noise direction. This may indicate that the predicted filters are attending to the speech signal. Besides, the responses in high-speech-energy regions are generally lower than others, which suggests an automatic gain control effect of the predicted filters.

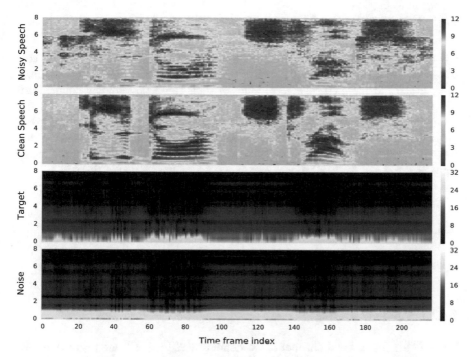

Fig. 5.7 Visualizations of the predicted beamformer responses at different frequencies (Y-axis) across time (X-axis) at the target speech direction (3rd spectrogram) and interfering noise direction (4th) with the noisy (1st) and clean (2nd) speech spectrograms

Table 5.8 Comparison between two-channel factored and adaptive models

	WER (%)		Param	MultAdd
Model	CE	Seq.	(M)	(M)
Factored	20.4	17.1	18.9	35.1
NAB	20.5	17.2	24.0	28.8

5.5.3 Results Summary

Finally, to conclude this section, we show the results after sequence training compared to the factored model. Since the NAB model was trained without frequency convolution (i.e., LDNN), we did the same for the factored model. Table 5.8 shows that while the factored model can potentially handle different directions by enumerating many look directions in the spatial filtering layer, the adaptive model can achieve similar performance with much less computational complexity, as measured by both the parameters and the multiplies and additions (M+A) of the model, as shown in the table.

5.6 Filtering in the Frequency Domain

Up to now, we have presented three multichannel models in the time domain. However, it is well known that convolution between two time-domain signals is equivalent to the elementwise product of their frequency-domain counterparts [3, 5]. A benefit of operating in the complex FFT space is that elementwise products are much faster to compute compared to convolutions, particularly when the convolution filters and input size are large as in our multichannel raw-waveform models. In this section, we describe how we can implement both the factored model from Sect. 5.4 and the NAB model from Sect. 5.5, in the frequency domain.

5.6.1 Factored Model

In this section, we describe the factored model in the frequency domain.

5.6.1.1 Spatial Filtering

For a frame index l and channel c, we denote by $X_c[l] \in \mathbb{C}^K$ the result of an M-point FFT of $x_c[t]$, and by $H_c^p \in \mathbb{C}^K$ the FFT of h_c^p. Note that we ignore negative frequencies because the time-domain inputs are real, and thus our frequency-domain representation of an M-point FFT contains only $K = M/2 + 1$ unique complex-valued frequency bands. The spatial convolution layer in (5.2) can be represented by (5.6) in the frequency domain, where $\cdot$ denotes an elementwise product. We denote the output of this layer as $Y^p[l] \in \mathbb{C}^K$ for each look direction p:

$$Y^p[l] = \sum_{c=0}^{C} X_c[l] \cdot H_{c-1}^p. \tag{5.6}$$

There are many different algorithms for implementing the "spectral filtering" layer in the frequency domain, some of which are presented here [31]. Just to give readers a high-level overview of "spectral filtering", in this chapter we choose to describe only the complex linear projection [38] method.

5.6.1.2 Spectral Filtering: Complex Linear Projection

It is straightforward to rewrite the convolution in (5.3) as an elementwise product in frequency, for each filter f and look direction p:

$$W_f^p[l] = Y^p[l] \cdot G_f. \tag{5.7}$$

In the above equation, $W_f^p[l] \in \mathbb{C}^K$ and $G_f \in \mathbb{C}^K$ is the FFT of the time-domain filter g_f in (5.3). There is no frequency-domain equivalent to the maxpooling operation in the time domain. Therefore to mimic maxpooling exactly requires taking the inverse FFT of $W_f^p[l]$ and performing the pooling operation in the time domain, which is computationally expensive to do for each look direction p and filter output f.

As an alternative, [38] recently proposed the complex linear projection (CLP) model, which performs average pooling in the frequency domain and results in similar performance to a single-channel raw-waveform model. Similarly to the waveform model, the pooling operation is followed by a pointwise absolute-value nonlinearity and log compression. The one-dimensional output for look direction p and filter f is given by

$$Z_f^p[l] = \log \left| \sum_{k=1}^{N} W_f^p[l,k] \right|. \tag{5.8}$$

5.6.2 NAB Model

In the frequency-domain NAB setup, we have an LSTM which predicts complex FFT (CFFT) inputs for both channels. Given a 512-point FFT input, this amounts to predicting 4×257 frequency points with real and imaginary components for two channels, which is much more than the predicted filter size in the time domain (i.e., $1.5\,\text{ms} = 25$ taps). After the complex filters are predicted for each channel, an elementwise product is done with the FFT of the input for each channel, mimicking the convolution in (5.4) in the frequency domain. The output of this is given to a single-channel LDNN in the frequency domain, which does spectral decomposition using the CLP method, and then acoustic modeling.

5.6.3 Results: Factored Model

5.6.3.1 Performance

First, we explore the performance of the frequency-domain factored model. Note this model does not have any frequency convolution layer. We explore this for a similar setting to most efficient raw-waveform factored setups [31], namely $P = 5$ look directions in the spatial layer and $F = 128$ filters in the spectral layer. The input is 32 ms instead of 35 ms like the raw waveform, as this allows us to take a $D = 512$-point FFT without zero-padding at a sampling rate of 16 kHz. A 35 ms input would have required us to take a 1024-point FFT, and we have not found any big difference in performance between 32 and 35 ms inputs for the raw waveform.

Table 5.9 Frequency-
domain factored-model
performance

Model	Spatial M+A	Spectral M+A	Total M+A	WER CE	WER ST
Time	525.6K	15.71M	35.1M	20.4	17.1
CLP	10.3K	655.4K	19.6M	20.5	17.2

Table 5.10 Results with a
64 ms window size

Feature	Spatial M+A	Spectral M+A	Total M+A	WER ST
Time	906.1K	33.81M	53.6M	17.1
CLP	20.5K	1.3M	20.2M	17.1

Table 5.9 shows the performance of the time- and frequency-domain factored models, as well as the total number of multiplication and addition operations (M+A) for different layers of the model. The table shows that the CLP factored model reduces the number of operations by a factor of 1.9× over the best waveform model, with a small degradation in WER.

However, given that the frequency models are more computationally efficient, we explored improving WER by increasing the window size (and therefore computational complexity) of the factored models. Specifically, since longer windows typically help with localization [39], we explored using 64 ms input windows for both models. With a 64 ms input, the frequency models require a 1024-point FFT. Table 5.10 shows that the frequency models improve the WER over using a smaller 32 ms input, and still perform roughly the same. However, the frequency model now has an even larger computational-complexity saving of 2.7× compared to the time-domain model.

5.6.3.2   Comparison Between Learning in Time vs. Frequency

Figure 5.8a shows the spatial responses (i.e., beam patterns) for both the time- and frequency-domain spatial layers. The beam patterns show the magnitude response in dB as a function of frequency and direction of arrival, i.e., each horizontal slice of the beam pattern corresponds to the filter's magnitude response for a signal coming from a particular direction. In each frequency band (vertical slice), lighter shades indicate that sounds from those directions are passed through, while darker shades indicate directions whose energy is attenuated. The figure shows that the spatial filters learned in the time domain are band-limited, unlike those learned in the frequency domain. Furthermore, the peaks and nulls are aligned well across frequencies for the time-domain filters.

The differences between these models can be seen further in the magnitude responses of the spectral-layer filters, as well as in the outputs of the spectral layers from different look directions plotted for an example signal. Figure 5.8b illustrates that the magnitude responses in both the time and the CLP models look qualitatively similar, and learn bandpass filters with increasing center frequency.

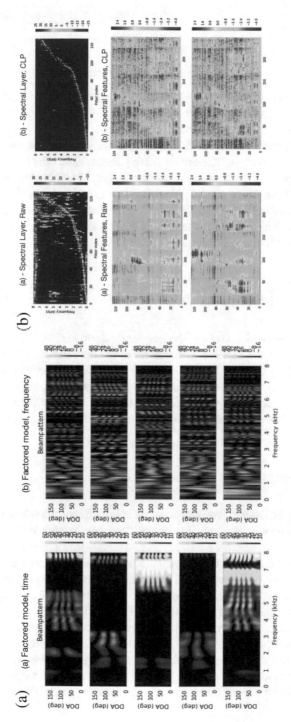

Fig. 5.8 (a) Beam patterns of time and frequency models. (b) Time and frequency domain spatial responses

Table 5.11 Comparison
between time and frequency
NAB models

Model	WER (%)	Param (M)	MultAdd (M)
Raw	20.5	24.6	35.3
CLP	21.0	24.7	25.1

However, because the spatial layers in time and frequency are quite different, we
see that the spectral-layer outputs in time are much more diverse in different spatial
directions compared to the CLP model.

At some level, time-domain and frequency-domain representations are inter-
changeable, but they result in networks that are parametrized very differently. Even
though the time and frequency models all learn different spatial filters, they all seem
to have similar WERs. There are roughly 18M parameters in the LDNN model that
sits above the spatial/spectral layers, which accounts for over 90% of the parameters
in the model. Any differences between the spatial layers in time and frequency are
likely accounted for in the LDNN part of the network.

5.6.4 Results: Adaptive Model

Next, we explore the performance of the frequency-domain NAB model. Table 5.11
shows the WER and computational complexity of the raw-waveform and CLP NAB
models. While using CLP features greatly reduces computational complexity, the
performance is worse than the raw-waveform model. One hypothesis we have is
that frequency-domain processing requires predicting a higher-dimensional filter,
which we can see from the table leads to a degradation in performance.

5.7 Final Comparison, Rerecorded Data

Finally, we also evaluated the performance of different multichannel models
presented in this chapter on a real "rerecorded" test set. Reverberation-I was when
the microphone was placed on a coffee table, whereas Reverberation-II was when
the mic was placed on a TV stand. Since this set contained a circular microphone
geometry but our models were trained on a linear microphone geometry, we only
report results with two channels to form a linear array with a 7.5 cm spacing. The
models, however, were trained with a 14 cm spacing.

Table 5.12 shows the results with different multichannel models. All raw-
waveform models were trained with 35 ms inputs and 128 spectral decomposition
filters. The factored model has 5 look directions. The CLP factored model was
trained with a 64 ms input, 5 look directions, and 256 spectral decomposition filters.
All front ends used an LDNN architecture in the upper layers of the network.

Table 5.12 WER on "rerecorded" set

Model	Rev.-I	Rev.-II	Rev.-I noisy	Rev.-II noisy	Ave
1 channel raw	18.6	18.5	27.8	26.7	22.9
2 channel raw, unfactored	17.9	17.6	25.9	24.7	21.5
2 channel raw, factored	17.1	16.9	24.6	24.2	20.7
2 channel CLP, factored	17.4	17.1	25.7	24.4	21.2
2 channel raw, NAB	17.8	18.1	27.1	26.1	22.3

Notice that the two-channel raw factored model gives a 13% relative improvement over a single channel, with larger improvements on noisier test sets, which is to be expected. In addition, the CLP factored model performs slightly worse than the raw factored model on this test set. One hypothesis is that the CLP factored model captures much less spatial diversity than the raw-waveform model, as shown in Fig. 5.8. Finally, the NAB model performs much worse than the factored model. Perhaps because the NAB model learns a set of adaptive filters, it is more sensitive to mismatches between training and test conditions compared to the other models.

5.8   Conclusions and Future Work

In this chapter, we introduced a methodology to do multichannel enhancement and acoustic modeling jointly within a neural network framework. First, we developed a unfactored raw-waveform multichannel model, and showed that this model performed as well as a model given oracle knowledge of the true location. Next, we introduced a factored multichannel model to separate out spatial and spectral filtering operations, and found that this offered an improvement over the unfactored model. Next, we introduced an adaptive beamforming method, which we found to match the performance of the multichannel model with far fewer computations. Finally, we showed that we can match the performance of the raw-waveform factored model, with far fewer computations, with a frequency-domain factored model. Overall, the factored model provides a 5–13% relative improvement over single-channel and traditional signal-processing techniques, on both simulated and rerecorded sets.

References

1. Allen, J.B., Berkley, D.A.: Image method for efficiently simulation room-small acoustics. J. Acoust. Soc. Am. **65**(4), 943–950 (1979)
2. Benesty, J., Chen, J., Huang, Y.: Microphone Array Signal Processing. Springer, Berlin (2009)

3. Bengio, Y., Lecun, Y.: Scaling Learning Algorithms Towards AI. Large Scale Kernel Machines. MIT press, Cambridge (2007)
4. Bengio, S., Vinyals, O., Jaitly, N., Shazeer, N.: Scheduled sampling for sequence prediction with recurrent neural networks. In: Advances in Neural Information Processing Systems, pp. 1171–1179 (2015)
5. Bracewell, R.: The Fourier Transform and Its Applications, 3rd edn. McGraw-Hill, New York (1999)
6. Brandstein, M., Ward, D.: Microphone Arrays: Signal Processing Techniques and Applications. Springer, Berlin (2001)
7. Chen, Z., Watanabe, S., Erdoğan, H., Hershey, J.R.: Speech enhancement and recognition using multi-task learning of long short-term memory recurrent neural networks. In: Proceedings of Interspeech, pp. 3274–3278. ISCA (2015)
8. Chung, J., Gulcehre, C., Cho, K., Bengio, Y.: Gated feedback recurrent neural networks. arXiv preprint. arXiv:1502.02367 (2015)
9. Dean, J., Corrado, G., Monga, R., Chen, K., Devin, M., Le, Q., Mao, M., Ranzato, M., Senior, A., Tucker, P., Yang, K., Ng, A.: Large scale distributed deep networks. In: Proceedings of NIPS (2012)
10. Delcroix, M., Yoshioka, T., Ogawa, A., Kubo, Y., Fujimoto, M., Ito, N., Kinoshita, K., Espi, M., Hori, T., Nakatani, T., Nakamura, A.: Linear prediction-based dereverberation with advanced speech enhancement and recognition technologies for the REVERB challenge. In: REVERB Workshop (2014)
11. Dieleman, S., Schrauwen, B.: End-to-end learning for music audio. In: 2014 IEEE International Conference on Acoustics, Speech and Signal Processing (ICASSP), pp. 6964–6968. IEEE (2014)
12. Giri, R., Seltzer, M.L., Droppo, J., Yu, D.: Improving speech recognition in reverberation using a room-aware deep neural network and multi-task learning. In: Proceedings of ICASSP, pp. 5014–5018. IEEE (2015)
13. Glorot, X., Bengio, Y.: Understanding the difficulty of training deep feedforward neural networks. In: Proceedings of AISTATS (2014)
14. Griffiths, L.J., Jim, C.W.: An alternative approach to linearly constrained adaptive beamforming. IEEE Trans. Antennas Propag. 30(1), 27–34 (1982)
15. Hain, T., Burget, L., Dines, J., Garner, P., Grezl, F., Hannani, A., Huijbregts, M., Karafiat, M., Lincoln, M., Wan, V.: Transcribing meetings with the AMIDA systems. IEEE Trans. Audio Speech Lang. Process. 20(2), 486–498 (2012)
16. Heigold, G., McDermott, E., Vanhoucke, V., Senior, A., Bacchiani, M.: Asynchronous stochastic optimization for sequence training of deep neural networks. In: Proceedings of ICASSP (2014)
17. Hershey, J.R., Roux, J.L., Weninger, F.: Deep unfolding: model-based inspiration of novel deep architectures. CoRR abs/1409.2574 (2014)
18. Hoshen, Y., Weiss, R.J., Wilson, K.W.: Speech acoustic modeling from raw multichannel waveforms. In: Proceedings of ICASSP (2015)
19. Jaitly, N., Hinton, G.: Learning a better representation of speech soundwaves using restricted Boltzmann machines. In: Proceedings of ICASSP (2011)
20. Knapp, C.H., Carter, G.C.: The generalized correlation method for estimation of time delay. IEEE Trans. Acoust. Speech Signal Process. 24(4), 320–327 (1976)
21. Li, B., Sainath, T.N., Weiss, R.J., Wilson, K.W., Bacchiani, M.: Neural network adaptive beamforming for robust multichannel speech recognition. In: Proceedings of Interspeech (2016)
22. Liu, Y., Zhang, P., Hain, T.: Using neural network front-ends on far-field multiple microphones based speech recognition. In: Proceedings of ICASSP (2014)
23. Mohamed, A., Hinton, G., Penn, G.: Understanding how deep belief networks perform acoustic modelling. In: Proceedings of ICASSP (2012)
24. Palaz, D., Collobert, R., Doss, M.: Estimating phoneme class conditional probabilities from raw speech signal using convolutional neural networks. In: Proceedings of Interspeech (2014)

25. Sainath, T.N., Kingsbury, B., Mohamed, A., Dahl, G., Saon, G., Soltau, H., Beran, T., Aravkin, A., Ramabhadran, B.: Improvements to deep convolutional neural networks for LVCSR. In: Proceedings of ASRU (2013)
26. Sainath, T.N., Li, B.: Modeling time–frequency patterns with LSTM vs. convolutional architectures for LVCSR tasks. In: Proceedings of Interspeech (2016)
27. Sainath, T.N., Kingsbury, B., Sindhwani, V., Arisoy, E., Ramabhadran, B.: Low-rank matrix factorization for deep neural network training with high-dimensional output targets. In: Proceedings of ICASSP (2013)
28. Sainath, T.N., Vinyals, O., Senior, A., Sak, H.: Convolutional, long short-term memory, fully connected deep neural networks. In: Proceedings of ICASSP (2015)
29. Sainath, T.N., Weiss, R.J., Wilson, K.W., Narayanan, A., Bacchiani, M., Senior, A.: Speaker localization and microphone spacing invariant acoustic modeling from raw multichannel waveforms. In: Proceedings of ASRU (2015)
30. Sainath, T.N., Weiss, R.J., Wilson, K.W., Senior, A., Vinyals, O.: Learning the speech front-end with raw waveform CLDNNs. In: Proceedings of Interspeech (2015)
31. Sainath, T.N., Narayanan, A., Weiss, R.J., Wilson, K.W., Bacchiani, M., Shafran, I.: Reducing the computational complexity of multimicrophone acoustic models with integrated feature extraction. In: Proceedings of Interspeech (2016)
32. Sainath, T.N., Weiss, R.J., Wilson, K.W., Narayanan, A., Bacchiani, M.: Factored spatial and spectral multichannel raw waveform CLDNNs. In: Proceedings of ICASSP (2016)
33. Sak, H., Senior, A., Beaufays, F.: Long short-term memory recurrent neural network architectures for large scale acoustic modeling. In: Proceedings of Interspeech (2014)
34. Seltzer, M., Raj, B., Stern, R.M.: Likelihood-maximizing beamforming for robust handsfree speech recognition. IEEE Trans. Audio Speech Lang. Process. 12(5), 489–498 (2004)
35. Stolcke, A., Anguera, X., Boakye, K., Çetin, O., Janin, A., Magimai-Doss, M., Wooters, C., Zheng, J.: The SRI-ICSI Spring 2007 meeting and lecture recognition system. In: Multimodal Technologies for Perception of Humans. Lecture Notes in Computer Science, vol. 2, pp. 450–463. Springer, Berlin (2008)
36. Swietojanski, P., Ghoshal, A., Renals, S.: Hybrid acoustic models for distant and multichannel large vocabulary speech recognition. In: Proceedings of ASRU (2013)
37. Tüske, Z., Golik, P., Schlüter, R., Ney, H.: Acoustic modeling with deep neural networks using raw time signal for LVCSR. In: Proceedings of Interspeech (2014)
38. Variani, E., Sainath, T.N., Shafran, I.: Complex linear projection (CLP): a discriminative approach to joint feature extraction and acoustic modeling. In: Proceedings of Interspeech (2016)
39. Veen, B.D., Buckley, K.M.: Beamforming: a versatile approach to spatial filtering. IEEE ASSP Mag. 5(2), 4–24 (1988)
40. Xiao, X., Watanabe, S., Erdogan, H., Lu, L., Hershey, J., Seltzer, M.L., Chen, G., Zhang, Y., Mandel, M., Yu, D.: Deep beamforming networks for multi-channel speech recognition. In: Proceedings of ICASSP (2016)
41. Xiao, X., Zhao, S., Zhong, X., Jones, D.L., Chng, E.S., Li, H.: A learning-based approach to direction of arrival estimation in noisy and reverberant environments. In: 2015 IEEE International Conference on Acoustics, Speech and Signal Processing (ICASSP), pp. 2814–2818. IEEE (2015)
42. Zhang, Y., Chuangsuwanich, E., Glass, J.R.: Extracting deep neural network bottleneck features using low-rank matrix factorization. In: ICASSP, pp. 185–189 (2014)

Chapter 6
Novel Deep Architectures in Speech Processing

John R. Hershey, Jonathan Le Roux, Shinji Watanabe, Scott Wisdom, Zhuo Chen, and Yusuf Isik

Abstract Model-based methods and deep neural networks have both been tremendously successful paradigms in machine learning. In model-based methods, problem domain knowledge can be built into the constraints of the model. In addition, unsupervised inference tasks such as adaptation and clustering are handled in a natural way. However, these benefits typically come at the expense of difficulties during inference. In contrast, deterministic deep neural networks are constructed in such a way that inference is straightforward, and discriminative training is relatively easy. However, their typically generic architectures often make it unclear how to incorporate specific problem knowledge or to perform flexible tasks such as unsupervised inference. This chapter introduces frameworks to provide the advantages of both approaches. To do so, we start with a model-based approach and an associated inference algorithm, and reinterpret inference iterations as layers in a deep network, while generalizing the parametrization to create a more powerful network. We show how such frameworks yield new understanding of conventional networks, and how they can result in novel networks for speech processing, including networks based on nonnegative matrix factorization, complex Gaussian microphone array signal processing, and a network inspired by efficient spectral clustering. We then discuss what has been learned in recent work and provide a prospectus for future research in this area.

J.R. Hershey (✉) • J. Le Roux • S. Watanabe
Mitsubishi Electric Research Laboratories (MERL), Cambridge, MA, USA
e-mail: hershey@merl.com

S. Wisdom
University of Washington, Seattle, WA, USA

Z. Chen
Columbia University, New York, NY, USA

Y. Isik
Sabanci University, Istanbul, Turkey

© Springer International Publishing AG 2017
S. Watanabe et al. (eds.), *New Era for Robust Speech Recognition*,
DOI 10.1007/978-3-319-64680-0_6

6.1 Introduction

Two of the most successful frameworks in machine learning are model-based methods and deep neural networks (DNNs). Each offers important well-known advantages and disadvantages. The goal of this chapter is to provide a general strategy to obtain the advantages of both approaches while avoiding many of their disadvantages. The general idea can be summarized as follows: given a model-based approach that requires an iterative inference method, we *unfold* the iterations into a layerwise structure analogous to a neural network. We then *untie* the model parameters across layers to obtain novel neural-network-like architectures that can easily be trained discriminatively using gradient-based methods. The resulting formula combines the expressive power of a conventional deep network with the internal structure of the model-based approach, while allowing inference to be performed in a fixed number of layers that can be optimized for best performance. This approach was introduced in [19] as *deep unfolding*.

One advantage of generative model-based approaches, such as probabilistic graphical models, is that they allow us to use prior knowledge and intuition to reason at the *problem level* in devising inference algorithms, or what David Marr called the "computational theory" level of analysis [18, 36]. Important assumptions about problem constraints can often be incorporated into a model-based approach in a straightforward way. Examples include constraints from the world, such as the linear additivity of signals, visual occlusion, and three-dimensional geometry, as well as more subtle statistical assumptions such as conditional independence, latent variable structure, sparsity, low-rank covariances, and so on. Of course, getting the assumptions wrong will limit performance, but by hypothesizing and testing different problem-level constraints, insight into the nature of the problem can be gained and used as inspiration to improve the modeling assumptions [18].

Unfortunately, inference in complex probabilistic models can be both mathematically and computationally intractable. Approximate methods, such as belief propagation and variational approximations, allow us to derive iterative algorithms to infer the latent variables of interest. However, the approximations further weaken the constraints of the model, and iterative methods are still often too slow for time-sensitive applications. In such cases, rigorous discriminative optimization of such models can be challenging because they may involve bilevel optimization, where the parameter optimization depends in turn on an iterative inference algorithm [6].

Deterministic deep neural networks, which have recently become the state of the art in many applications, are formulated such that the inference is computed via a closed-form expression, organized into *layers*, which are typically executed in sequence. Discriminative training of the networks can be used to optimize the speed-versus-accuracy trade-off, and has become indispensable in producing systems that perform very well in a particular application. However, a disadvantage is that conventional DNNs are closer to "black-box" mechanisms than problem-level formulations, and it can be difficult to incorporate prior knowledge about the problem. Even when one has a working DNN system, it is not clear how it actually achieves its results, and so discovering how to modify its architecture to achieve better results could be considered as much an art as a science.

In this chapter we present a general framework that addresses these problems by bringing the problem-level formulation of model-based methods to the task of designing deep-neural-network architectures. Each step of the deep-unfolding framework uses well-known methods: deriving iterative inference methods for a given probabilistic model follows a long tradition that makes use of many standard tools, and unfolding the iterations and applying the chain rule for gradient-based training is also straightforward. We first show how conventional sigmoid neural networks can be understood as an application of deep unfolding to mean-field inference in Markov random fields (MRFs). Substituting belief propagation for mean-field inference exemplifies how deep unfolding can lead to alternative neural network architectures.

The remainder of the chapter focuses on specific generative models that embody the problem-level assumptions encountered in audio, such as linear mixing and reverberation, and discusses how to derive from them deep learning architectures for source separation. We first apply deep unfolding to nonnegative matrix factorization (NMF) [31, 47, 58]. NMF has no closed-form solution and relies on iterative inference methods, typically formulated as multiplicative updates. We unfold these iterations, resulting in a novel nonnegative deep network architecture introduced in [30] that can be more powerful than NMF, while still incorporating its basic additivity assumptions. We also apply deep unfolding to a generative model for channel and source estimation introduced in [61]. Finally, we show how we can unfold a clustering algorithm to enable end-to-end training of a speech separation system known as *deep clustering* [21, 25].

6.1.1 Relationship to the Literature

Some recent work has addressed the idea of unfolding inference algorithms and using gradient descent to optimize them in the context of a variety of models and inference methods. Both sparse coding [16, 49] and nonnegative matrix factorization [50, 63] have been addressed using unfolding and back-propagation or other optimization methods. In [51], gradient-based optimization of loopy belief propagation was applied to binary pairwise Markov random fields. In [7, 8], tree-reweighted belief propagation and mean-field inference were unrolled and trained via gradient descent. In [14], inference in a graphical model was implemented via an ensemble of unfolded inference algorithms trained to predict one held-out variable given the others. In all of this work, unfolding was done without untying parameters, so only an approximation to the original model was optimized.

In our view, the untying of the parameters is an important step in creating new deep architectures that can be competitive with conventional deep networks. Some recent work has begun to address the untying of parameters for Markov random field inference algorithms. Belief-propagation-style inference was learned in [43], using logistic regression with untied parameters. Simultaneously with our work, [32] introduced unfolded mean-field inference and untied parameters. However, in both of these, only conventional sigmoid networks resulted from the untying.

6.2 General Formulation of Deep Unfolding

In the general setting, we consider generative models for which inference is an optimization problem. One example is variational inference, where a lower bound on the data likelihood is optimized to estimate approximate posterior probabilities, which can then be used to compute conditional expected values of hidden variables. Here, we present a general formulation based on a model, determined by *parameters* θ, that specifies the relationships between *hidden* quantities of interest y_i and the *observed* variables x_i for each data instance i. The parameter set, θ, contains all parameters used in the model: for Markov random fields, θ contains the potential functions, while for Gaussian-based models, it contains the means and variances, and for basis expansion models, it contains the basis functions. The quantities of interest, y_i, are typically estimates of latent variables important for a particular task. For example, in a scene-labeling task, y_i might be the labels of the pixels; in denoising, y_i might be the posterior mean of the latent clean signal. At test time, estimating these quantities of interest involves optimizing an inference objective function $\mathscr{F}_\theta(x_i, \phi_i)$, where ϕ_i are intermediate variables (considered as vectors) from which y_i can be computed:

$$\hat{\phi}_i(x_i|\theta) = \arg\min_{\phi_i} \mathscr{F}_\theta(x_i, \phi_i), \quad \hat{y}_i(x_i|\theta) = g_\theta(x_i, \hat{\phi}_i(x_i|\theta)), \qquad (6.1)$$

where g_θ is an estimator for y_i. For many interesting cases, this optimization cannot be easily done and leads to an iterative inference algorithm. In probabilistic generative models, $\mathscr{F}$ might be an approximation to the negative log-likelihood, y_i could be taken to represent hidden variables, and ϕ_i to represent an estimate of their posterior distribution. For example, in variational inference algorithms, ϕ_i could be taken to be the variational parameters. In sum–product loopy belief propagation, the ϕ_i would be the posterior marginal probabilities. On the other hand, for the nonprobabilistic formulation of NMF, ϕ_i can be taken as the activation coefficients of the basis functions that are updated at inference time. Note that the x_i, y_i can all be sequences or have other underlying structure, but here for simplicity we ignore their structure.

At training time, we may optimize the parameters θ using a discriminative objective function,

$$\mathscr{E}_\theta \stackrel{\text{def}}{=} \sum_i \mathscr{D}(y_i^*, \hat{y}_i(x_i|\theta)), \qquad (6.2)$$

where $\mathscr{D}$ is a loss function and y_i^* a reference value. In the general case, minimization of (6.2) is a *bilevel* optimization problem, since $\hat{y}_i(x_i|\theta)$ is itself determined by an optimization problem (6.1) that depends on the parameters θ.

We assume that the intermediate variables ϕ_i in (6.1) can be optimized iteratively using update steps $k \in \{1, \ldots, K\}$ of the form[1]

$$\phi_i^k = f_\theta(x_i, \phi_i^{k-1}), \tag{6.3}$$

beginning with ϕ_i^0. Consider optimizing the parameters θ with respect to our loss using gradient-based methods such as stochastic gradient descent. Efficiently computing the gradient entails back-propagation: the intermediate value for each iteration is stored in memory in the forward pass, and derivatives are computed at the stored values in the backward pass using the chain rule. Thus we arrive at a neural-network-like architecture with one layer per iteration. The intermediate variables $\phi^1, \ldots, \phi^K$ are the nodes of layers 1 to K and (6.3) determines the transformation and activation function between layers. Finally, the y_i^K are the nodes of the output layer, and are obtained by $y_i^K = g_\theta(x_i, \phi_i^K)$.

If the parameters θ are the same for all layers, then this amounts to discriminative optimization of the original model, under a fixed number of iterations. Treating each iteration as a neural network layer, this can be viewed as a recurrent network in the layer-to-layer direction. Such networks are known as *deep recursive networks* (DRNs) to distinguish them from ordinary recurrent neural networks (RNNs). Generally, deep networks do not have this structure, and it has been noted that DRNs can be more difficult to learn [27]. Moreover, nonrecursive deep networks seem to function via a progressive refinement of representations, from primitive sensory representations in early layers to more sophisticated and abstract ones in later layers.

In the deep-unfolding framework, we hypothesize that allowing the individual layers/iterations to differ may allow the network to implement more complex inference procedures. We therefore consider *untying* the parameters across layers. To formulate this untying, we define parameters $\theta \overset{\text{def}}{=} \{\theta^k\}_{k=0}^K$ for each layer, so that $\phi_i^k = f_{\theta^{k-1}}(x_i, \phi_i^{k-1})$ and $y_i^K = g_{\theta^K}(x_i, \phi_i^K)$. Then we can compute the derivatives recursively as in back-propagation,

$$\frac{\partial \mathscr{E}}{\partial \phi_i^K} = \frac{\partial \mathscr{D}}{\partial y_i^K} \frac{\partial y_i^K}{\partial \phi_i^K}, \qquad \frac{\partial \mathscr{E}}{\partial \theta^K} = \sum_i \frac{\partial \mathscr{D}}{\partial y_i^K} \frac{\partial y_i^K}{\partial \theta^K}, \tag{6.4}$$

$$\frac{\partial \mathscr{E}}{\partial \phi_i^k} = \frac{\partial \mathscr{E}}{\partial \phi_i^{k+1}} \frac{\partial \phi_i^{k+1}}{\partial \phi_i^k}, \qquad \frac{\partial \mathscr{E}}{\partial \theta^k} = \sum_i \frac{\partial \mathscr{E}}{\partial \phi_i^{k+1}} \frac{\partial \phi_i^{k+1}}{\partial \theta^k}, \tag{6.5}$$

---

[1] Indices k in superscript always refer to the iteration index (and similarly for l defined later as the source index).

where $k < K$, and we sum over all the intermediate indices of the derivatives. The specific derivations will of course depend on the form of f, g, and $\mathscr{D}$, for which we give examples below.

6.3 Unfolding Markov Random Fields

It is easy to show that conventional sigmoid networks can be obtained by unfolding and untying mean-field inference on discrete-state pairwise Markov random fields. Although generic MRFs are not a good example of incorporating problem-level knowledge, it is instructive to consider them, both in order to understand conventional networks in terms of unfolding MRFs, and to generalize conventional deep networks by changing either the model or the inference algorithm prior to unfolding.

Here, we first review how mean-field updates can lead to conventional sigmoid networks. Then, we show how belief propagation leads to a different deep architecture. Finally, we unify the two architectures using a general power mean formulation.

For simplicity we restrict discussion to pairwise MRFs. More general MRFs with higher-order factors can be easily expressed as pairwise MRFs by creating an auxiliary random variable for each higher-order factor. We first give a general formulation with arbitrary state spaces, and then discuss the special case of binary MRFs which lead to sigmoid networks when unfolded. Also, for simplicity, we partition the variables into hidden and observed random variables, and omit connections between observed variables, since these do not affect inference.

A pairwise MRF is represented here by an undirected graph whose vertices index hidden random variables h_i taking values in $\mathscr{H}_i$ for i in $\mathscr{I}_h = \{1, \ldots, N_h\}$ and observed variables v_l taking values in $\mathscr{V}_l$ for l in $\mathscr{I}_v = \{1, \ldots, N_v\}$. We abuse notation by using h_i, v_l to refer to both random variables and their values and by omitting their ranges in summations. The factors of the probability distribution are associated with edges of the graph. Edges between hidden variables are identified by unordered pairs of indices $(i, j) \equiv (j, i)$ in the edge set $\mathscr{E}_{hh} \overset{\text{def}}{=} \{(i, j) : i$ and j are connected$\}$. The set of edges (i, l) between hidden and observed variables is $\mathscr{E}_{hv} \overset{\text{def}}{=} \{(i, l) : i$ and l are connected$\}$. The neighborhoods of node i are $\mathscr{N}_i^{hh} \overset{\text{def}}{=} \{j | (i, j) \in \mathscr{E}_{hh}\}$ for hidden nodes, and $\mathscr{N}_i^{hv} = \{l | (i, l) \in \mathscr{E}_{hv}\}$ for visible nodes. The edge factors between hidden variables are parametrized by log potential functions $\Psi(h_i, h_j) \overset{\text{def}}{=} \Psi_{h_i, h_j}(h_i, h_j)$, and the hidden-to-visible potentials by $\Psi(h_i, v_l) \overset{\text{def}}{=} \Psi_{h_i, v_l}(h_i, v_l)$, where we again abuse notation by indexing the functions using their arguments.

The MRF posterior probability distribution can then be written

$$p(h|v) = \frac{1}{z(\Psi, v)} \prod_{(i,j) \in \mathscr{E}_{\mathsf{hh}}} e^{\Psi(h_i, h_j)} \prod_{(i,l) \in \mathscr{E}_{\mathsf{hv}}} e^{\Psi(h_i, v_l)} \tag{6.6}$$

$$\propto \exp\left(\sum_{(i,j) \in \mathscr{E}_{\mathsf{hh}}} \Psi(h_i, h_j) + \sum_{(i,l) \in \mathscr{E}_{\mathsf{hv}}} \Psi(h_i, v_l)\right), \tag{6.7}$$

where $z(\Psi, v) = \sum_h p(h, v)$ is a normalizer that depends on both the parameters and the combination of visible states. For discrete h_i and v_l, the log potential functions are typically represented using scalar parameters for each combination of values taken by their arguments, and the MRF can be formulated as an exponential-family model using indicator functions as features. It is worth noting that computing $z(\Psi, v)$ is generally intractable in fully connected MRFs due to the need to evaluate an exponential number of hidden state combinations; hence the need for approximate inference methods.

6.3.1 Mean-Field Inference

In variational methods, of which the *mean field* (MF) approximation is a special case, we perform a tractable approximate inference by minimizing the Kullback–Leibler (KL) divergence between an approximate posterior q_{h} and the true posterior $p_{\mathsf{h}|\mathsf{v}}$. Equivalently, we maximize a lower bound on the likelihood obtained via Jensen's inequality:

$$\arg\min_{q_{\mathsf{h}}} D_{\mathrm{KL}}(q_{\mathsf{h}} \| p_{\mathsf{h}|\mathsf{v}}) = \arg\max_{q_{\mathsf{h}}} \mathscr{L}(q_{\mathsf{h}}, p_{\mathsf{h},\mathsf{v}}), \tag{6.8}$$

$$\mathscr{L}(q_{\mathsf{h}}, p_{\mathsf{h},\mathsf{v}}) \stackrel{\text{def}}{=} \sum_h q(h) \log \frac{p(h, v)}{q(h)} \leq \log p(v). \tag{6.9}$$

In the mean-field approximation, the posterior is fully factorized over the variables so that $q(h) = \prod_{i \in \mathscr{I}_{\mathsf{h}}} q(h_i)$, which is a product of marginal posteriors. This leads to bound-preserving update equations of the form

$$q(h_i) \propto \exp\left(\sum_{j \in \mathscr{N}_i^{\mathsf{hh}}} \sum_{h_j} q(h_j)\Psi(h_i, h_j) + \sum_{l \in \mathscr{N}_i^{\mathsf{hv}}} \Psi(h_i, v_l^*)\right), \tag{6.10}$$

where $\sum_{h_i} q(h_i) = 1$, and v_l^* is the observed value of v_j. Normalizing $q(h_i)$ leads to a multivariate logistic or "sigmoid" function,

$$q(h_i) = \frac{\exp\left(\sum_{j\in\mathcal{N}_i^{hh}} \sum_{h_j} q(h_j)\Psi(h_i, h_j) + \sum_{l\in\mathcal{N}_i^{hv}} \Psi(h_i, v_l^*)\right)}{\sum_{h_i'} \exp\left(\sum_{j\in\mathcal{N}_i^{hh}} \sum_{h_j} q(h_j)\Psi(h_i', h_j) + \sum_{l\in\mathcal{N}_i^{hv}} \Psi(h_i', v_l^*)\right)}. \tag{6.11}$$

Here we formulate the updates in terms of messages to facilitate comparison with belief propagation:

$$q(h_i) \propto \exp\left(\sum_{j\in\mathcal{N}_i^{hh}} \log m_{j\rightarrow i}(h_i) + \sum_{l\in\mathcal{N}_i^{hv}} \Psi(h_i, v_l^*)\right), \tag{6.12}$$

where messages $m_{j\rightarrow i}(h_i)$ from j to i at value h_i are given by

$$m_{j\rightarrow i}(h_i) \propto \exp\left(\sum_{h_j} q(h_j)\Psi(h_i, h_j)\right). \tag{6.13}$$

The two updates (6.12) and (6.13) together constitute the activation function for one layer in the unfolded MRF network. In order to maintain the variational bound, the updates must be done according to an update schedule that avoids synchronous updates of directly interdependent q functions. However, in the context of discriminative training with an unfolded model, maintaining the bound may not be necessary. Nevertheless, the specific ordering of updates may have a strong effect on the rate of convergence of inference, and can be optimized along with the model parameters as in [19].

To compare with conventional sigmoid neural networks, we consider an MRF with binary random variables, $h_i, v_l \in \{0, 1\} = \mathcal{H}_i = \mathcal{V}_i$. The MRF posterior distribution can be written

$$p(h|v) \propto \exp\left(\sum_{i,j\in\mathcal{I}_h} \frac{1}{2}a_{i,j}h_ih_j + \sum_{i\in\mathcal{I}_h} b_ih_i + \sum_{i\in\mathcal{I}_h, l\in\mathcal{I}_v} c_{i,l}h_iv_l^*\right), \tag{6.14}$$

with suitable choices for $a_{i,j}$, b_i, and $c_{i,l}$, derived from Ψ. The factor of $1/2$ comes from the fact that $a_{i,j} = a_{j,i}$, and each edge potential is counted twice in the sum. In matrix notation, with $A = \{a_{i,j}\}_{i,j\in\mathcal{I}_h}$, where $a_{i,j} = 0$ for $(i, j) \notin \mathcal{E}_{hh}$, and similarly for matrix C and vector $b = \{b_i\}_{i\in\mathcal{I}_h}$, we can write the desired posterior as

$$p(h|v) \propto \exp\left(\tfrac{1}{2}h^TAh + h^Tb + h^TCv^*\right). \tag{6.15}$$

Note that $a_{i,i} = 0$ in the original model since there are no self-edges. Unfolding and untying parameters and using synchronous updates then leads to a conventional sigmoid network structure,

$$\mu^k = \text{logistic}\left(A^k \mu^{k-1} + b^k + C^k v^*\right), \tag{6.16}$$

where μ^k is the vector $[\mu_i^k]_{i \in \mathscr{I}_h}$ of activations of layer k, and $\mu_i^k \overset{\text{def}}{=} q^k(h_i = 1)$. This can be recognized as a sigmoid network having a special structure in which inputs are connected to all the layers. This structure is a consequence of unfolding a model in which any hidden variables may be directly connected to observations. However, as we can untie the parameters in any way we please, to emulate the conventional case where the first layer depends only on the inputs and each subsequent layer depends only on the previous one, we can allow $c_{i,l}^k$ to be nonzero only in the first frame, $k = 0$. The initial distribution $\mu_i^{k=0}$, as well as the associated weights $a_{i,j}^{k=1}$, can be set to zero. We can also relax the constraint $a_{i,i}^k = 0$ from the original model to reach the full generality of the conventional sigmoid network.

It is worth noting that conventional feedforward sigmoid networks can also be derived, more simply, by starting with a deep, layerwise binary MRF, and performing a single forward pass of mean-field updates starting with the input and ending with the last layer. This corresponds to a special case of the MRF structure and update schedule in our framework. When looking at a given conventional neural network, then, we may be able to interpret it in two different ways, either as an approximate MF inference in an MRF with the same structure as the neural network, or as a deep unfolding of a model with a more compact structure.

The main point of all of this is that once we have a model and inference algorithm corresponding to a given neural network, we can consider changing the inference algorithm or model structure in order to generate alternative neural network architectures. For example, instead of using mean-field inference, one could unfold the model using belief propagation.

6.3.2   Belief Propagation

Belief propagation (BP) is an algorithm for computing posterior probabilities, which leads to an exact solution for tree-structured graphical models [39]. When applied to graphs with loops it is known as *loopy belief propagation*. It can be interpreted as a fixed-point algorithm for the stationary points of the Bethe free energy [64], which in turn can be seen as an approximation to the Kullback–Leibler divergence between the approximated posterior and the true posterior distribution. Algorithms in the style of belief propagation have been thought to produce better results on general Markov random field problems [57], and hence there is a motivation to investigate deep network architectures based on BP. Some previous work has explored unfolding of BP without untying the parameters [7, 8], which focused on an extension to loopy BP based on tree-reweighted BP approaches [55], but for simplicity we begin with the standard sum–product version of BP.

In BP as in MF methods, the update equations are formulated in terms of marginal posteriors, known as beliefs, based on messages:

$$q(h_i) \propto \prod_{j \in \mathcal{N}_i^{hh}} m_{j \to i}(h_i) \prod_{l \in \mathcal{N}_i^{hv}} e^{\Psi(h_i, v_l^*)}, \qquad (6.17)$$

where $\sum_{h_i} q(h_i) = 1$, with messages defined by

$$m_{j \to i}(h_i) \propto \sum_{h_j} \frac{q(h_j)}{m_{i \to j}(h_j)} e^{\Psi(h_i, h_j)}. \qquad (6.18)$$

As in the MF updates, normalization of messages is optional. However, in contrast to MF, in which the beliefs have to be normalized in each iteration, the normalization of beliefs in BP is optional and can be done whenever desired for numerical reasons, or to compute output predictions. For comparison with the MF equations, we formulate (6.17) as

$$q(h_i) \propto \exp\left(\sum_{j \in \mathcal{N}_i^{hh}} \log m_{j \to i}(h_i) + \sum_{l \in \mathcal{N}_i^{hv}} \Psi(h_i, v_l^*) \right), \qquad (6.19)$$

and see that (6.19) is identical to (6.12), so that only the messages differ between MF (6.13) versus BP (6.18).

For tree-structured graphs, the exclusion of the incoming message $m_{i \to j}(h_j)$ in (6.18) prevents "feedback" by ensuring that each message is only incorporated once into a given belief, and the updates yield exact marginals, from which the full posterior can be computed.

In the general case where MRFs may have cycles, the exclusion of incoming messages no longer completely prevents feedback, and the approximate marginals are no longer guaranteed to converge to the true marginals. However, loopy BP works well in practice for some problems, with an appropriate message-passing schedule, which can also be optimized as part of the model as described in [19]. As in the MF case, we can obtain similar updates by starting with a layerwise graph structure, with an update schedule that passes sequentially through the layers, in the manner of a feedforward neural network. In this case, the incoming messages $m_{i \to j}(h_j)$ in (6.18) can be considered uniform and can be ignored, leading to even simpler messages, $m_{j \to i}^k(h_i) \propto \sum_{h_j} q^{k-1}(h_j) e^{\Psi^{k-1}(h_i, h_j)}$.

In [19], generalized messages are derived to yield an architecture that can encompass both the MF and BP messages as special cases, along with formulas that generalize sum–product and max-product varieties of BP. Formulating the latter in the log domain, with $u(h_i) = \log q(h_i)$, with a soft-max parameter κ, yields

$$u^k(h_i) = \sum_{j \in \mathcal{N}_i^{hh}} \frac{1}{\kappa} \log\left(\sum_{h_j} \frac{1}{N_{h_j}} \exp\left(\kappa u^{k-1}(h_j) + \kappa \Psi^k(h_i, h_j) \right) \right) + \sum_{l \in \mathcal{N}_i^{hv}} \Psi^k(h_i, v_l^*).$$

$$(6.20)$$

This is similar in spirit to *softmaxout* [66]. The max-product messages ($\kappa \to \infty$) yield a particularly simple and tractable form,

$$u^k(h_i) = \sum_{j \in \mathcal{N}_i^{hh}} \max_{h_j} \left(u^{k-1}(h_j) + \Psi^k(h_i, h_j)\right) + \sum_{l \in \mathcal{N}_i^{hv}} \Psi^k(h_i, v_l^*), \qquad (6.21)$$

which appears to be similar in spirit to maxout [15].

In the end, we arrive at a variety of different nonlinear activation functions that would have been difficult to derive by any other means. In initial proof-of-concept experiments, we found that architectures in this family gave performance on MNIST comparable to the state of the art. However, we leave experiments on generic MRFs for other work, and in the rest of this chapter we turn to models that incorporate specific problem domain knowledge.

6.4 Deep Nonnegative Matrix Factorization

While discrete MRFs are an interesting general case, one point of this work is to incorporate problem-level knowledge into a novel deep architecture. To that end, here we apply the proposed deep-unfolding framework to the nonnegative matrix factorization model, which can be applied to any nonnegative signal. NMF [31] is a popular algorithm commonly used for challenging single-channel audio source separation tasks, such as speech enhancement in the presence of difficult nonstationary noise (e.g., music and other speech). The NMF model encompasses the simple problem-level assumptions that the power spectra of different sources approximately add together. The basic idea is to represent the features of the sources via sets of basis functions and their activation coefficients, one set per source. Mixtures of signals are then analyzed using the concatenated sets of basis functions, and each source is reconstructed using its corresponding activations and basis set.

However, the training-time and test-time objectives of NMF differ: the parameters are optimized to best represent single sources, but at test time NMF is used to separate mixtures. Training the NMF parameters to improve the separation performance, termed *discriminative NMF*, involves a generally difficult bilevel optimization, where the top-level optimization seeks the best basis function parameters, and the bottom-level optimization seeks the best NMF activations given the basis for each of the training examples. This is challenging because evaluating the effect of changing the top-level parameters requires optimizing the bottom level. One approach to this problem involves first finding the optimum of the bottom level, and then implicitly differentiating at the solution with respect to the parameters [50]. Here we show how deep unfolding leads to a different solution that can be interpreted as a novel deep network architecture.

NMF operates on a matrix of F-dimensional nonnegative spectral features, usually the power or magnitude spectrogram of the mixture, $\mathbf{M} = [\mathbf{m}_1, \ldots, \mathbf{m}_T]$, where T is the number of frames and $\mathbf{m}_t \in \mathbb{R}_+^F$, $t = 1, \ldots, T$, are obtained by short-time Fourier transformation of the time-domain signal. With L sources, each source $l \in \{1, \ldots, L\}$ is represented using a matrix containing R_l nonnegative basis column vectors $\mathbf{W}^l = [\mathbf{w}_r^l]_{r=1}^{R_l}$, multiplied by a matrix of activation column vectors $\mathbf{H}^l = [\mathbf{h}_t^l]_{t=1}^T$, for each time t. The rth row of $\mathbf{H}^l$ contains the activations for the corresponding basis $\mathbf{w}_r^l$ at each time t. A columnwise normalized $\widetilde{\mathbf{W}}^l$ can be used to avoid scaling indeterminacy. The basic assumptions can then be written as

$$\mathbf{M} \approx \sum_l \mathbf{S}^l \approx \sum_l \widetilde{\mathbf{W}}^l \mathbf{H}^l = \widetilde{\mathbf{W}} \mathbf{H}, \tag{6.22}$$

where $\widetilde{\mathbf{W}} = [\widetilde{\mathbf{W}}^1, \ldots, \widetilde{\mathbf{W}}^L]$ and $\mathbf{H}^\mathsf{T} = [\mathbf{H}^{1\mathsf{T}}, \ldots, \mathbf{H}^{L\mathsf{T}}]^\mathsf{T}$. The β-divergence, D_β, is an appropriate cost function for this approximation [12], which casts inference as an optimization of $\hat{\mathbf{H}}$,

$$\hat{\mathbf{H}} = \arg \min_{\mathbf{H}} D_\beta(\mathbf{M} \mid \widetilde{\mathbf{W}} \mathbf{H}) + \mu |\mathbf{H}|_1. \tag{6.23}$$

For $\beta = 1$, D_β is the generalized KL divergence, whereas $\beta = 2$ yields the squared error. An L1 sparsity constraint with weight μ favors solutions where few basis vectors are active at a time.

The following multiplicative updates minimize (6.23) subject to nonnegativity constraints [12]:

$$\mathbf{H}^k = \mathbf{H}^{k-1} \circ \frac{\widetilde{\mathbf{W}}^T \left(\mathbf{M} \circ \left(\widetilde{\mathbf{W}} \mathbf{H}^{k-1} \right)^{\beta-2} \right)}{\widetilde{\mathbf{W}}^T \left(\widetilde{\mathbf{W}} \mathbf{H}^{k-1} \right)^{\beta-1} + \mu}, \tag{6.24}$$

for iteration $k \in \{1, \ldots, K\}$, where $\circ$ denotes elementwise multiplication, the matrix quotient is elementwise, and $\mathbf{H}^0$ is initialized randomly.

After K iterations, to reconstruct each source, typically a Wiener-filtering-like approach is used, which enforces the constraint that all the source estimates $\widetilde{\mathbf{S}}^{l,K}$ sum to the mixture:

$$\widetilde{\mathbf{S}}^{l,K} = \frac{\widetilde{\mathbf{W}}^l \mathbf{H}^{l,K}}{\sum_{l'} \widetilde{\mathbf{W}}^{l'} \mathbf{H}^{l',K}} \circ \mathbf{M}. \tag{6.25}$$

While in general NMF bases are trained independently on each source before being combined, the combination is not trained discriminatively for good separation performance on a mixture. Recently, discriminative methods have been applied to sparse dictionary-based methods to achieve better performance in particular tasks [34]. In a similar way, we can discriminatively train NMF bases for source separation. The following optimization problem for training bases, termed *discriminative*

NMF (DNMF), was proposed in [50, 58]:

$$\hat{\mathbf{W}} = \arg\min_{\mathbf{W}} \sum_l \gamma_l D_{\beta_2}\left(\mathbf{S}^l \mid \mathbf{W}^l \hat{\mathbf{H}}^l(\mathbf{M}, \mathbf{W})\right), \qquad (6.26)$$

$$\hat{\mathbf{H}}(\mathbf{M}, \mathbf{W}) = \arg\min_{\mathbf{H}} D_{\beta_1}\left(\mathbf{M} \mid \widetilde{\mathbf{W}}\mathbf{H}\right) + \mu |\mathbf{H}|_1, \qquad (6.27)$$

where β_1 controls the divergence used in the bottom-level analysis objective, and β_2 controls the divergence used in the top-level reconstruction objective. The weights γ_l account for the application-dependent importance of source l; for example, in speech denoising, we focus on reconstructing the speech signal. The first part, (6.26), minimizes the reconstruction error given $\hat{\mathbf{H}}$. The second part ensures that $\hat{\mathbf{H}}$ is the activations that arise from the test-time inference objective. Given the bases $\mathbf{W}$, the activations $\hat{\mathbf{H}}(\mathbf{M}, \mathbf{W})$ are uniquely determined, due to the convexity of (6.27). Nonetheless, the above remains a difficult bilevel optimization problem, since the bases $\mathbf{W}$ occur in both levels.

In [50], the bilevel problem was approached by directly solving for the derivatives of the lower-level problem after convergence.

Here, based on our framework, we unfold the entire model as a deep nonnegative neural network, and we untie the parameters across layers as $\mathbf{W}^k$ for $k = 1, \dots, K$. We call this new model *deep NMF*. In addition, (6.25) is incorporated into the discriminative criteria as

$$\hat{\mathbf{W}} = \arg\min_{\mathbf{W}} \sum_l \gamma_l D_{\beta_2}\left(\mathbf{S}^l \mid \widetilde{\mathbf{S}}^{l,K}(\mathbf{M}, \mathbf{W})\right). \qquad (6.28)$$

In NMF, multiplicative updates are often derived using a heuristic approach which splits the gradient into a positive and a negative part and uses their ratio as a multiplication factor to update the value of the variable of interest. Here, we use a similar approach to train the unfolded network while respecting the nonnegativity constraints:

$$\mathbf{W}^k \Leftarrow \mathbf{W}^k \circ \frac{[\nabla_{\mathbf{W}^k}\mathscr{E}]_-}{[\nabla_{\mathbf{W}^k}\mathscr{E}]_+}. \qquad (6.29)$$

We need to back-propagate a split between the positive and negative parts of the gradient, which can be done through the chain rule. For example, we can use

$$\left[\frac{\partial \mathscr{E}}{\partial h^k}\right]_- = \left[\frac{\partial \mathscr{E}}{\partial h^{k+1}}\right]_+ \left[\frac{\partial h^{k+1}}{\partial h^k}\right]_- + \left[\frac{\partial \mathscr{E}}{\partial h^{k+1}}\right]_- \left[\frac{\partial h^{k+1}}{\partial h^k}\right]_+,$$

where $a = [a]_+ - [a]_-$ is a split of $a \in \mathbb{R}$ with $[a]_\pm \geq 0$.

Results for deep unfolding of sparse NMF (SNMF) [10] were reported in [30] on the 2nd CHiME Speech Separation and Recognition Challenge corpus [54]. The task was speaker-independent speech enhancement in reverberated noisy mixtures.

Speech from the Wall Street Journal (WSJ-0) corpus of read speech was mixed with mostly nonstationary noise sources recorded in a home environment, at six signal-to-noise ratios (SNRs) from -6 to $9\,dB$. The deep NMF architecture was compared with two baselines, sparse NMF and standard feedforward sigmoid DNNs. The input features consisted of $T = 9$ consecutive frames of left context, ending with the target frame, obtained as short-time Fourier spectral magnitudes, as described in [30]. For the DNNs, magnitude spectra were replaced with logarithmic magnitude spectra. The DNN output was a masking function trained such that, when the masking function was applied to the mixture, it best reconstructed the clean speech. The DNNs were trained on the CHiME training set, using back-propagation, stochastic gradient descent with momentum, and discriminative layerwise pretraining. Early stopping based on cross-validation with the CHiME development set, and Gaussian input noise (standard deviation 0.1) were used to prevent overfitting on the training set.

Deep unfolding, with parameter untying on the last two layers, improved performance in terms of signal-to-distortion ratio (SDR) [53] from $9.2\,dB$ for SNMF to $10.2\,dB$ for deep NMF using 4.8M parameters. The best DNN in these experiments achieved $9.6\,dB$ using over 5M parameters. However, in later experiments with improved optimization, the DNN performance was improved to $11.2\,dB$ SDR using 4.1M parameters. Close on its heels, deep NMF with 2M parameters achieved $10.7\,dB$ SDR, and another example with 5M parameters achieved $10.9\,dB$ SNR in subsequent experiments.

The experiments are not conclusive with respect to DNNs versus deep NMF, but deep NMF is clearly superior to SNMF. Moreover, the generative-model formalism provides some additional benefits. For example, it is straightforward to perform inference with missing data without retraining the model, in the same way as with ordinary NMF. Of course, DNNs can also be seen as deep unfolding of a probabilistic model, so we can also derive the appropriate test-time missing-data formulation for DNNs from the deep-unfolding framework. In both cases, deep unfolding allows us to consider operations which are easy in the generative-modeling framework, and transfer them in the appropriate way to the corresponding deep-network architectures.

6.5 Multichannel Deep Unfolding

Whereas NMF is a particularly simple model for single-channel acoustic data, here we consider the application of deep unfolding to a more complex model for multichannel source separation. It is well known that exploiting multiple microphones can greatly improve speech enhancement and recognition performance in the presence of noise, other speakers, and reverberation. Multiple microphones enable the use of beamforming [17], multichannel filtering [48], and clustering of spatial features [9, 35].

In this section, we follow [61] in unfolding a multichannel Gaussian mixture model (MCGMM), resulting in a deep MCGMM computational network that directly processes complex-valued frequency-domain multichannel audio and has an architecture defined explicitly by a generative model, thus combining the advantages of deep networks and model-based approaches.

Conventional speech acoustic models have previously been used to optimize beamformers for example by maximizing likelihood [44]. However, DNN speech models have recently been very successful for single-channel speech enhancement [11, 23, 38, 59] and recognition [33, 45]. Their combination with multichannel methods is not as straightforward due to the absence of a likelihood function, but there have been a few steps in this direction. Swietojanski et al. [52] proposed a convolutional neural network (CNN) architecture for automatic speech recognition (ASR) using multichannel audio, where different microphone channels were pooled together. Hoshen et al. [22] used a CNN-DNN for acoustic modeling on raw time-domain multichannel audio. However, while DNN-based methods are effective, they require empirical exploration to determine the best network architecture. Furthermore, it is difficult to directly incorporate domain knowledge into generic networks.

As an alternative to such approaches, we consider deriving a network architecture starting with a generative model by Attias [1]. We show how unfolding inference in this model results in improved source separation performance for multichannel mixtures of two simultaneous speakers. The resulting deep MCGMM computational network directly processes complex-valued frequency-domain multichannel audio and has an architecture defined explicitly by a generative model. We further extend the deep MCGMM by modeling states as an MRF, whose unfolded mean-field inference updates contribute additional context.

6.5.1 Source Separation Using Multichannel Gaussian Mixture Model

We assume that J acoustic sources x^j are recorded by I microphones. Let $Y_{f,t} \in \mathbb{C}^I$ be the complex-valued short-time Fourier transform (STFT) coefficients of the I microphones at frame $t \in [0, T-1]$ and frequency $f \in [0, F-1]$. The STFT window and FFT lengths are both taken to be $N_w = 2(F-1)$. The ith microphone signal is given by

$$Y_{f,t}^i = \sum_j B_f^{i,j} X_{f,t}^j + V_{f,t}^i, \tag{6.30}$$

where $X_{f,t}^j$ is the STFT coefficient of the jth source, $V_{f,t}^i$ is additive, zero-mean, circular, complex-valued noise, and $B_f^{i,j}$ is the value at frequency f of the FFT of the channel $b^{i,j}$ from source j to microphone i, where we assume a narrowband channel model: that is, the channel impulse response $b^{i,j}$ is shorter than the analysis window length N_{w}. By using a narrowband assumption, the effect of the channel is a complex-valued gain $B_f^{i,j}$ in each frequency bin f for each microphone–source pair (i,j).

We model each source as a mixture of zero-mean, circular, complex-valued Gaussians with mixture states $z_t^j \in [1, Z]$:

$$X_{f,t}^j | z_t^j \sim \mathcal{N}_{\mathbb{C}}(0, 1/\gamma_f^{j,z}), \tag{6.31}$$

where $\gamma_f^{j,z}$ are state-dependent precisions. Each channel is assumed to have a small amount of additive, independent, zero-mean, circular, complex-valued Gaussian noise. The observations are thus distributed as

$$Y_{f,t}^i | X_{f,t}^{1:J} \sim \mathcal{N}_{\mathbb{C}}\left(\sum_j B_f^{i,j} X_{f,t}^j, 1/\psi_f^i \right), \tag{6.32}$$

where ψ_f^i is a precision for the additive sensor noise $V_{f,t}^i$. The states z^j for source j have priors $\pi^{j,z} := p(z^j = z)$, where z is a value in $[1, Z]$. The channel model B_f is here considered a parameter. A graph of the model is shown in Fig. 6.1.

Exact inference in this model is intractable because the E-step requires summing over an exponential number of terms $(\mathcal{O}(Z^J))$ in the marginalization over states.

Fig. 6.1 Graphical model of MCGMM

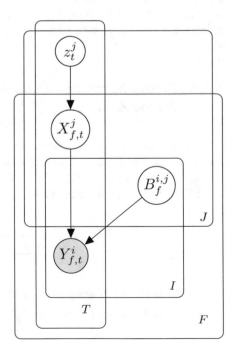

However, an approximate variational algorithm [26] can be derived, which was done by Attias [1]. The approximate inference algorithm uses the variational approximation

$$q(X_{f,t}^{1:J}, z_t^{1:J}) = \left[\prod_f \prod_j q(X_{f,t}^j | z_t^j)\right]\left[\prod_j q(z_t^j)\right], \tag{6.33}$$

where $q(X_{f,t}^j | z_{f,t}^j) = \mathcal{N}_{\mathbb{C}}(X_{f,t}^j; \bar{\mu}_{f,t}^{j,z}, \bar{\gamma}_f^{j,z})$, and $q(z_{f,t}^j) = \bar{\pi}_t^{j,z}$. In this variational approximation, $\bar{\mu}_{f,t}^{j,z}$ is the state-dependent variational posterior mean and $\bar{\gamma}_f^{j,z}$ is the state-dependent variational posterior precision of source j at time–frequency (t, f). The variational updates are given in Attias [1, eqs. (10)–(15)].

The deep-unfolding framework is applied to the variational expectation maximization (EM) updates for MCGMM. A potential challenge is that several updates in the complex-valued unfolded MCGMM involve nonholomorphic functions of complex-valued variables. Because of these nonholomorphic functions, the usual complex gradient is not sufficient to perform gradient descent. One possible approach is to take derivatives of the real and imaginary parts separately. However, these real–imaginary derivatives can be cumbersome algebraically, and furthermore they do not correspond to the standard complex-derivative definition for holomorphic functions [29]. Fortunately, we can sidestep these issues by using a generalization of the complex gradient defined using Wirtinger calculus [29].

6.5.2 Unfolding the Multichannel Gaussian Mixture Model

The variational inference in the MCGMM makes use of the following updates [60, 61]. For each iteration k, the E-step consists of the following updates, which are done independently at all times t:

$$\bar{\gamma}_f^{j,z,(k)} \leftarrow [B_f^{(k-1)}]_{:,j}^H \psi_f [B_f^{(k-1)}]_{:,j} + \gamma_f^{j,z,(k)}, \tag{6.34}$$

$$\bar{\mu}_{f,t}^{j,z,(k)} \leftarrow \frac{[B_f^{(k-1)}]_{:,j}^H \psi_f}{\bar{\gamma}_f^{j,z,(k)}}\left(Y_{f,t} - [B_f^{(k-1)}]_{:,\backslash j} \hat{X}_{f,t}^{\backslash j,(k-1)}\right), \tag{6.35}$$

$$L_t^{j,z,(k)} \leftarrow \log \pi^{j,z} + \sum_f \log \frac{\gamma_f^{j,z,(k)}}{\bar{\gamma}_f^{j,z,(k)}} + \sum_f \bar{\gamma}_f^{j,z,(k)} \left|\bar{\mu}_f^{j,z,(k)}\right|^2, \tag{6.36}$$

$$\bar{\pi}_t^{j,z,(k)} \leftarrow \text{softmax}\left(L_t^{j,1:Z,(k)}\right), \tag{6.37}$$

$$\hat{X}_{f,t}^{j,(k)} \leftarrow \sum_z \bar{\pi}_t^{j,z,(k)} \bar{\mu}_{f,t}^{j,z,(k)}. \tag{6.38}$$

The M-step then updates the time-invariant channel parameters $B_f^{(k)}$:

$$\hat{\Sigma}_f^{YX} \leftarrow \left\langle Y_{f,t}\left(\hat{X}_{f,t}^{(k)}\right)^H \right\rangle_t, \tag{6.39}$$

$$\left[\hat{\Sigma}_f^{\hat{X}\hat{X}}\right]_{j,j} \leftarrow \left\langle \sum_z \bar{\pi}_t^{j,z,(k)}\left(\frac{1}{\bar{\gamma}_f^{j,z,(k)}} + \left|\bar{\mu}_{f,t}^{j,z,(k)}\right|^2\right)\right\rangle_t, \tag{6.40}$$

$$B_f^{(k)} \leftarrow \hat{\Sigma}_f^{Y\hat{X}}\left(\hat{\Sigma}_f^{\hat{X}\hat{X}}\right)^{-1}. \tag{6.41}$$

In order to unfold the variational EM algorithm of Attias [1], we make a few simplifications. For example, the algorithm requires solving a $J \times J$ linear system of equations in each iteration to preserve the variational bound on the log-likelihood. To avoid this, we can individually update each of the J independent state-dependent posterior source means, at the cost of doing sequential updates to maintain the variational bound. Here, we elect to perform the individual updates synchronously, which breaks the bound. In practice, we have not observed degradation of the separation performance, as long as the synchronous updates are preceded by at least a few iterations of the original variational updates.

A computational graph of the resulting network is shown in Fig. 6.2. Note that this results in a somewhat complicated computational architecture that is radically different from a conventional neural network, yet retains some similar layer types such as softmax and linear computations.

6.5.3 MRF Extension of the MCGMM

We would like to improve the deep MCGMM network's ability to estimate the correct state for each source at the output. One way to accomplish this is to add feedback to the network such that the estimated posterior log-likelihoods $L_t^{j,z,(k)}$ of the states in layer k (6.36) use information about the estimated posterior state likelihoods $\pi_t^{j,z,(k-1)}$ (6.37) in the previous layer, $k-1$.

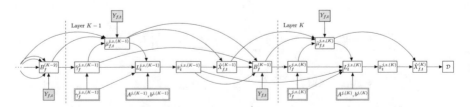

Fig. 6.2 Last two layers of the unfolded deep MCGMM. *Boxes with double lines* are the discriminatively trained source parameters, and *shaded boxes* represent the observed data

In the model, we can incorporate structure into the states by replacing the mixture model with an MRF. In Sect. 6.3 we showed that unfolding mean-field inference in a binary MRF leads to a deep feedforward sigmoid network. Given an MRF with M hidden binary random variables s_m, log potentials Ψ_{ss}, and the log-likelihood of the observed data L_{obs}, the posterior distribution can be written as

$$p(s|v) \propto \exp\left(\frac{1}{2}s^T As + s^T b + s^T L_{\text{obs}}\right), \tag{6.42}$$

where $s := s_{1:M}$, $A \in \mathbb{R}^{M \times M}$, $A_{m,m} = 0$ for all m, $A_{m_1,m_2} = A_{m_2,m_1}$ for $m_1 \neq m_2$, $b \in \mathbb{R}^M$ are derived from the log potentials Ψ_{ss}, and $L_{\text{obs}} \in \mathbb{R}^M$

The variational posterior probability $\bar{\pi}^{(k)} := \{q^{(k)}(s_m)\}_{m=1:M}$ in iteration k of the mean-field inference algorithm is then

$$\bar{\pi}^{(k)} = \sigma\left(A\bar{\pi}^{(k-1)} + b + L_{\text{obs}}\right), \tag{6.43}$$

where σ is the sigmoid function. Notice that A and $b + L_{\text{obs}}$ define an affine transformation, and if these parameters are untied across layers, $A^{(k)}$ and $b^{(k)}$, then (6.43) is equivalent to one layer of a deep feedforward sigmoid network. Discriminatively training the $A^{(k)}$ and $b^{(k)}$ in each layer is equivalent to finding a different set of log potential functions for the MRF for each iteration, such that the result of K iterations of inference minimizes the discriminative cost function. The expression $A^{(k)}\bar{\pi}^{(k-1)} + b^{(k)}$ is essentially a prior on the state log-likelihoods that varies from iteration to iteration, with feedback from the previously estimated state likelihoods $\bar{\pi}^{(k-1)}$.

To apply this in our model, we can replace the multinomial state $z_t^j \in [1, Z]$ of a source with an MRF as in the above to make the deep MCGMM more powerful. To do this, let each multinomial state z_t^j be mapped to Z binary random variables $s_t^{j,z}$ in a fully connected MRF, where $s_t^{j,1:Z}$ is constrained to be one-hot. We use the variational approximation $q\left(s_t^{j,1:Z}\right) = \prod_z \bar{\pi}_t^{j,z}$ for the binary random variables $s_t^{j,z}$, with variational probabilities $\bar{\pi}_t^{j,z} := q\left(s_t^{j,z} = 1, s_t^{j,z'} = 0, \forall z' \neq z\right)$. Rather than using the usual mean-field distribution for a binary random variable, here we constrain the variational posterior to behave like our multinomial Gaussian-mixture-model (GMM) states. So, instead of being the variational probability of a multinomial, $\bar{\pi}_f^{j,z}$ is the variational probability that the zth element of $s_t^{j,1:Z}$ is set to 1 and the other elements are set to 0. Then, if we unfold the mean-field inference for the hidden binary states $s_t^{j,z}$, we replace the multinomial prior $\log \pi^{j,z}$ in the update (6.36) with

$$L_{\text{prior},t}^{j,z,(k)} = A^{(k)}\bar{\pi}_t^{j,z,(k-1)} + b^{(k)}, \tag{6.44}$$

where the parameters $A^{(k)} \in \mathbb{R}^{Z \times Z}$ and $b^{(k)} \in \mathbb{R}^Z$ can be layer-dependent. When $A^{(k)} = 0$ and $b^{(k)} = \log \pi^{j,z}$ for all k, the new update (6.45) simplifies to the original variational update for the MCGMM.

The new update for $L_t^{j,z,(k)}$ that replaces (6.36) is thus

$$L_t^{j,z,(k)} \leftarrow L_{\text{prior},t}^{j,z,(k)} + \alpha L_{\text{acoustic},t}^{j,z,(k)}, \tag{6.45}$$

with

$$L_{\text{acoustic},t}^{j,z,(k)} = \sum_f \log \frac{\gamma_f^{j,z,(k)}}{\bar{\gamma}_f^{j,z,(k)}} + \sum_f \bar{\gamma}_f^{j,z,(k)} \left| \bar{\mu}_f^{j,z,(k)} \right|^2. \tag{6.46}$$

Equation (6.46) is the part of the log-likelihood corresponding to acoustic information and α is an "acoustic weight" that expresses the importance of the acoustic evidence over the prior. Thus we arrive at a hybrid model that has a standard sigmoid neural network as a subcomponent, derived from a coherent graphical model framework. This makes it easy, for example, to add temporal context by connecting MRFs across time, and to form either convolutional networks or recurrent networks depending on the message-passing schedule we use for inference, by experimenting with different probabilistic relations and inference algorithms to obtain a family of related deep networks.

6.5.4 Experiments and Discussion

We used a modified version of the SimData and multicondition training (mcTrain) data components of the REVERB challenge dataset [28]. Each signal consisted of a single-channel speech utterance from the WSJCAM0 dataset [41] reverberated using measured 8-channel room impulse responses (RIRs) in different rooms. SimData uses RIRs from three different rooms, and mcTrain uses RIRs from six different rooms. Stationary noise that was recorded in each particular room was added at 20 dB SNR. To create a dataset of overlapping speech, we added a second speech signal to each signal that had been reverberated using a measured RIR that corresponded to a different position in the same room. No normalization of the power of the reverberated speech sources was performed, in order to test realistic conditions. The power ratio between speaker 1 and speaker 2 ranged from about -10 to $+10$ dB.

The initial source precisions $\gamma_f^{j,z,(0)}$ were trained on a gender-specific split of the WSJCAM0 training set. That is, two separate 256-component GMMs were trained for male and female speakers. Then these gender-specific GMMs were concatenated into a 512-component GMM. A GMM was first trained on the log-magnitude STFTs. Then, using the labels ℓ from the result, the GMM precisions γ_f^z

were set to be $1/\sum_{t:\ell(t)=z}|X_{f,t}|^2$. The MRF parameters were initialized as $A^{(0)} = 0$ and $b^{(0)} = \log \pi^z$. Both sources were initialized with the same source model.

Since our main interest here is to observe the performance improvement of the deep MCGMM over the conventional MCGMM, we used an oracle least-squares initialization for the channel model for each file:

$$B_f^{ij,(0)} = \hat{\Sigma}_f^{YX} \left(\hat{\Sigma}_f^{XX}\right)^{-1}, \tag{6.47}$$

where $\hat{\Sigma}_f^{YX}$ is the frequency-domain cross-covariance between the microphone observations $Y_{f,t}$ and reference sources $X_{f,t}$, and $\hat{\Sigma}_f^{XX}$ is the covariance between the reference sources $X_{f,t}$.

For each file, ten iterations of variational updates, as described in Sect. 6.5.1, were run. The output of these iterations was fed to a network of $K = 5$ simplified update layers, as described in Sect. 6.5.2. The parameters $\Theta^{(k)} = \left\{A^{(k)}, b^{(k)}, \gamma_f^{j,z,(k)}\right\}$ were untied between layers and discriminatively trained. We used an "error-to-source" (ESR) cost function given by

$$\mathcal{D}_{\mathrm{ESR}}(X_{f,t}, \hat{X}_{f,t}^{(K)}) = \sum_j \frac{\sum_{f,t}\left|\hat{X}_{f,t}^j - X_{f,t}^j\right|^2}{\sum_{f,t}\left|X_{f,t}^j\right|^2}, \tag{6.48}$$

where $\hat{X}_{f,t}^{(K)}$ are the estimated source STFT coefficients from the last (Kth) layer and $X_{f,t}$ are the clean single-channel references. By minimizing (6.48), the signal-to-noise ratio of both sources was maximized. Since many of the updates contained nonholomorphic functions of the complex variables, we used Wirtinger calculus to derive generalized gradients. Refer to the supplementary materials [60] for a detailed description of the gradients and their derivation. To ensure the GMM source precisions $\gamma_f^{j,z,(k)}$ remained nonnegative, we optimized $\lambda_f^{j,z,(k)} := \log \gamma_f^{j,z,(k)}$, and replaced all instances of $\gamma_f^{j,z,(k)}$ in the updates with $\exp \lambda_f^{j,z,(k)}$. Stochastic gradient descent was used for back-propagation, and a gradient step was taken using one mixture signal at a time. The initial learning rate was set to $\eta = 0.02$, and an annealing schedule was used such that the learning rate for the nth signal was

$$\eta^{(n)} = \frac{\eta^{(0)}}{1 + dn}, \tag{6.49}$$

where d is a constant that determines the rate of decay. For our experiments, we set $d = 1/(20 \cdot 780)$. We used a momentum of 0.9. A validation set was built from 65 randomly selected files from the SimData development set, and its error was measured after every 78 gradient steps.

MATLAB was used to implement the MCGMM variational inference algorithm, the forward pass of the deep MCGMM, and the gradient computations for discrim-

Table 6.1 Source separation
results for the deep MCGMM

MCGMM var. EM layers	DMCGMM layers	Trained layers	SNR (dB)
No proc.	–	–	−0.78
10	0	0	4.33
15	0	0	4.31
10	1	0/1	4.33/4.47
10	2	0/2	4.57/4.75
10	3	0/3	4.20/4.59
10	4	0/4	4.30/4.70

inative training. All computations were performed on an Nvidia Titan X graphics
processing unit (GPU) using the MATLAB Parallel Processing Toolbox. Using this
implementation, for a 10 s audio file it took about 5 s to perform the MCGMM
variational algorithm and about 10 s to perform a deep MCGMM forward pass and
back-propagation gradient computation.

Table 6.1 shows the resulting SNRs of the sources, averaged across the validation
set, for different numbers of discriminatively trained deep MCGMM layers and
amounts of training data, where the SNR for the time-domain estimate $\hat{x}$ with
reference x is defined as

$$\mathrm{SNR}(\hat{x}, x) = 10 \log_{10} \frac{\sum_n x_n^2}{\sum_n (\hat{x}_n - x_n)^2}. \tag{6.50}$$

We can see that the SNR improvement after discriminative training increases
as the number of trained layers increases. In future work, we will explore other
enhancements and generalizations of this network, including incorporation of
recurrence and long short-term memories (LSTMs), more sophisticated versions
and extensions of the model, other types of cost functions such as cross-entropy on
the source states, and combination with automatic speech recognition systems.

6.6 End-to-End Deep Clustering

An especially challenging problem is that of separating multiple speakers when their
individual characteristics are not known. This so-called *cocktail party* problem [5]
has proven extremely challenging for computers, and separating and recognizing
speech in such conditions has been the holy grail of speech processing for more
than 50 years.

Deep-learning approaches have recently been applied to simpler enhancement
tasks [24, 56, 59, 62]. However, these methods treat the mask inference as a
classification problem, and so were thought to be inadequate when the sources were
of the same class, since there is then an arbitrary ambiguity about which output
belongs to which target signal. We call this the *permutation problem*: there are

multiple valid output masks that differ only by a permutation of the order of the sources, so a global decision is needed to choose a permutation.

In the baseline method of [21], the permutation problem was addressed for a mask estimation network by using a permutation-invariant training method, in which the best one-to-one assignment of network outputs to reference signals was chosen during training. Although the initial attempts in [21] failed, this approach was subsequently shown to work with the addition of a signal estimation objective function [65].

Deep clustering [21, 25], however, solves the permutation problem by using a representation that is independent of permutation of the source labels. It produces an embedding for each time–frequency element in the spectrogram, such that the pairwise affinities between the embeddings of different time–frequency bins represent the desired segmentation. Clustering the embeddings then produces the segmentation, which can be used as a mask to extract each of the sources. Because of the embedding-based representation it can flexibly represent any number of sources, allowing the number of inferred sources to be decided at test time.

In this section we show how the deep-clustering model can be extended to allow end-to-end training for signal estimation, as in [25]. The original deep-clustering system was intended only to recover a binary mask for each source, leaving recovery of the missing features for subsequent stages. In [25], enhancement layers were incorporated to refine the signal estimate. Here we show how deep unfolding is applied to the iterations of the soft clustering inference algorithm. This allows us to train the entire system *end-to-end*, training jointly through the deep-clustering embeddings, and the clustering and enhancement stages. Thus we can use a more direct signal approximation objective instead of the original mask-based objective.

6.6.1 Deep-Clustering Model

Here we review the deep-clustering formalism introduced in [20, 21]. We define as x a raw input signal and as $X_i = g_i(x), i \in \{1, \dots, N\}$, a feature vector indexed by an element i. In audio signals, i is typically a time–frequency (TF) index (t, f), where t indexes a frame of the signal, f indexes the frequency, and $X_i = X_{t,f}$ is the value of the complex spectrogram at the corresponding TF bin. We assume that the TF bins can be partitioned into sets of TF bins in which each source dominates. Once estimated, the partition for each source serves as a TF mask to be applied to X_i, yielding the TF components of each source that are uncorrupted by other sources. The STFT can then be inverted to obtain estimates of each isolated source. The target partition in a given mixture is represented by the indicator $Y = \{y_{i,c}\}$, mapping each element i to each of C components of the mixture, so that $y_{i,c} = 1$ if element i is in cluster c. Then $A = YY^T$ is a binary affinity matrix that represents the cluster assignments in a permutation-independent way: $A_{i,j} = 1$ if i and j belong to the same cluster and $A_{i,j} = 0$ otherwise, and $(YP)(YP)^T = YY^T$ for any permutation matrix P.

To estimate the partition, we seek D-dimensional embeddings $V = f_\theta(x) \in \mathbb{R}^{N \times D}$, parametrized by θ, such that clustering the embeddings yields a partition of $\{1, \ldots, N\}$ that is close to the target. In [21] and this work, $V = f_\theta(X)$ is based on a deep neural network that is a global function of the entire input signal X. Each embedding $v_i \in \mathbb{R}^D$ has unit norm, i.e., $|v_i|^2 = 1$. We consider the embeddings V to implicitly represent an $N \times N$ estimated affinity matrix $\hat{A} = VV^T$, and we optimize the embeddings such that, for an input X, $\hat{A}$ matches the ideal affinities A. This is done by minimizing, with respect to $V = f_\theta(X)$, the training cost function

$$\mathscr{C}_Y(V) = \|\hat{A} - A\|_F^2 = \|VV^T - YY^T\|_F^2 \tag{6.51}$$

summed over training examples, where $\| \cdot \|_F^2$ is the squared Frobenius norm. Due to its low-rank nature, the objective and its gradient can be formulated so as to avoid operations on all pairs of elements, leading to an efficient implementation.

At test time, the embeddings $V = f_\theta(X)$ are computed on the test signal X, and the rows $v_i \in \mathbb{R}^D$ are clustered using K-means. The resulting cluster assignments $\bar{Y}$ are used as binary masks on the complex spectrogram of the mixture, to estimate the sources.

6.6.2 Optimizing Signal Reconstruction

Deep clustering solves the difficult problem of segmenting the spectrogram into regions dominated by each source. It does not, however, solve the problem of recovering the sources in regions strongly dominated by other sources. We propose to use a second-stage enhancement network to obtain better source estimates, in particular for the missing regions. For each source c, the enhancement network first processes the concatenation of the amplitude spectrogram x of the mixture and the $\hat{s}_c$ of the deep clustering estimate through a bidirectional long short-term memory (BLSTM) layer and a feedforward linear layer, to produce an output z_c. Sequence-level mean and variance normalization is applied to the input, and the network parameters are shared for all sources. A soft-max is then used to combine the outputs z_c across sources, forming a mask $m_{c,i} = e^{z_{c,i}} / \sum_{c'} e^{z_{c',i}}$ at each TF bin i. This mask is applied to the mixture, yielding the final estimate $\tilde{s}_{c,i} = m_{c,i} x_i$. During training, we optimize the enhancement cost function $\mathscr{C}_E = \min_{\pi \in \mathscr{P}} \sum_{c,i} (s_{c,i} - \tilde{s}_{\pi(c),i})^2$, where $\mathscr{P}$ is the set of permutations on $\{1, \ldots, C\}$. Since the enhancement network is trained to directly improve the signal reconstruction, it may improve upon deep clustering, especially in regions where the signal is dominated by other sources.

6.6.3 End-to-End Training

In order to consider end-to-end training in the sense of jointly training the deep clustering with the enhancement stage, we need to compute gradients of the clustering step. In [21], hard K-means clustering was used to cluster the embeddings. The resulting binary masks cannot be directly optimized to improve signal fidelity, because the optimal masks are generally continuous, and because the hard clustering is not differentiable. Here we propose a soft K-means algorithm that enables us to directly optimize the estimated speech for signal fidelity.

In [21], clustering was performed with equal weights on the TF embeddings, although weights were used in the training objective in order to train only on TF elements with significant energy. Here we introduce similar weights w_i for each embedding v_i to focus the clustering on TF elements with significant energy. The goal is mainly to avoid clustering silence regions, which may have noisy embeddings, and for which mask estimation errors are inconsequential.

The soft weighted K-means algorithm can be interpreted as a weighted EM algorithm for a Gaussian mixture model with tied circular covariances. It alternates between computing the assignment of every embedding to each centroid, and updating the centroids:

$$\gamma_{i,c} = \frac{e^{-\alpha|v_i - \mu_c|^2}}{\sum_{c'} e^{-\alpha|v_i - \mu_{c'}|^2}}, \qquad \mu_c = \frac{\sum_i \gamma_{i,c} w_i v_i}{\sum_i \gamma_{i,c} w_i}, \qquad (6.52)$$

where μ_c is the estimated mean of cluster c, and $\gamma_{i,j}$ is the estimated assignment of embedding i to the cluster c. The parameter α controls the hardness of the clustering. As the value of α increases, the algorithm approaches K-means.

The weights w_i may be set in a variety of ways. A reasonable choice could be to set w_i according to the power of the mixture in each TF bin. Here we set the weights to 1, except in silence TF bins where the weight is set to 0. Silence is defined using a threshold on the energy relative to the maximum of the mixture.

End-to-end training is performed by *unfolding* the steps of (6.52), and treating them as layers in a clustering network, according to the general framework known as deep unfolding [19]. The gradients of each step are thus passed to the previous layers using standard back-propagation.

It is also interesting to compare the unfolded clustering algorithm to attention and segmentation models [2, 3, 37, 42]. The $\gamma_{i,c}$ in (6.52) correspond to the attentional mask, and the μ_c can be considered a bank of attention vectors that define this mask. Here the μ_c are recomputed as an average, but could alternately be produced by the network as, for example, in [3]. One advantage of the deep-unfolding approach is that we can consider a richer class of existing clustering models to extend our architecture, for example, by adding various priors [4, 13] or by using a pairwise MRF [46], which could be unfolded in a similar way to yield alternative architectures.

Table 6.2 SDR/Magnitude SNR improvements (dB) and word error rate (WER) with enhancement network

Model	Same-gender	Different-gender	Overall	WER (%)
dpcl	8.6/8.9	11.7/11.4	10.3/10.2	87.9
dpcl + enh	9.1/10.7	11.9/13.6	10.6/12.3	32.8
End-to-end	9.4/11.1	12.0/13.7	10.8/12.5	30.8

6.6.4   Experiments

End-to-end deep clustering was evaluated on a single-channel speaker-independent speech separation task, considering mixtures of two and three speakers with all gender combinations. The data were mixtures derived from the Wall Street Journal (WSJ0) corpus, generated by randomly selecting utterances by different speakers from the WSJ0 training set si_tr_s, and mixing them at various SNRs randomly chosen between 0 and 10 dB. Details of the experimental setup are given in [25].

A second-stage enhancement network was used on top of the baseline deep clustering ("dpcl") model. The enhancement network featured two BLSTM layers with 300 units in each LSTM layer, with one instance per source followed by a soft-max layer to form a masking function. We first trained the enhancement network separately ("dpcl + enh"), followed by end-to-end fine-tuning by unfolding iterations of the clustering algorithm, in combination with the dpcl model ("end-to-end"). Table 6.2 shows the improvement in SDR as well as the *magnitude SNR* the (SNR computed from the magnitude spectrograms).

The magnitude SNR is insensitive to phase estimation errors introduced by using the noisy phases for reconstruction, whereas the SDR might get worse as a result of phase errors, even if the amplitudes are accurate. Speech recognition uses features based on the amplitudes, and hence the improvements in magnitude SNR seem to predict the improvements in WER due to the enhancement and end-to-end training. Figure 6.3 shows that the SDR improvements of the end-to-end model are consistently good on nearly all of the two-speaker test mixtures.

6.6.4.1   ASR Performance

We evaluated the ASR performance (WER) with GMM-based clean-speech WSJ models obtained by a standard Kaldi recipe [40]. The noisy baseline result on the mixtures was 89.1%, while the result on the clean speech was 19.9%. The raw output from dpcl did not work well, despite good perceptual quality, possibly due to the effect of near-zero values in the masked spectrum, which is known to degrade ASR performance. However, the enhancement networks significantly mitigated the degradation, and finally obtained 30.8% with the end-to-end network.

Fig. 6.3 Scatter plot of the input SDRs and the corresponding improvements. *Color* indicates density

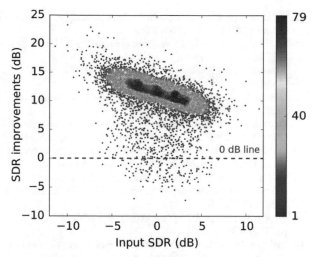

6.7 Conclusion

In conclusion, a general framework was introduced that allows model-based approaches to guide the exploration of the space of deep-network architectures, which would otherwise be difficult to navigate. We have shown how conventional sigmoid networks could be seen as unfolded mean-field inference in Markov random fields, leading to possible generalizations to other inference algorithms such as belief propagation and its variants. We demonstrated how model-based problem constraints of nonnegative matrix factorization can be incorporated via deep unfolding into a novel deep architecture. We implemented a novel, complex microphone array adaptation network by discriminatively training a generative-model inference algorithm, and extending it in novel ways. Finally we showed how a unfolding a clustering algorithm can enable end-to-end training of a speech separation algorithm.

By reasoning at the problem level with the model-based approach, our methodology allowed us to derive architectures and training methods that otherwise would be difficult to obtain. We hope that this framework will help realize some of the benefits of probabilistic models, such as the ability to incorporate problem domain knowledge, in the context of deep networks.

References

1. Attias, H.: New EM algorithms for source separation and deconvolution with a microphone array. In: Proceedings of ICASSP, vol. 5, pp. 297–300 (2003)
2. Ba, J., Mnih, V., Kavukcuoglu, K.: Multiple object recognition with visual attention (2014). arXiv:1412.7755

3. Bahdanau, D., Cho, K., Bengio, Y.: Neural machine translation by jointly learning to align and translate (2014). arXiv:1409.0473
4. Blei, D.M., Jordan, M.I.: Variational inference for Dirichlet process mixtures. Bayesian Anal. **1**(1), 121–144 (2006)
5. Bregman, A.S.: Auditory Scene Analysis: The Perceptual Organization of Sound. MIT Press, Cambridge (1990)
6. Colson, B., Marcotte, P., Savard, G.: An overview of bilevel optimization. Ann. Oper. Res. **153**(1), 235–256 (2007)
7. Domke, J.: Parameter learning with truncated message-passing. In: IEEE Conference on Computer Vision and Pattern Recognition (CVPR), pp. 2937–2943 (2011)
8. Domke, J.: Learning graphical model parameters with approximate marginal inference. IEEE Trans. Pattern Anal. Mach. Intell. **35**(10), 2454 (2013)
9. Duong, N., Vincent, E., Gribonval, R.: Under-determined reverberant audio source separation using a full-rank spatial covariance model. IEEE Trans. Audio Speech Lang. Process. **18**(7), 1830–1840 (2010)
10. Eggert, J., Körner, E.: Sparse coding and NMF. In: Proceedings of Neural Networks, vol. 4, pp. 2529–2533 (2004)
11. Erdogan, H., Hershey, J.R., Watanabe, S., Le Roux, J.: Phase-sensitive and recognition-boosted speech separation using deep recurrent neural networks. In: Proceedings of ICASSP (2015)
12. Févotte, C., Bertin, N., Durrieu, J.L.: Nonnegative matrix factorization with the Itakura–Saito divergence: with application to music analysis. Neural Comput. **21**(3), 793–830 (2009)
13. Figueiredo, M.A.T., Jain, A.K.: Unsupervised learning of finite mixture models. IEEE Trans. Pattern Anal. Mach. Intell. **24**(3), 381–396 (2002)
14. Goodfellow, I.J., Mirza, M., Courville, A., Bengio, Y.: Multi-prediction deep Boltzmann machines. In: Advances in Neural Information Processing Systems, pp. 548–556 (2013)
15. Goodfellow, I.J., Warde-Farley, D., Mirza, M., Courville, A., Bengio, Y.: Maxout networks (2013). arXiv:1302.4389
16. Gregor, K., LeCun, Y.: Learning fast approximations of sparse coding. In: ICML, pp. 399–406 (2010)
17. Habets, E., Benesty, J., Cohen, I., Gannot, S., Dmochowski, J.: New insights into the MVDR beamformer in room acoustics. IEEE Trans. Audio Speech Lang. Process. **18**(1), 158–170 (2010)
18. Hershey, J.R.: Perceptual inference in generative models. Ph.D. thesis, University of California, San Diego (2005)
19. Hershey, J.R., Le Roux, J., Weninger, F.: Deep unfolding: model-based inspiration of novel deep architectures (2014). arXiv:1409.2574
20. Hershey, J.R., Chen, Z., Le Roux, J., Watanabe, S.: Deep clustering: discriminative embeddings for segmentation and separation (2015). arXiv:1508.04306
21. Hershey, J.R., Chen, Z., Le Roux, J., Watanabe, S.: Deep clustering: discriminative embeddings for segmentation and separation. In: Proceedings of ICASSP (2016)
22. Hoshen, Y., Weiss, R.J., Wilson, K.W.: Speech acoustic modeling from raw multichannel waveforms. In: Proceedings of ICASSP (2015)
23. Huang, P.S., Kim, M., Hasegawa-Johnson, M., Smaragdis, P.: Deep learning for monaural speech separation. In: Proceedings of ICASSP, pp. 1562–1566 (2014)
24. Huang, P.S., Kim, M., Hasegawa-Johnson, M., Smaragdis, P.: Joint optimization of masks and deep recurrent neural networks for monaural source separation (2015). arXiv:1502.04149
25. Isik, Y., Le Roux, J., Chen, Z., Watanabe, S., Hershey, J.R.: Single-channel multi-speaker separation using deep clustering. In: Proceedings of ISCA Interspeech (2016)
26. Jordan, M.I., Ghahramani, Z., Jaakkola, T.S., Saul, L.K.: An introduction to variational methods for graphical models. Mach. Learn. **37**(2), 183–233 (1999)
27. Kaiser, L., Sutskever, I.: Neural GPUs learn algorithms (2015). arXiv:1511.08228

28. Kinoshita, K., Delcroix, M., Yoshioka, T., Nakatani, T., Habets, E., Haeb-Umbach, R., Leut-nant, V., Sehr, A., Kellermann, W., Maas, R.: The REVERB challenge: a common evaluation framework for dereverberation and recognition of reverberant speech. In: Proceedings of WASPAA (2013)

29. Kreutz-Delgado, K.: The complex gradient operator and the CR-calculus (2009). arXiv:0906.4835

30. Le Roux, J., Hershey, J.R., Weninger, F.J.: Deep NMF for speech enhancement. In: Proceedings of ICASSP (2015)

31. Lee, D.D., Seung, H.S.: Algorithms for non-negative matrix factorization. In: NIPS, pp. 556–562 (2001)

32. Li, Y., Zemel, R.: Mean field networks. In: Learning Tractable Probabilistic Models (2014)

33. Li, J., Deng, L., Gong, Y., Haeb-Umbach, R.: An overview of noise-robust automatic speech recognition. IEEE/ACM Trans. Audio Speech Lang. Process. **22**(4), 745–777 (2014)

34. Mairal, J., Bach, F., Ponce, J.: Task-driven dictionary learning. IEEE Trans. Pattern Anal. Mach. Intell. **34**(4), 791–804 (2012)

35. Mandel, M.I., Weiss, R.J., Ellis, D.P.: Model-based expectation-maximization source separa-tion and localization. IEEE Trans. Audio Speech Lang. Process. **18**(2), 382–394 (2010)

36. Marr, D.: Vision: A Computational Investigation into the Human Representation and Process-ing of Visual Information. H. Freeman, San Francisco (1982)

37. Mnih, V., Heess, N., Graves, A., et al.: Recurrent models of visual attention. In: Advances in Neural Information Processing Systems, pp. 2204–2212 (2014)

38. Narayanan, A., Wang, D.: Ideal ratio mask estimation using deep neural networks for robust speech recognition. In: Proceedings of ICASSP, pp. 7092–7096 (2013)

39. Pearl, J.: Probabilistic Reasoning in Intelligent Systems: Networks of Plausible Inference. Morgan Kaufmann, San Francisco (1988)

40. Povey, D., Ghoshal, A., Boulianne, G., Burget, L., Glembek, O., Goel, N., Hannemann, M., Motlicek, P., Qian, Y., Schwarz, P., Silovsky, J., Stemmer, G., Vesely, K.: The Kaldi speech recognition toolkit. In: Proceedings of ASRU (2011)

41. Robinson, T., Fransen, J., Pye, D., Foote, J., Renals, S.: WSJCAM0: a British English speech corpus for large vocabulary continuous speech recognition. In: Proceedings of ICASSP, pp. 81–84 (1995)

42. Romera-Paredes, B., Torr, P.H.: Recurrent instance segmentation (2015). arXiv:1511.08250

43. Ross, S., Munoz, D., Hebert, M., Bagnell, J.A.: Learning message-passing inference machines for structured prediction. In: IEEE Conference on Computer Vision and Pattern Recognition (CVPR), pp. 2737–2744 (2011)

44. Seltzer, M.L., Raj, B., Stern, R.M.: Likelihood-maximizing beamforming for robust hands-free speech recognition. IEEE Trans. Audio Speech Process. **12**(5), 489–498 (2004)

45. Seltzer, M.L., Yu, D., Wang, Y.: An investigation of deep neural networks for noise robust speech recognition. In: Proceedings of ICASSP, pp. 7398–7402 (2013)

46. Shental, N., Zomet, A., Hertz, T., Weiss, Y.: Pairwise clustering and graphical models. In: Advances in Neural Information Processing Systems, pp. 185–192 (2004)

47. Smaragdis, P., Raj, B., Shashanka, M.: Supervised and semi-supervised separation of sounds from single-channel mixtures. In: Proceedings of ICA, pp. 414–421 (2007)

48. Souden, M., Araki, S., Kinoshita, K., Nakatani, T., Sawada, H.: A multichannel MMSE-based framework for speech source separation and noise reduction. IEEE Trans. Audio Speech Lang. Process. **21**(9), 1913–1928 (2013)

49. Sprechmann, P., Litman, R., Yakar, T.B., Bronstein, A.M., Sapiro, G.: Supervised sparse analysis and synthesis operators. In: NIPS, pp. 908–916 (2013)

50. Sprechmann, P., Bronstein, A.M., Sapiro, G.: Supervised non-Euclidean sparse NMF via bilevel optimization with applications to speech enhancement. In: Proceedings of HSCMA (2014)

51. Stoyanov, V., Ropson, A., Eisner, J.: Empirical risk minimization of graphical model parame-ters given approximate inference, decoding, and model structure. In: International Conference on Artificial Intelligence and Statistics, pp. 725–733 (2011)

52. Swietojanski, P., Ghoshal, A., Renals, S.: Convolutional neural networks for distant speech recognition. IEEE Signal Process. Lett. **21**(9), 1120–1124 (2014)
53. Vincent, E., Gribonval, R., Févotte, C.: Performance measurement in blind audio source separation. IEEE Trans. Audio Speech Lang. Process. **14**(4), 1462–1469 (2006)
54. Vincent, E., Barker, J., Watanabe, S., Le Roux, J., Nesta, F., Matassoni, M.: The second 'CHiME' speech separation and recognition challenge: datasets, tasks and baselines. In: Proceedings of ICASSP, pp. 126–130 (2013)
55. Wainwright, M.J., Jaakkola, T.S., Willsky, A.S.: A new class of upper bounds on the log partition function. IEEE Trans. Inf. Theory **51**(7), 2313–2335 (2005)
56. Wang, Y., Narayanan, A., Wang, D.: On training targets for supervised speech separation. IEEE/ACM IEEE Trans. Audio Speech Lang. Process. **22**(12), 1849–1858 (2014)
57. Weiss, Y.: Comparing the mean field method and belief propagation for approximate inference in MRFs. In: Advanced Mean Field Methods Theory and Practice, pp. 229–240 (2001)
58. Weninger, F., Le Roux, J., Hershey, J.R., Watanabe, S.: Discriminative NMF and its application to single-channel source separation. In: Proceedings of ISCA Interspeech (2014)
59. Weninger, F., Erdogan, H., Watanabe, S., Vincent, E., Le Roux, J., Hershey, J.R., Schuller, B.: Speech enhancement with LSTM recurrent neural networks and its application to noise-robust ASR. In: Latent Variable Analysis and Signal Separation (LVA), pp. 91–99 (2015)
60. Wisdom, S., Hershey, J.R., Le Roux, J., Watanabe, S.: Deep unfolding for multichannel source separation: supplementary materials. http://www.merl.com/demos/deep-MCGMM (2015)
61. Wisdom, S., Hershey, J., Le Roux, J., Watanabe, S.: Deep unfolding for multichannel source separation. In: Proceedings of ICASSP, pp. 121–125 (2016)
62. Xu, Y., Du, J., Dai, L.R., Lee, C.H.: An experimental study on speech enhancement based on deep neural networks. IEEE Signal Process. Lett. **21**(1), 65–68 (2014)
63. Yakar, T.B., Litman, R., Sprechmann, P., Bronstein, A., Sapiro, G.: Bilevel sparse models for polyphonic music transcription. In: Proceedings of ISMIR (2013)
64. Yedidia, J.S., Freeman, W.T., Weiss, Y.: Constructing free-energy approximations and generalized belief propagation algorithms. IEEE Trans. Inf. Theory **51**(7), 2282–2312 (2005)
65. Yu, D., Kolbæk, M., Tan, Z.H., Jensen, J.: Permutation invariant training of deep models for speaker-independent multi-talker speech separation (2016). arXiv:1607.00325
66. Zhang, X., Trmal, J., Povey, D., Khudanpur, S.: Improving deep neural network acoustic models using generalized maxout networks. In: Proceedings of ICASSP (2014)

Chapter 7
Deep Recurrent Networks for Separation and Recognition of Single-Channel Speech in Nonstationary Background Audio

Hakan Erdogan, John R. Hershey, Shinji Watanabe, and Jonathan Le Roux

Abstract We investigate the use of deep neural networks and deep recurrent neural networks for separation and recognition of speech in challenging environments. Mask prediction networks received considerable interest recently for speech separation and speech enhancement problems where the background signals are nonstationary and challenging. Initial signal-level enhancement with deep neural networks has also been shown to be useful for noise-robust speech recognition in these environments. We consider using various loss functions for training the networks and illustrate differences among them. We compare the performance of deep computational architectures with conventional statistical techniques as well as variants of nonnegative matrix factorization, and establish that one can achieve impressively superior results with deep-learning-based techniques on this problem.

7.1 Introduction

Speech enhancement is a classical signal-processing research area which aims to denoise and possibly dereverberate speech signals corrupted with noise and reverberation [1, 17]. We can find publications in this area that date back to the 1970s. Classical learning-free approaches used statistical modeling and estimation of noise parameters for a given utterance and used those models to enhance noisy speech. The Wiener filter and spectral subtraction are probably the earliest examples of speech enhancement methods.

This work was largely completed when the first author was on sabbatical leave at MERL from his faculty position at Sabanci University, Istanbul.

H. Erdogan (✉)
Microsoft Research, Redmond, WA, USA
e-mail: hakan.erdogan@microsoft.com

J.R. Hershey • S. Watanabe • J. Le Roux
Mitsubishi Electric Research Laboratories (MERL), Cambridge, MA, USA

© Springer International Publishing AG 2017
S. Watanabe et al. (eds.), *New Era for Robust Speech Recognition*,
DOI 10.1007/978-3-319-64680-0_7

165

Source separation, however, is a newer research area which tries to solve the "cocktail party" problem of audio signals, i.e., to separate individual signals recorded using single or multiple microphones. Separation of speech from background noise can be referred to as speech enhancement or speech–background separation, whereas separation of speech from another speech signal can be referred to as speech segregation, speech–speech separation, or simply speech separation.

We focus on single-channel mixtures of speech and background noise in this chapter. Newer methods for speech–background separation make use of learning techniques to learn the characteristics of speech and noise from a set of training data and use this information at test time. One of the earliest techniques that makes use of training data is nonnegative matrix factorization (NMF) [16]. In NMF, one can train dictionaries separately from speech and noise data and combine them to form a concatenated dictionary that is used at test time [23]. After performing matrix factorization on the test data, one can obtain an estimate of each source.

More recently, deep learning models have been used to build speech–background separation systems with remarkable success [18–20, 30, 31, 35]. Deep learning is used simply as a type of denoising autoencoder, where we feed in noisy data as input and expect the network to output enhanced speech. For this purpose, we provide clean speech as a target during training.

Recurrent neural networks (RNNs) are arguably best suited for time sequence data since they can remember past events that are relevant to prediction at the current time. This efficient use of contextual information is beneficial to obtaining better predictions. While deep neural networks require explicitly providing contextual information by feeding data from neighboring frames by splicing feature vectors from neighbors together, in recurrent nets, such explicit feeding is unnecessary since past inputs are already used in predictions. In bidirectional RNNs, we also take inputs from future neighbors, which can further improve predictions.

In this chapter, we review speech–background separation methods with a focus on neural networks and report our experimental results on the CHiME-2 data [26].

7.2   Problem Description

The single-channel speech–background separation problem is illustrated in Fig. 7.1 and can be explained simply as follows: given an observed mixed signal $y[n] = x[n] + u[n]$, estimate the original speech signal $x[n]$ from it. Due to the underdetermined nature of this problem, it is not easy to estimate $x[n]$ from $y[n]$. We require to have training data for both $x[n]$ and $u[n]$ such that we can get some idea about the sources and separate them from the observed mixed signal. In practice, the sources will also be reverberated, which means there is an underlying clean signal which passes through a filter before being observed in the mixture. However, in this chapter, we do not wish to dereverberate the signals and we wish only to obtain the sources as accurately as possible, not caring about whether the sources have been reverberated or not. The reason for this is that moderate levels of reverberation

Fig. 7.1 Speech–background noise separation problem

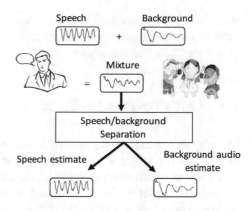

are not harmful for both human and machine recognition of speech. However, removal of noise is important for improving automatic speech recognition (ASR) performance.

The problem can be stated in the short-time Fourier transform (STFT) domain as follows:

$$Y(t,f) = X(t,f) + U(t,f), \qquad (7.1)$$

where $Y(t,f)$, $X(t,f)$, and $U(t,f)$ are the STFTs of the mixed signal, speech signal, and noise, respectively. We define the STFT as

$$Y(t,f) = \sum_{n=0}^{N-1} y[n + tL]w_a[n]e^{-j2\pi nf/N}. \qquad (7.2)$$

The analysis window function $w_a[n]$ is of length N and the signal is shifted by L samples for each frame t. Here $t = 0, 1, \ldots, N_t - 1$ and $f = 0, 1, \ldots, N - 1$ are integers representing the frame index and frequency index, and we can interpret the integer frequency index f as the continuous frequency $(f/N)f_s$ where f_s is the sampling rate.

Typically, we try to obtain an estimate $\hat{X}(t,f)$ from observed data $Y(t,f)$ and use the inverse STFT to go back to the time domain. The Inverse STFT involves inverse Fourier transform and overlap-add operations as follows:

$$\hat{x}_t[n] = \frac{1}{N} \sum_{f=0}^{N-1} X(t,f)e^{j2\pi nf/N}, \qquad (7.3)$$

$$\hat{x}[n] = \sum_{t=0}^{N_t-1} w_s[n - tL]\hat{x}_t[n - tL], \qquad (7.4)$$

where $w_s[n]$ is the synthesis window. We can find analysis and synthesis window pairs that result in perfect reconstruction when the inverse STFT is applied to the unmodified STFT of any signal.

7.3 Learning-Free Methods

Learning-free methods for speech enhancement generally operate as shown in Fig. 7.2. Noise variance is estimated from the utterance at hand. The critical parameter is this variance (or the a priori signal-to-noise ratio (SNR) which can be derived from it) and there have been various suggestions for its estimation, such as using minimum-energy frames [3] or initial or final frames of an utterance for which one assumes no speech is present.

The gain or mask parameter can be found using a Wiener filter or spectral subtraction. The criterion for enhancement can be the minimum mean squared error (MMSE) [4], the MMSE in the log magnitude domain [5], or other similar criteria. One of the best learning-free speech enhancement algorithms is the optimally modified log spectral amplitude (OMLSA) algorithm [2], which uses MMSE in the log magnitude domain with improved noise statistics prediction techniques. These studies do not make use of training data and try to operate within the confines of the given utterance.

Once the real gain parameter (or mask) $\hat{M}(t,f)$ is obtained, the enhanced signal is obtained by an inverse STFT on the pseudo-STFT $\hat{X}(t,f) = \hat{M}(t,f)Y(t,f)$. We use the term "pseudo-STFT" since it may not correspond to the STFT of a signal in the time domain [14]. However, one can still perform the inverse STFT to go back to the time domain to obtain an enhanced signal. Thus, we can see that one of the goals of speech enhancement methods is to obtain a "true" gain/mask function $M^*(t,f)$. Ideally, we can calculate ideal or oracle masks which would give almost perfect speech reconstruction from mixed signal data. We will touch upon these ideal masks in Sect. 7.5.2.1.

Learning-free methods do not make use of training data for enhancement, instead relying on statistical models that are estimated at test time only. However, it recently became clear that it would be beneficial to learn the characteristics of sources for better separation of mixed signals. We review next the use of NMF and deep learning architectures for solving the source separation problem.

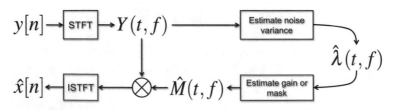

Fig. 7.2 Illustration of basic steps in learning-free methods for speech enhancement

7.4 Nonnegative Matrix Factorization

NMF has been extensively used for source separation or speech enhancement problems. A typical way of using NMF is to build dictionaries for each source type (e.g., speech and noise) and use these dictionaries at test time together to estimate each source from mixed data.

NMF is used to find nonnegative dictionaries with nonnegative coefficients that can explain away nonnegative data. Magnitude STFTs (or magnitude spectrograms) or power spectrograms can be considered as a data/observation matrix where each column corresponds to a frame and each row corresponds to a frequency. Then, we can find a nonnegative dictionary that explains each frame of observed data as

$$Y \approx BG,$$

where Y is the magnitude or power spectrogram matrix and B is a dictionary matrix which has much fewer columns. This is similar to PCA but with nonnegativity imposed on both the dictionary B and the gains or time-activations G [16]. NMF can be used to obtain both of the matrices B and G, but we can also work with fixed matrices for either of them since the updates are serialized for B and G. This decomposition is illustrated in Fig. 7.3. After training, we use the dictionary B as a model for the training data.

Let us assume that the first source is speech and the second source is background noise in a source separation problem. After training models B_1 and B_2 for two sources, we concatenate them at test time to obtain a new $B = [B_1\ B_2]$ [22]. This new dictionary is used to decompose the mixed-signal spectrogram, which enables its separation into two parts:

$$Y = \begin{bmatrix} B_1\ B_2 \end{bmatrix} \begin{bmatrix} G_1 \\ G_2 \end{bmatrix} = B_1 G_1 + B_2 G_2.$$

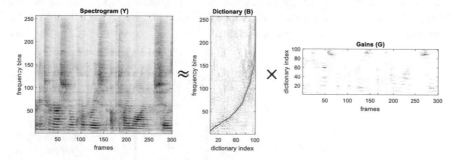

Fig. 7.3 Illustration of NMF for spectral dictionary learning where one hundred dictionary entries are sorted according to the location of their peak value and the spectrograms display third root of magnitude

These two parts can be interpreted as "projections" of the matrix Y onto convex cones nonnegatively spanned by B_1 and B_2, respectively, and each one can be used as an estimate of each source derived from the mixed signal through NMF. To get even better results [8], we can estimate a mask matrix from them as follows:

$$\hat{M} = \frac{B_1 G_1}{B_1 G_1 + B_2 G_2},$$
(7.5)

where division is coordinatewise (i.e., a Hadamard division). This mask matrix is always between 0 and 1 and can be used to reconstruct the first source by doing an inverse STFT on $\hat{M}(t,f)Y(t,f)$. This is similar in spirit to an adaptive Wiener filter, where we consider each component as power-spectral predictions of each source.

Many variants of NMF exist, such as sparse NMF [15, 27], exemplar-based NMF [7], discriminative NMF [32], and others.

7.5 Deep Learning for Source Separation

Deep learning is a booming area of research which seems to be useful in almost all learning-based problems. If there is plenty of training data available, it seems that deep learning can outperform any other learning technique. In source separation problems as well, given sufficient training data, deep learning techniques are likely to lead to systems with superior performance.

Computational networks can be used for prediction of sources from a mixed signal. The straightforward way to use a network is to use the mixed signal as an input and to expect the network to produce the source of interest at the output [12, 29, 35]. This requires training the network with simulated mixtures, since we require to know the clean sources during training.

The use of a deep network for enhancement is illustrated in Fig. 7.4. The parameter set of the network is denoted by the variable w. The number of parameters that need to be learned can be as high as several millions, especially for deeper networks. The output of the network can be the enhanced signal directly or a mask function. The mask function plays the same role as the one used in learning-free

Fig. 7.4 Using a deep neural network for enhancement

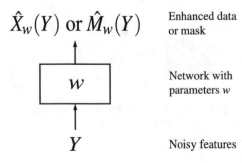

$$\hat{X}_w(Y) \text{ or } \hat{M}_w(Y)$$

Enhanced data or mask

w

Network with parameters w

Y

Noisy features

methods and the one found in (7.5) for NMF. The training is performed using simulated mixtures of relevant sources.

Another possible way to use deep learning for separation is to use a network as a classifier, which can be used to check the fidelity of source estimates as proposed in [9]. However, this requires solving an optimization problem at test time, which is slow and requires a good initialization to work well.

7.5.1 Recurrent and Long Short-Term Memory Networks

Deep recurrent networks are learning machines that can be used for learning from sequential data. In an RNN, we utilize hidden nodes, for which we calculate activations using inputs from both a lower layer and its own output value at a previous step of the sequence. This recursive nature allows the network to make use of past inputs to make decisions on current outputs. A reverse directional RNN makes use of future context in a similar way as a regular RNN makes use of past context. By combining the two, we obtain a bidirectional RNN. These RNN architectures are illustrated in Fig. 7.5.

Recurrent nets are hard to train due to the so-called vanishing- or exploding-gradient problem: the gradients tend to either die out or explode as they are back-propagated in time. This causes the network to fail to learn from data. One solution to this gradient back-propagation problem is to use long short-term memory (LSTM) networks [10]. These are special recurrent neural networks that make use of memory cells which have a temporal weight of one or zero, thus making good gradient propagation possible.

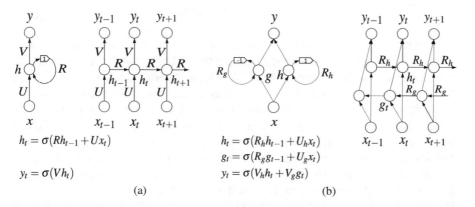

$$h_t = \sigma(Rh_{t-1} + Ux_t)$$
$$y_t = \sigma(Vh_t)$$
(a)

$$h_t = \sigma(R_h h_{t-1} + U_h x_t)$$
$$g_t = \sigma(R_g g_{t-1} + U_g x_t)$$
$$y_t = \sigma(V_h h_t + V_g g_t)$$
(b)

Fig. 7.5 Illustration of (**a**) RNN and (**b**) bidirectional RNN with input sequence $x_1, \ldots, x_T$ and output sequence $y_1, \ldots, y_T$, without explicit time expansion on the left and with explicit time expansion on the right

7.5.2 Mask Versus Signal Prediction

As we have mentioned in the beginning of Sect. 7.5, prediction of the speech signal can be achieved either directly or through a mask. When the network is used to infer a mask, its output can be multiplied by the noisy input to achieve signal prediction. Other alternative prediction targets for a network are the magnitude STFT, power spectrogram, and log power spectrogram of the clean signal.

7.5.2.1 Ideal Masks and Phase-Sensitive Mask

Since the source STFTs are complex-valued, to have perfect reconstruction of the sources we would need to predict the phases of the sources. However, phase prediction is notoriously hard and, in the problem of source separation, using the mixed signal's phase tends to work quite well. For example, in speech enhancement, it has been shown that the noisy phase is the MMSE estimator for the phase [4]. Intuitively, it makes sense as well, since when a time–frequency bin is dominated by one source, that bin's magnitude and phase will be close to that source's magnitude and phase. Hence, using the noisy phase makes sense since it will be quite close to the phase of the original source that dominates that time–frequency bin, and its value in other time–frequency bins will not have a big impact as the corresponding magnitude will be relatively small.

Among real masks, there is no unique definition of what the "optimal" or "ideal" mask is. Indeed, no real mask can reconstruct the speech signal exactly, due to phase mismatch. Dropping dependence on time–frequency to simplify the notation, we can write the mixture equation for each time–frequency bin as $Y = X + U$, or

$$|Y|e^{j\theta_y} = |X|e^{j\theta_x} + |U|e^{j\theta_u}. \tag{7.6}$$

In Table 7.1, we list various possible definitions of an "ideal" mask for this problem, each one ideal under certain assumptions. Almost all of these masks have been considered in the literature. The ideal ratio mask (IRM) is optimal if we assume the phases of both speech and noise are the same, but this is often false in practice. Similarly, an ideal amplitude filter (IAF) would predict the magnitude of the speech signal correctly but, since the phase of it would be wrong, the end result could be a bad prediction when we consider both the magnitude and the phase of the target speech signal. Considering these problems with other ideal real filters, we introduced in [6] the phase-sensitive filter (PSF), which takes into account the phase difference $\theta = \theta_y - \theta_x$ between the mixed signal and the source of interest. A PSF will not reconstruct the signal exactly, since the phase will still be wrong, but the error will be minimal as compared to other ideal real filters. We conjecture that the PSF is the best real mask that can be defined in terms of improving the SNR, or other similar signal-level metrics such as source-to-distortion ratio (SDR).

Table 7.1 Various masking functions M for computing a speech estimate $\hat{X} = MY$, their formulas in terms of M, and conditions for optimality

Target mask/filter	Formula	Optimality principle
Ideal binary mask (IBM)	$M^{\text{ibm}} = \delta(|X| > |U|)$	max SNR $M \in \{0, 1\}$
Ideal ratio mask (IRM)	$M^{\text{irm}} = \dfrac{|X|}{|X| + |U|}$	max SNR $\theta_x = \theta_u$,
Wiener-like	$M^{\text{wf}} = \dfrac{|X|^2}{|X|^2 + |U|^2}$	max SNR, expected power
Ideal amplitude	$M^{\text{iaf}} = |X|/|Y|$	Exact $|\hat{X}|$, max SNR $\theta_x = \theta_y$
Phase-sensitive filter	$M^{\text{psf}} = \dfrac{|X|}{|Y|}\cos(\theta)$	max SNR given $M \in \mathbb{R}$
Ideal complex filter	$M^{\text{icf}} = X/Y$	max SNR given $M \in \mathbb{C}$

In the IBM, $\delta(p)$ is 1 if the expression p is true and 0 otherwise

Fig. 7.6 Illustration of ideal amplitude filter (IAF), ideal ratio mask (IRM), and phase-sensitive filter (PSF) for three geometric arrangements of $Y = X + U$

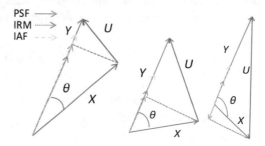

The PSF, IRM, and IAF filters are illustrated in Fig. 7.6 for several arrangements of Y, X and U. This figure clearly shows that using a PSF would provide a much lower error $|MY - X|$ in magnitude as compared to using either the IRM or the IAF.

7.5.2.2 Evaluating Ideal Masks

We evaluated each ideal filter on the CHiME-2 development set. Table 7.2 shows that we can obtain much better SDR values with a phase-sensitive oracle filter. The SDR is a blind source separation evaluation metric [25] and is closely related to the SNR. The phase-sensitive oracle filter can take any real value. Even when it is truncated to being between 0 and 1, we obtain a better SDR value using PSF as compared to other ideal filters.

During training time, since we train with simulated mixtures, we know the phase differences as well. So, we can use a loss function that considers the phase differences. This loss function, which calculates the error in the complex domain, is called the phase-sensitive approximation (PSA) loss function and will be introduced in the next section.

Table 7.2 SDR results (in dB) at various SNR levels on the left channel of the CHiME-2 development (dev) data using each oracle mask

dev	−6 dB	9 dB	Avg
IBM	14.56	20.89	17.59
IRM	14.13	20.69	17.29
Wiener-like	15.20	21.49	18.21
Ideal amplitude	13.97	21.35	17.52
Phase-sensitive filter	17.74	24.09	20.76
Truncated PSF	16.13	22.49	19.17

7.5.3 Loss Functions and Inputs

Several loss functions may be considered in training deep learning systems for source separation or speech enhancement. Let us define the outputs of a network as $\hat{X}_w$, $\hat{L}_w$, and $\hat{M}_w$ when the network is predicting the magnitude STFT, the log magnitude STFT, and the time–frequency mask, respectively. For direct prediction of magnitude spectra, one can use the squared error as a loss function,

$$D_{\text{MSE}}(w) = \sum_{t,f} \left| |X(t,f)| - \hat{X}_w(t,f) \right|^2, \tag{7.7}$$

or one can use the log-spectral error, which is the squared error in the log-magnitude-spectrum domain,

$$D_{\text{LMSE}}(w) = \sum_{t,f} \left| \log |X(t,f)| - \hat{L}_w(t,f) \right|^2. \tag{7.8}$$

For a mask prediction network, corresponding losses can be similarly defined. For example, given an ideal mask $M^*(t,f)$, we can define a squared-error loss in the mask domain, which we call the mask approximation (MA) loss,

$$D_{\text{MA}}(w) = \sum_{t,f} \left| M^*(t,f) - \hat{M}_w(t,f) \right|^2, \tag{7.9}$$

or we can define the error in the signal magnitude domain, which we call the magnitude spectrum approximation (MSA) loss,

$$D_{\text{MSA}}(w) = \sum_{t,f} \left| |X(t,f)| - \hat{M}_w(t,f)|Y(t,f)| \right|^2. \tag{7.10}$$

Predicting the mask may be easier than predicting the spectra or log spectra directly. One reason for that may be that we can use a sigmoid output layer which is always between 0 and 1 and suits a mask well, whereas we need a rectified linear output layer for the magnitude spectrum, which has infinite range. Another reason may be that when the mask is equal to 1 for a time–frequency bin, which

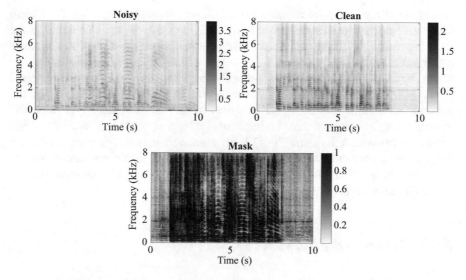

Fig. 7.7 Noisy and clean spectrograms and the corresponding ideal amplitude filter/mask for an example signal. Spectrograms display absolute value to the power 0.3

happens when that bin is dominated by speech, the network does not need to learn to reproduce the input itself at the output; it only needs to output a value of 1 for the mask. Finally, we can say that the mask is smoother and easier to predict than the magnitude spectrogram. We can see the difference between a clean spectrogram and the corresponding mask in Fig. 7.7.

7.5.4 Phase-Sensitive Approximation Loss Function

The corresponding loss function for the phase-sensitive ideal filter (the PSF) defined in Sect. 7.5.2.1 is called the PSA loss and defined as follows:

$$D_{\text{PSA}}(w) = \sum_{t,f} \left| X(t,f) - \hat{M}_w(t,f)Y(t,f) \right|^2. \tag{7.11}$$

Note that the error is defined using the complex STFT values of speech and mixed signals. This is equivalent to the following loss:

$$D_{\text{PSA}}(w) = \sum_{t,f} \left| |X(t,f)| \cos(\theta(t,f)) - \hat{M}_w(t,f)|Y(t,f)| \right|^2, \tag{7.12}$$

where $\theta(t,f) = \theta_y(t,f) - \theta_x(t,f)$ is the phase angle between $X(t,f)$ and $Y(t,f)$. Note that, with the PSA loss, the network still predicts a real mask $\hat{M}(t,f)$ and does not perform any phase prediction at all. It is only during training that the network makes use of the phase differences. The relation between the phase-sensitive ideal filter and the PSA loss function is that the PSF ideal filter is the minimizer of the PSA loss function among real masks when the magnitudes and phases of mixture components are known.

So, what is the network learning with the PSA loss function as compared to the MSA loss function if they are both predicting real masks? Basically, the network learns to shrink the mask estimates by an amount $\cos(\theta(t,f))$ (which is known during training but must be implicitly guessed at test time). Since the phase difference between $X(t,f)$ and $Y(t,f)$ is high when there is a high amount of noise $U(t,f)$, we can say that the network needs to assess whether the amount of noise is high enough, and if that is the case, it needs to shrink the mask estimate more than a (non-phase-sensitive) ratio mask would call for.

7.5.5 Inputs to the Network

Neural networks for separation or enhancement typically use features extracted from the mixed signal as input and aim to output the enhanced target signal. It is interesting to experiment with various features of the mixed signal that can be used as inputs.

7.5.5.1 Spectral Features

In neural networks for ASR, typically one uses 40-dimensional log mel filterbank features. For denoising, one can use the magnitude spectrogram as an input directly, or experiment with various log mel filterbank features. In [31], input features of full magnitude STFT log mel filterbanks with 40, 60, and 100 features were compared and it was found that 100 log mel filterbank features worked the best among the alternatives for separation of speech from background noise. We thus use 100 log mel filterbank features extracted from the mixed signal in this chapter. We use the shorthand "MFB" to refer to these 100 log mel filterbank features.

7.5.5.2 Speech-State Information

In addition to spectral features, one can add extra inputs from other information sources to improve performance. In [6], we introduced the idea of adding speech recognition states as inputs to the network. ASR systems make use of language models which incorporate word context information to help improve speech recognition accuracies. It is unlikely an RNN could infer word-level information from only acoustic data, whereas ASR language models are trained with large amounts of text data. ASR systems can be used for extracting a predicted speech state for each frame of input, and this extra information can potentially be exploited by a neural network to improve its enhancement capability. The basic hypothesis is that if the network knew what the speaker was saying, it could enhance noisy speech better. This intuition turns out to be correct, and we show that we can obtain improvements in performance by using extra input features derived from ASR hypotheses.

The input features from ASR hypotheses are obtained as follows. First, speech recognition is performed on the noisy data and a hypothesis is obtained. Then, the noisy signal is aligned with this hypothesis and an aligned state is determined for each frame. Finally, instead of feeding in the state information directly as a one-hot vector, we feed in the average of the log mel filterbank features aligned with that speech state in the training data. That is, the added feature vector has the same dimension as the original noisy feature vector and indicates the average feature value for that state in the training data. We use "SSI" to refer to these features in our experiments.

7.5.5.3   Enhanced Features

One other type of additional information is the spectral information from a previous round of enhancement. Hence, after enhancing the signal with a deep learning system in a first round, we can further use the enhanced signal as an additional input to improve the results as well. One interpretation here is that the network may detect uncertainties that can be derived from the previous round of enhancement and the noisy data and use information from these uncertainties. Another interpretation is that this may be a way of building a deeper network which can use information extracted from a lower-layer network as input. We have shown that performance can be improved by using this kind of additional information as well. This additional input also has the same dimension as the noisy data feature vector. We use the shorthand "ENH" to refer to these features in our experiments.

The various types of inputs to the enhancement network are illustrated in Fig. 7.8.

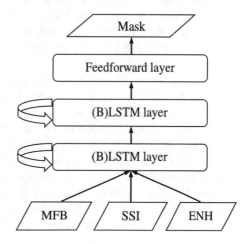

Fig. 7.8 Recurrent neural networks with various input features used for mask prediction

7.6 Experiments and Results

We performed experiments on the CHiME-2 dataset [26] to measure the effectiveness of recurrent neural networks for speech–background separation compared to other techniques introduced in the literature.

The CHiME-2 dataset consists of two-channel recordings of speech plus background audio. The speech part was simulated and the noise part was obtained from real recordings in a living room environment. The living room noises include kids playing and talking, TV, and other household noises. These noises were recorded with two microphones. The speech signal is an utterance from the Wall Street Journal speech database [21], reverberated using appropriate room impulse responses with the speaker assumed to be equidistant from the microphones [26]. CHiME-2 is challenging in that the background audio also contains speech (albeit kids' speech) and is very different from audio originating from a stationary noise source.

7.6.1 Neural Network Training

Several tricks of the trade exist for efficiently training recurrent nets. We used some of them here to obtain a suitable mask prediction neural network.

We performed a supervised version of layer-by-layer pretraining. We first trained a single-layer bidirectional LSTM ((B)LSTM), and then starting from the first-layer (B)LSTM and ignoring output layer weights, we added a new (B)LSTM layer, which improved results further. The inputs were mean and variance normalized over all the training data to have zero mean and unit variance. We added 0-mean Gaussian noise with variance 0.1 to the inputs to help improve generalization. We have not yet included dropout and batch normalization, which are newer techniques, but expect they may help improve performance in the future. We used stochastic gradient descent with momentum to train the neural networks we used in this chapter. The learning rate was 10^{-6} and the momentum coefficient was 0.9. Validation cost was used as the stopping criterion: when it stopped decreasing for ten epochs, the training was stopped.

We initially trained a mask-prediction network with a mask-domain squared-error loss function where the target mask was a 100 dimensional mel-transformed ideal ratio mask. Then, we added another layer on top of the previous network, which was initialized with the transpose or the pseudo-inverse of the mel transform matrix and mapped its inputs to the full-spectrum domain. Then, we continue training with a mask-domain squared error loss function, which we call the mask-approximation (MA) loss in the full-spectrum domain, again with an ideal ratio mask as the target. After these initial training steps, we switched to a signal-domain loss function such as the MSA loss function introduced in Sect. 7.5.3. This strategy turns out to result in better performance for mask-prediction networks [31].

For the PSA loss function, we always initialized from the network trained using the steps described above for the MSA loss function. We do not have to repeat the initial steps for every network we train if we already have a good network to initialize from. Note that it was recently shown in [34] that using the PSF as the target M^* for training with a mask-domain (MA) loss (see (7.9)) leads to better results than using the IRM. However, since we determined that the MSA loss yielded better results than the MA loss, we did not go back to the MA loss and modify it to use the PSF as the ideal mask target. Instead, we compared the performances of the MSA and PSA signal-domain losses directly.

In addition, when training a network with newly added input dimensions, we usually started the network from an earlier trained network, with additional weights corresponding to the new inputs set to zero. This way, we can make sure that the network starts out with the previous loss value and that we can further reduce that loss value using the new inputs.

7.6.2   Results on CHiME-2

We experimented with single-channel enhancement and two-channel enhancement with the CHiME-2 data. In the case of two channels, we just took the mean of the two microphones, since this corresponds to beamforming due to the speaker being assumed equidistant from the microphones. We applied enhancement to the averaged signal afterwards.

We used various measures to compare the performance of deep recurrent networks with other earlier proposals, some of which do not require any training data. Our initial results are shown in Table 7.3. The metrics used are the SDR [25], source-to-interference ratio (SIR) [25], perceptual evaluation of speech quality

Table 7.3 Evaluation results on the CHiME-2 evaluation dataset with left-channel audio only

Method	Loss	Input	SDR	SIR	PESQ	STOI	CEPD
No enh.			2.34	2.34	1.55	0.82	42.16
Log-MMSE			3.53	4.04	1.54	0.80	44.15
VTS			2.84	5.02	1.53	0.80	45.90
OMLSA			5.97	6.46	1.54	0.82	44.48
Sparse NMF			10.10	12.33	1.94	0.85	25.24
DNN 3×1024	MA	mag-STFT	11.43	14.17	2.28	0.88	20.34
DNN 3×1024	MSA	mag-STFT	12.16	15.53	2.36	0.89	17.44
DNN 3×1024	MA	MFB	12.02	14.89	2.31	0.88	20.01
DNN 3×1024	MSA	MFB	12.50	16.01	2.40	0.89	17.61
LSTM 2×256	MA	mag-STFT	13.19	16.50	2.59	0.90	15.61
LSTM 2×256	MSA	mag-STFT	13.59	17.46	2.63	**0.91**	14.27
LSTM 2×256	MA	MFB	13.69	17.62	2.60	**0.91**	15.45
LSTM 2×256	MSA	MFB	**13.91**	**17.97**	**2.67**	**0.91**	**13.95**

(PESQ) [11], short-time objective intelligibility (STOI) [24], and cepstral distance (CEPD) [13]. For the first four measures, higher values indicate better performance, whereas for the cepstral distance, the lower the better. Bold values indicate the best values in each column.

These initial results show that learning-based systems can perform much better than their learning-free counterparts on the CHiME-2 data. This is probably because the background noise in CHiME-2 is not stationary and has speech-like elements. Conventional methods cannot cope with this nonstationary and speech-like nature of the noise and produce much lower performance. In comparison with sparse NMF, neural-network-based systems perform much better on this task. We can also clearly see that LSTM networks with 2 layers and 256 nodes in each layer perform better than deep-neural-network (DNN) systems with 3 layers and 1024 nodes in each layer, probably because they make use of contextual data in a better way. The number of layers and nodes in each layer was partly optimized on the validation data [31]. The DNN systems used ten contextual concatenated frame features as input, whereas the LSTM systems used single frame features as input. In addition, it is clear that the signal-domain loss function (MSA) leads to better results than the mask-domain loss function (MA). Finally, we can see that using log mel filterbank features with 100 filters is better than directly using magnitude STFT features.

Altogether, a 2-layer LSTM network, using real mask prediction with the magnitude spectrum approximation loss function and 100 log mel filterbank features as input, already gives extremely good results on this task. There is, however, still room for improvement, as we investigate next.

The improvements we consider are (1) using a bidirectional LSTM, (2) using a PSA loss function, and (3) additionally using speech state information (SSI) as input. For the BLSTM network, we increased the total number of nodes per layer from 256 to 384 since there were forward and backward directional nodes, and we obtained better results with 384 as compared to 256 nodes. The results including these improvements are provided in Table 7.4. They indicate that using the phase-sensitive loss function is helpful to improve signal-level measures such as SDR and SIR. It may not always be useful for improving the PESQ, STOI, and CEPD measures, but it does not make them worse either.

We show the spectrograms of signals enhanced using various methods for an example utterance in Fig. 7.9. It is clear that learning-based methods, especially

Table 7.4 Evaluation results on the CHiME-2 evaluation dataset with left-channel audio only, using further improvements on enhancement networks

Method	Loss	Input	SDR	SIR	PESQ	STOI	CEPD
No enh.			2.34	2.34	1.55	0.82	42.16
LSTM 2 × 256	MSA	MFB	13.91	17.97	2.67	0.91	13.95
LSTM 2 × 256	PSA	MFB	14.14	19.20	2.64	0.91	13.85
BLSTM 2 × 384	PSA	MFB	14.51	19.78	2.78	0.91	12.77
BLSTM 2 × 384	PSA	MFB+SSI	**14.75**	**20.45**	**2.86**	**0.92**	**12.52**

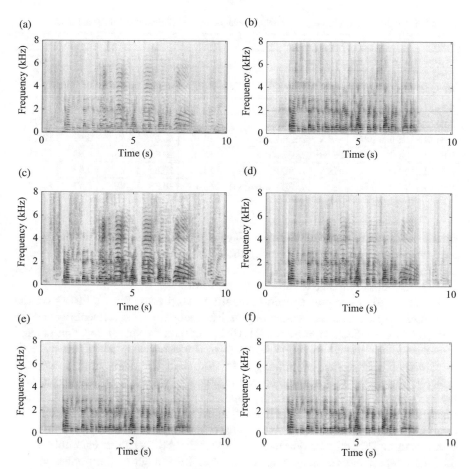

Fig. 7.9 Spectrograms for an example utterance at −6 dB SNR. (**a**) Noisy, (**b**) clean spectrograms. Enhanced spectrograms obtained with (**c**) OMLSA, (**d**) NMF, (**e**) LSTM-MSA, (**f**) BLSTM-PSA-SSI methods. Each spectrogram image is normalized individually

LSTM- and BLSTM-based ones, yield much better results as compared to the alternatives.

We also performed some experiments with two-channel data. The results are given in Table 7.5. In this table, in addition to earlier approaches, we also consider adding a third type of input to the network, which is the enhanced features (ENH) obtained from a previous round of enhancement (this enhancement is the best among earlier ones, and uses BF+BLSTM, PSA, and MFB+SSI inputs). Note that the results are not directly comparable to the previous two tables since their baseline is different. The clean signal was formed by taking the average of two channels of reverberated clean signals.

We observe that the PSA loss function improves the SDR, SIR, PESQ, and CEPD metrics as compared to the MSA loss, even though the improvement may not be so

Table 7.5 Evaluation results on the CHiME-2 evaluation dataset using two-channel data

Method	Loss	Input	SDR	SIR	PESQ	STOI	CEPD
Beamforming (BF)			1.74	1.74	1.53	0.83	41.83
BF+LSTM 2 × 256	MSA	MFB	14.17	18.03	2.63	0.92	13.55
BF+LSTM 2 × 256	PSA	MFB	14.49	19.66	2.67	0.92	12.77
BF+BLSTM 2 × 384	MSA	MFB	14.46	18.31	2.73	**0.93**	12.50
BF+BLSTM 2 × 384	PSA	MFB	14.88	20.23	2.80	**0.93**	11.78
BF+BLSTM 2 × 384	MSA	MFB+SSI	14.67	18.61	2.82	**0.93**	12.26
BF+BLSTM 2 × 384	PSA	MFB+SSI	15.07	20.40	2.86	**0.93**	11.46
BF+BLSTM 2 × 384	MSA	MFB+SSI+ENH	14.71	18.66	2.82	**0.93**	12.20
BF+BLSTM 2 × 384	PSA	MFB+SSI+ENH	**15.13**	**20.55**	**2.90**	**0.93**	**11.28**

significant for PESQ and CEPD but significant for the SIR metric. The STOI metric does not seem to change much among the advanced methods, since intelligibility is already quite high in the data. We observe that using ENH features improves the results further as compared to the earlier best result.

In Table 7.6, we report speech recognition word error rates (WERs) on the CHiME-2 development and evaluation sets. The recognition experiments were done using a DNN + hidden-Markov-model (HMM) system for speech recognition, and sequence-discriminative training was performed without adaptation. The systems were trained with enhanced training data. Although our main purpose was not directly to improve ASR accuracies, we observe that we can significantly improve the recognition accuracies using the proposed speech enhancement methods. The phase-sensitive approximation loss function is not always the clear winner in the case of WER. Its performance is close to that of the magnitude signal approximation loss function. This may be due to the fact that the ASR systems care only about magnitude spectrum accuracy, and it may thus be enough to train enhancement networks to focus on that only. These results show that when the background noise is noticeably removed from the speech signal, the recognition rates are also significantly improved. If we have speech recognition in mind, we can build networks that can attempt to reconstruct clean features, or even directly minimize speech recognition losses such as cross-entropy followed by sequence-discriminative losses as well. However, due to the benefit of masking, we believe it would help to first train an enhancement network and then continue to train a joint neural network for recognition purposes. These ideas are left as future work.

One of our earlier papers showed that we can further reduce the WER to 13.76% using discriminative training with noisy and enhanced stacked features [33]. Since then, it has been shown in [28] that using a masking/feature-extraction/recognition joint network trained with sequence-discriminative training and speaker adaptation can improve the WER to 11.23% using only jointly trained mel-filterbank-like features, and a WER of 10.63% can be achieved after feature-level combination using multiple robust features.

Table 7.6 WER on the CHiME-2 dataset with DNN-HMM acoustic models using stereo training (predicting clean HMM states from noisy data) and sequence-discriminative training, using enhanced speech features as input

Method	Loss	Input	WER (dev) Avg	WER (eval) Input SNR (dB)						
				−6	−3	0	3	6	9	Avg
Single-channel systems										
None			29.39	40.31	30.00	23.37	17.88	15.02	13.86	23.41
NMF-SA [32]			28.38	37.57	28.88	22.23	16.25	14.55	12.63	22.02
LSTM 2 × 256	MSA	MFB	23.99	30.92	23.26	18.72	14.35	12.85	11.68	18.63
LSTM 2 × 256	PSA	MFB	23.72	30.90	22.34	18.77	14.12	12.40	11.34	18.31
BLSTM 2 × 384	PSA	MFB	22.87	29.20	23.11	17.11	13.99	11.75	11.26	17.74
BLSTM 2 × 384	PSA	MFB+SSI	**21.54**	**28.04**	**20.03**	**16.05**	**13.04**	**11.38**	**10.97**	**16.58**
Two-channel systems										
BF			25.64	35.55	26.88	21.60	16.61	13.90	12.16	21.12
2 ch-NMF			25.13	32.19	23.05	20.04	15.54	13.19	12.72	19.46
BF+LSTM 2 × 256	MSA	MFB	19.03	24.86	17.65	15.11	11.41	10.20	9.68	14.82
BF+LSTM 2 × 256	PSA	MFB	19.20	24.15	17.63	14.91	11.73	9.75	9.58	14.63
BF+BLSTM 2 × 384	MSA	MFB	18.35	23.76	17.92	14.48	11.58	9.86	9.19	14.47
BF+BLSTM 2 × 384	MSA	MFB+SSI	18.41	24.38	**16.74**	14.80	11.06	**9.23**	9.32	14.25
BF+BLSTM 2 × 384	PSA	MFB+SSI	18.19	23.97	16.81	14.42	11.19	9.64	9.40	14.24
BF+BLSTM 2 × 384	MSA	MFB+SSI+ENH	**18.16**	**23.03**	17.21	14.16	**10.61**	9.25	9.45	**13.95**
BF+BLSTM 2 × 384	PSA	MFB+SSI+ENH	18.28	23.54	16.81	**14.07**	10.78	9.32	**9.17**	**13.95**

7.6.3 Discussion of Results

Our experiments show that recurrent networks, especially long short-term memory variants, are extremely effective for single-channel source separation problems as compared to earlier approaches including NMF. We evaluated the enhancement results with various metrics and, in all of them, we see large improvements in separation quality. The phase-sensitive loss function is extremely effective in improving the SDR and especially the SIR metrics. In terms of word error rates, phase-sensitive and magnitude signal-domain losses yield close results.

7.7 Conclusion

We experimented with single-channel source separation in the context of separating speech from background noise, where the background noises were recorded in a real-life living room. The mixing was simulated to enable objective measures to be computed. Future work should address real mixing and find ways to evaluate methods based on real mixtures. It would be desirable to develop metrics that do not require ground truth references for this purpose. When ASR accuracies are the final goal, we can use WER as the metric, but the goal of speech separation is not always limited to ASR. The final goal may be to improve perceptual quality and/or intelligibility for human–human communications or to improve separation for hearing aids and potentially other purposes.

References

1. Benesty, J., Makino, S., Chen, J.: Speech Enhancement. Springer Science & Business Media, New York (2005)
2. Cohen, I.: Optimal speech enhancement under signal presence uncertainty using log-spectral amplitude estimator. IEEE Signal Process. Lett. 9(4), 113–116 (2002)
3. Cohen, I., Berdugo, B.: Noise estimation by minima controlled recursive averaging for robust speech enhancement. IEEE Signal Process. Lett. 9(1), 12–15 (2002)
4. Ephraim, Y., Malah, D.: Speech enhancement using a minimum-mean square error short-time spectral amplitude estimator. IEEE Trans. Acoust. Speech Signal Process. 32(6), 1109–1121 (1984)
5. Ephraim, Y., Malah, D.: Speech enhancement using a minimum mean-square error log-spectral amplitude estimator. IEEE Trans. Acoust. Speech Signal Process. 33(2), 443–445 (1985)
6. Erdogan, H., Hershey, J.R., Watanabe, S., Le Roux, J.: Phase-sensitive and recognition-boosted speech separation using deep recurrent neural networks. In: Proceedings of the IEEE International Conference on Acoustics, Speech, and Signal Processing (ICASSP), Brisbane (2015)
7. Gemmeke, J.F., Virtanen, T., Hurmalainen, A.: Exemplar-based sparse representations for noise robust automatic speech recognition. IEEE Trans. Audio Speech Lang. Process. 19(7), 2067–2080 (2011)

8. Grais, E.M., Erdogan, H.: Single channel speech music separation using nonnegative matrix factorization and spectral masks. In: Proceedings of the International Conference on Digital Signal Processing (DSP), pp. 1–6 (2011)
9. Grais, E.M., Sen, M.U., Erdogan, H.: Deep neural networks for single channel source separation. In: Proceedings of the IEEE International Conference on Acoustics, Speech, and Signal Processing (ICASSP), Florence (2014)
10. Hochreiter, S., Schmidhuber, J.: Long short-term memory. Neural Comput. 9(8), 1735–1780 (1997)
11. Hu, Y., Loizou, P.C.: Evaluation of objective quality measures for speech enhancement. IEEE Trans. Audio Speech Lang. Process. 16(1), 229–238 (2008)
12. Huang, P.S., Kim, M., Hasegawa-Johnson, M., Smaragdis, P.: Deep learning for monaural speech separation. In: Proceedings of the IEEE International Conference on Acoustics, Speech, and Signal Processing (ICASSP), Florence, pp. 1581–1585 (2014)
13. Kubichek, R.: Mel-cepstral distance measure for objective speech quality assessment. In: IEEE Pacific Rim Conference on Communications, Computers and Signal Processing, vol. 1, pp. 125–128 (1993)
14. Le Roux, J., Vincent, E., Mizuno, Y., Kameoka, H., Ono, N., Sagayama, S.: Consistent Wiener filtering: generalized time–frequency masking respecting spectrogram consistency. In: Proceedings of the International Conference on Latent Variable Analysis and Signal Separation, pp. 89–96 (2010)
15. Le Roux, J., Weninger, F.J., Hershey, J.R.: Sparse NMF – half-baked or well done? Technical Report, TR2015-023, Mitsubishi Electric Research Laboratories (MERL), Cambridge, MA (2015)
16. Lee, D.D., Seung, H.S.: Algorithms for non-negative matrix factorization. In: Advances in Neural Information Processing Systems (NIPS), pp. 556–562 (2001)
17. Loizou, P.C.: Speech Enhancement: Theory and Practice. CRC Press, Boca Raton, FL (2013)
18. Lu, X., Tsao, Y., Matsuda, S., Hori, C.: Speech enhancement based on deep denoising autoencoder. In: Proceedings of the Interspeech, Lyon, pp. 3444–3448 (2013)
19. Maas, A.L., O'Neil, T.M., Hannun, A.Y., Ng, A.Y.: Recurrent neural network feature enhancement: the 2nd CHiME challenge. In: Proceedings of the CHiME Workshop on Machine Listening in Multisource Environments, Vancouver, pp. 79–80 (2013)
20. Narayanan, A., Wang, D.: Ideal ratio mask estimation using deep neural networks for robust speech recognition. In: Proceedings of the IEEE International Conference on Acoustics, Speech, and Signal Processing (ICASSP), Vancouver, pp. 7092–7096 (2013)
21. Paul, D.B., Baker, J.M.: The design for the Wall Street Journal-based CSR corpus. In: Proceedings of the Workshop on Speech and Natural Language, pp. 357–362 (1992)
22. Schmidt, M.N., Olsson, R.K.: Single-channel speech separation using sparse non-negative matrix factorization. In: Proceedings of the Interspeech, Pittsburgh, PA, pp. 1652–55 (2006)
23. Smaragdis, P.: Convolutive speech bases and their application to supervised speech separation. IEEE Trans. Audio Speech Lang. Process. 15(1), 1–14 (2007)
24. Taal, C.H., Hendriks, R.C., Heusdens, R., Jensen, J.: An algorithm for intelligibility prediction of time–frequency weighted noisy speech. IEEE Trans. Audio Speech Lang. Process. 19(7), 2125–2136 (2011)
25. Vincent, E., Gribonval, R., Févotte, C.: Performance measurement in blind audio source separation. IEEE Trans. Audio Speech Lang. Process. 14(4), 1462–1469 (2006)
26. Vincent, E., Barker, J., Watanabe, S., Le Roux, J., Nesta, F., Matassoni, M.: The second "CHiME" speech separation and recognition challenge: datasets, tasks and baselines. In: Proceedings of the IEEE International Conference on Acoustics, Speech, and Signal Processing (ICASSP), Vancouver, pp. 126–130 (2013)
27. Virtanen, T.: Monaural sound source separation by nonnegative matrix factorization with temporal continuity and sparseness criteria. IEEE Trans. Audio Speech Lang. Process. 15(3), 1066–1074 (2007)
28. Wang, Z.Q., Wang, D.: A joint training framework for robust automatic speech recognition. IEEE/ACM Trans. Audio Speech Lang. Process. 24(4), 796–806 (2016)

29. Wang, Y., Narayanan, A., Wang, D.: On training targets for supervised speech separation. IEEE/ACM Trans. Audio Speech Lang. Process. **22**(12), 1849–58 (2014)
30. Weninger, F., Geiger, J., Wöllmer, M., Schuller, B., Rigoll, G.: The Munich feature enhancement approach to the 2013 CHiME challenge using BLSTM recurrent neural networks. In: Proceedings of the 2nd CHiME Speech Separation and Recognition Challenge held in conjunction with ICASSP 2013, Vancouver, pp. 86–90 (2013)
31. Weninger, F., Hershey, J.R., Le Roux, J., Schuller, B.: Discriminatively trained recurrent neural networks for single-channel speech separation. In: Proceedings of the IEEE Global Conference on Signal and Information Processing (GlobalSIP), pp. 577–581 (2014)
32. Weninger, F., Le Roux, J., Hershey, J., Watanabe, S.: Discriminative NMF and its application to single-channel source separation. In: Proceedings of the Interspeech, Singapore (2014)
33. Weninger, F., Erdogan, H., Watanabe, S., Vincent, E., Le Roux, J., Hershey, J.R., Schuller, B.: Speech enhancement with LSTM recurrent neural networks and its application to noise-robust ASR. In: Proceedings of the International Conference on Latent Variable Analysis and Signal Separation (LVA/ICA) (2015)
34. Williamson, D.S., Wang, Y., Wang, D.: Complex ratio masking for monaural speech separation. IEEE/ACM Trans. Audio Speech Lang. Process. **24**(3), 483–492 (2016)
35. Xu, Y., Du, J., Dai, L.R., Lee, C.H.: An experimental study on speech enhancement based on deep neural networks. Signal Process. Lett. **21**(1), 65–68 (2014)

Chapter 8
Robust Features in Deep-Learning-Based Speech Recognition

Vikramjit Mitra, Horacio Franco, Richard M. Stern, Julien van Hout,
Luciana Ferrer, Martin Graciarena, Wen Wang, Dimitra Vergyri,
Abeer Alwan, and John H.L. Hansen

Abstract Recent progress in deep learning has revolutionized speech recognition research, with Deep Neural Networks (DNNs) becoming the new state of the art for acoustic modeling. DNNs offer significantly lower speech recognition error rates compared to those provided by the previously used Gaussian Mixture Models (GMMs). Unfortunately, DNNs are data sensitive, and unseen data conditions can deteriorate their performance. Acoustic distortions such as noise, reverberation, channel differences, etc. add variation to the speech signal, which in turn impact DNN acoustic model performance. A straightforward solution to this issue is training the DNN models with these types of variation, which typically provides quite impressive performance. However, anticipating such variation is not always possible; in these cases, DNN recognition performance can deteriorate quite sharply.

V. Mitra (✉) • H. Franco • J. van Hout • M. Graciarena • W. Wang • D. Vergyri
Speech Technology and Research (STAR) Lab., SRI International, 333 Ravenswood Ave.,
Menlo Park, CA 94025-3493, USA
e-mail: vikramjitmitra@gmail.com; horacio.franco@sri.com; julien.vanhout@sri.com;
martin.graciarena@sri.com; wen.wang@sri.com; dimitra.vergyri@sri.com

R.M. Stern
Department of Electrical and Computer Engineering and Language Technologies Institute,
Carnegie Mellon University, 5000 Forbes Avenue, Pittsburgh, PA 15213, USA
e-mail: rms@cmu.edu

L. Ferrer
Instituto de Investigación en Ciencias de la Computación, CONICET-UBA, Oficina 15,
Pabellón I, Ciudad Universitaria, C1428EGA Ciudad de Buenos Aires, Argentina
e-mail: lferrer@dc.uba.ar

A. Alwan
Department of Electrical Engineering, University of California, Los Angeles, 405 Hilgard Ave.,
Los Angeles, CA 90095, USA
e-mail: alwan@ee.ucla.edu

J.H.L. Hansen
Center for Robust Speech Systems (CRSS), University of Texas at Dallas, 800 W Campbell Road,
Richardson, TX 75080-3021, USA
e-mail: John.Hansen@utdallas.edu

© Springer International Publishing AG 2017
S. Watanabe et al. (eds.), *New Era for Robust Speech Recognition*,
DOI 10.1007/978-3-319-64680-0_8

To avoid subjecting acoustic models to such variation, robust features have traditionally been used to create an invariant representation of the acoustic space. Most
commonly, robust feature-extraction strategies have explored three principal areas:
(a) enhancing the speech signal, with a goal of improving the perceptual quality
of speech; (b) reducing the distortion footprint, with signal-theoretic techniques
used to learn the distortion characteristics and subsequently filter them out of the
speech signal; and finally (c) leveraging knowledge from auditory neuroscience and
psychoacoustics, by using robust features inspired by auditory perception.

In this chapter, we present prominent robust feature-extraction strategies
explored by the speech recognition research community, and we discuss their
relevance to coping with data-mismatch problems in DNN-based acoustic modeling.
We present results demonstrating the efficacy of robust features in the new paradigm
of DNN acoustic models. And we discuss future directions in feature design for
making speech recognition systems more robust to unseen acoustic conditions. Note
that the approaches discussed in this chapter focus primarily on single channel data.

8.1 Introduction

Before the advent of deep learning, Gaussian-mixture-model (GMM)-based hidden
Markov models (HMMs) were the state-of-the-art acoustic models for automatic
speech recognition (ASR) systems. However, GMM-HMM systems are susceptible
to background noise and channel distortions, and a small mismatch between training
and testing conditions can make speech recognition a futile effort. To counter
this issue, the speech research community undertook significant efforts to reduce
the mismatch between training and testing conditions by processing the speech
signal, either through speech enhancement [106, 115] or by using robust signal-
processing techniques [28, 62, 77, 112]. Studies also explored making acoustic
models more robust by either using data augmentation or introducing a reliability
mask [18, 29, 65].

The emergence of the deep-neural-network (DNN) architecture has significantly
boosted speech recognition performance. Several studies [66, 85, 102] demonstrated
significant improvement in speech recognition performance from DNNs compared
to their GMM-HMM counterparts. Recent studies [23, 103] showed that DNNs
work quite well for noisy speech and again significantly improve performance under
these conditions compared to GMM-HMM systems. Given the versatility of DNN
systems, it has been stated [120] that speaker normalization techniques, such as
vocal tract length Normalization (VTLN) [122], do not significantly improve speech
recognition accuracy, as the DNN architecture's rich multiple projections through
multiple hidden layers enable it to learn a speaker-invariant data representation.

State-of-the-art DNN architectures also deviate from using traditional cepstral
representation by instead employing simpler spectral representations. While GMM-
HMM architectures necessitated uncorrelated observations due to their widely used
diagonal covariance design (which in turn required that the observations undergo
a decorrelation step using the popular discrete cosine transform (DCT)), DNN

architectures suffer from no such requirement. Rather, the neural network architectures are known to benefit from cross-correlations [76] and hence demonstrate similar or better performance when using spectral features rather than their cepstral counterparts [103].

The convolutional neural network (CNN) [1, 50] is often found to outperform fully connected DNN architectures [2]. CNNs are also expected to be noise robust [1], especially when noise or distortion is localized in the spectrum. Speaker normalization techniques, such as VTLN [122], are also found to have less impact on speech recognition accuracy for CNNs as compared to DNNs. With CNNs, the localized convolution filters across frequency tend to normalize the spectral variations in speech arising from vocal tract length differences, enabling the CNNs to learn speaker-invariant data representations. Recent results [80–82] confirm that CNNs are more robust to noise and channel degradation than DNNs. Typically, for speech recognition, a single layer of convolution filters is used on the input-contextualized feature space to create multiple feature maps that, in turn, are fed to fully connected DNNs. However, in [96], adding multiple convolution layers (usually up to two) was shown to improve the performance of CNN systems beyond their single-layer counterparts. A recent study [74] observed that performing convolution across the time and frequency dimensions gives better performance than that provided by the CNN counterparts, especially for reverberated speech. Temporal processing through the use of time delay neural networks (TDNNs) provided impressive results when dealing with reverberated speech [91].

DNN models can be quite sensitive to data mismatches, and changes in the background acoustic conditions can result in catastrophic failure of such models. Further, any unseen distortion introduced at the input layers of the DNN results in a chain reaction of distortion propagation through the DNN. Typically, deeper neural nets offer better speech recognition performance for seen data conditions than shallower neural nets, while shallower nets are relatively robust to unseen data conditions [75]. This observation is a direct consequence of distortion propagation through the hidden layers of the neural nets, where deeper neural nets typically have more distorted information at their output-activation level compared to shallower ones, as shown by Bengio [11]. The literature reports that data augmentation to match the evaluation condition [61, 90] improves the robustness of DNN acoustic models and combats data mismatch. All such conditions assume that we have a priori knowledge about the kind of distortion that the model will see, which is often quite difficult, if not impossible, to achieve. For example, ASR systems deployed in the wild encounter unpredictable and highly dynamic acoustic conditions that are unique and hence difficult to augment.

A series of speech recognition challenges (MGB [9], CHiME-3 [6], ASpIRE [45], REVERB-2014 [67], and many more) revealed the vulnerability of DNN systems to realistic acoustic conditions and variations, and has resulted in innovative ways of making DNN-based acoustic models more robust to unseen data conditions. Typically, robust acoustic features are used to improve acoustic models when dealing with noisy and channel-degraded acoustic data [80–82]. A recent study [97] showed that instead of performing ad hoc signal processing, as typically is done for robust feature generation, one can directly use the raw signal and employ a long short-term memory (LSTM) neural net to perform the signal processing for a DNN acoustic model where the feature-extraction-step parameters are jointly optimized with the acoustic-model parameters. Although such an approach is intriguing for future speech recognition research, it is unknown both whether the limited training data would impact acoustic-model behavior and how well such systems generalize to unseen data conditions, where feature transforms learned in a data-driven way may not generalize well for out-of-domain acoustic data.

Model adaptation is another alternative for dealing with unseen acoustic data. Several studies have explored novel ways of performing unsupervised adaptation of DNN acoustic models [88, 98, 118], where techniques based on maximum likelihood linear regression (MLLR) transforms, i-vectors, etc. have shown impressive performance gains over unadapted models. Supervised adaptation with a limited set of transcribed target domain data is typically found to be helpful [7], and such approaches mostly involve updating the DNN parameters with the supervised adaptation data with some regularization. The effectiveness of such approaches is usually proportional to the volume of available adaptation data; however, such systems are typically found to digress away from the original training data and to learn the details of the target adaptation data. A solution for coping with this issue was proposed in [121], where a Kullback–Leibler divergence (KLD) regularization was proposed for DNN adaptation, which differs from the typically used L2 regularization [68] in the sense that it constrains the model parameters themselves rather than the output probabilities.

In the next sections, we first provide a brief historical background on acoustic features as used in ASR systems, present some of the prominent robust-feature-extraction strategies that have been used in the literature, and discuss how some of those features have been used in the current DNN-based acoustic models.

8.2   Background

The study of speech technologies began during the second half of the eighteenth century, with an attempt to create machines that imitated the process of human speech production [59]. Acoustic phonetics dominated the early years of modern speech recognition research, with analysis of acoustic realizations of phonetic elements in spoken utterances being the primary focus. Vocal-tract resonances or formant structures in speech in sustained vowel contexts was widely researched, and the vowel space with respect to formant frequency values was defined [20, 40].

In the late 1960s, linear predictive coding (LPC) [3, 56] was introduced, which enabled estimating the vocal-tract response from speech waveforms. The introduction of LPC, in turn, enabled designing pattern recognition methodologies that recognized speech using LPC-based information [55, 93]. In 1980, Davis and Mermelstein [19] first introduced mel-frequency cepstral coefficients (MFCCs), which have since served as the acoustic feature of choice across all speech applications. The steps involved in MFCC feature computation consist of (1) short-time Fourier analysis using Hamming windows; (2) weighting of the short-time magnitude spectrum by a series of triangularly shaped filterbanks with peaks that are equally spaced in frequency according to the mel scale; (3) computation of the log of the total energy in the weighted spectrum; and (4) computation of a relatively small number of coefficients of the inverse DCT of the log power coefficients for each channel. The mel-filterbank crudely mimics human auditory filtering; the log-compression mimics the nonlinear psychophysical transfer function for intensity; and the inverse DCT provides a low-pass Fourier series representation of the frequency-warped log spectrum, where the fine structure corresponding to source information is filtered out, retaining mostly the phonetic content of speech.

The perceptual linear prediction (PLP) feature is somewhat different from the MFCC feature, but the motivating principles behind both features are similar. The steps involved in PLP feature extraction are as follows: (1) short-time Fourier analysis using Hamming windows (as in MFCC processing); (2) weighting of the power spectrum by a set of asymmetrical functions that are spaced according to the Bark scale, and are based on the auditory masking curves of [99]; (3) pre-emphasis to simulate the equal-loudness curve suggested by Makhoul and Cosell [70], to model the loudness contours of Fletcher and Munson; (4) a power-law nonlinearity with exponent 0.33 as suggested by Stevens et al. [107] to describe the intensity transfer function; (5) a smoothed approximation to the frequency response obtained by all-pole modeling; and (6) application of a linear recursion that converts the coefficients of the all-pole model to cepstral coefficients.

In this section, we have briefly discussed how speech science evolved and the motivation behind conventional feature extraction techniques, and described the steps involved. Next, we describe the various facets of speech-signal processing that have been explored to improve the performance and robustness of automatic speech recognition systems.

8.3 Approaches

Since the introduction of ASR systems, a tremendous effort has been made toward understanding the problem of speech recognition and making such systems more robust, with the reliability of ASR systems under realistic background conditions being a critical research topic. Digitized audio signals serve as input to ASR systems, and therefore signal-processing methodologies have been exhaustively investigated for coping with background conditions, with the aim of producing invariant speech representations that least impact ASR acoustic-model performance and thus speech recognition quality. The study of robust features explores different signal-processing techniques that produce reliable and invariant speech representations, where the phonetic classes are more easily recognizable, and the background distortions are minimized. In this section, we discuss robust-feature-extraction techniques that have been investigated in the ASR research literature.

8.3.1 Speech Enhancement

Speech enhancement has received a tremendous amount of attention over the previous few decades. A detailed exploration of the different speech enhancement techniques can be found in [10]. Most speech enhancement techniques aim to modify the short-time spectral amplitude (STSA) of noisy speech signals. Subtractive-type speech enhancement techniques assume that background noise is locally stationary, such that the noise characteristics can be estimated from the speech-absent/pause regions. Since the introduction of the spectral subtraction algorithm [13], several variants/enhancements of subtractive algorithms have been proposed [8, 44]. In [115], a detailed analysis of the various subtraction parameters was explored, and a generalized spectral subtraction algorithm that adapts its parameters based on the masking properties of the human auditory system was presented.

In [31], the robustness of ASR systems was investigated, where speech in additive noise conditions was considered. The ETSI (European Telecommunications Standards Institute) basic [27] and advanced [28] front ends have been proposed for distributed speech recognition (DSR). Such front ends perform speech enhancement to attenuate background noise, before extracting the spectral features for acoustic-model training. The ETSI advanced end has two stages, where the first stage consists of voice activity detection (VAD) to detect speech-absent regions for estimating the noise spectral characteristics necessary for speech enhancement, and the second stage performs speech enhancement followed by acoustic feature extraction. The ETSI advanced front end is typically found to offer better performance than the ETSI basic front end in noisy conditions [31].

Auditory scene analysis (ASA) is usually considered to be a key factor behind the human ability to robustly perceive speech in varying acoustic environments [14]. ASA helps human listeners to organize the audio mixture into streams [14] that correspond to the different sound sources in the mixture. A feature-based Computational auditory scene analysis (CASA) system was proposed in [105], which makes weak assumptions about the various sound sources in the mixture. In [116], an ideal binary time–frequency mask was proposed as a major computational goal of CASA, where the binary mask is constructed from a priori knowledge of the target and interference. Using time–frequency masks is motivated by the phenomenon of auditory masking, where a weaker signal is masked by a stronger one within a critical band [86]. Soft-mask-based approaches have been successfully applied to noise-robust ASR on small- and large-vocabulary tasks. In [113], a technique called log-spectral enhancement (LSEN) was proposed, in which the variability caused by noise in the log-spectra is reduced while preserving the variability from the speech energy. First, a signal-to-noise-ratio-based soft-decision mask is computed in the mel-spectral domain as an indicator of speech presence. Then, the known time–frequency correlation of speech is exploited by treating this mask as an image and performing median filtering and blurring to remove the outliers and to smooth the decision regions. Finally, log-spectral flooring is applied to the lifted spectra of both clean and noisy speech, so as to match their respective dynamic ranges and to emphasize the information in the spectral peaks.

8.3.2   Signal-Theoretic Techniques

Signal-theoretic approaches use signal characteristics to perform filtering or transformation of the speech signal to generate robust feature representations that can improve the robustness of speech recognition systems against varying acoustic conditions.

Typically, acoustic features are expected to demonstrate different distributions for different phonetic units. It is well known that noise, channels, reverberation, etc. result in significant deviation from the usual distributions for the different acoustic units. Such deviation typically results in acoustic-condition mismatch, in which the training data and the test data statistics do not match, resulting in significant errors during test data decoding.

The most direct robustness techniques are based on normalization of various statistics of the features. The simplest such approach is cepstral mean normalization (CMN), in which the mean of the cepstra in an utterance is subtracted frame by frame on a sentence-by-sentence basis, in both training and testing a speech recognition system. CMN has been so successful that it (or a similar technique) is used invariably in speech recognition. The success of CMN can be understood from two points of view. First, if a speech signal undergoes unknown linear filtering, showing that the filter imposes an additive shift on the cepstral coefficients is easy, provided the filter's impulse response is briefer than the analysis window's duration. Hence, subtracting the mean cepstral values eliminates any effects introduced by stationary linear filtering. Second, and more prosaically, equating the features' utterance-level means reduces variability between the features representing the training and testing data. This principle is easily extended to the features' other attributes, such as in mean variance normalization (MVN, in which the means are typically set to zero and the variances set to one in training and testing) and histogram equalization (HEQ, in which the values of the features are warped monotonically to match a standard distribution of their values) [46]. MVN and HEQ typically provide some additional robustness to many types of distortion, again by reducing the statistical disparities between the training and testing samples.

8.3.3   Perceptually Motivated Features

It is well known that human speech-processing capabilities surpass the capabilities of current automatic speech recognition and related technologies. Since the early 1980s, this observation has motivated development of feature extraction approaches for speech recognition systems that are based on auditory physiology and perception. Influential early examples include the auditory models of Seneff [104], Lyon [69], Ghitza [37], and Cohen [17]. Typically, these features provide little or no benefit for the recognition of clean speech, but they tend to be helpful in recognizing degraded speech.

Researchers have had differing opinions concerning which aspects of auditory processing are the most important to preserve in feature extraction schemes. The most successful auditory-modeling schemes have included some of the following components:

- Peripheral frequency selectivity, which typically includes a bank of filters that mimic the shape of the frequency-selective response of individual fibers of the auditory nerve and more central structures. The gammatone filterbank [89] is frequently used to implement this stage of processing.
- Rate-level response, which typically takes the form of an S-shaped function (such as a sigmoid or inverse tangent) that relates the signal intensity in a given frequency channel to the output level, as opposed to the strictly logarithmic relationship between input and output in MFCC and similar representations.
- Synchrony with low-frequency fine structure, in the form of a mechanism that responds in synchrony with the fine structure of the low-frequency components of sound. (This component is believed by some to improve recognition accuracy in noisy environments.)
- Emphasis of onsets and suppression of steady state-components, as in RelAltive SpecTrA (RASTA) processing. In effect, this enhances temporal contrast and improves recognition accuracy in reverberant environments.
- Lateral suppression, which enhances contrast in signal content with respect to frequency. This is believed by some to be useful especially in distinguishing components of complex sound fields.
- Modulation-spectrum analysis, which can be useful for separating speech and nonspeech components in noisy environments.

In recent years, advances in computation and statistical modeling of the features produced by auditory models have enabled much more practical use of physiologically and perceptually motivated features. Examples of successful systems include RASTA-PLP, TRAPS, PNCC, FDLP, MHEC, NMC, DOC, and many more.

Temporal processing plays a key role in human speech perception and ASR [26, 60]. For example, short-time spectral features, such as MFCCs, are routinely concatenated with their first- and second-order temporal derivatives. The delta (δ) and double delta (δ^2) feature coefficients capture temporal dynamics of the acoustic features, and are overwhelmingly used in speech tasks such as speech recognition, speaker recognition, language identification, etc. Temporal information has also been incorporated through extracting temporal amplitude modulation (AM) of speech spectra or cepstra. A widely popular and frequently used modulation-based acoustic feature is the RASTA-processed PLP feature [48], which uses an infinite impulse response (IIR) filter that emphasizes AM frequencies between 1 and 12 Hz. The goal of RASTA processing is retaining the perceptually relevant modulation bands that correspond to linguistically meaningful information while filtering out extrinsic information [26, 60]. RASTA processing effectively applies a bandpass filter to the compressed spectral amplitudes in the intermediate stages of the PLP features, with the intention of modeling the emphasis in the transient portions of incoming signals, which is considered to be an attribute of human auditory processing.

PLP features [47] were developed with the intent of obtaining a representation that was similar to MFCC features, but implemented in a manner that was attentive to more detailed attributes of peripheral auditory physiology and perception. Details regarding PLP processing are provided in Sect. 8.2. Many researchers have obtained better recognition accuracy with PLP features than with MFCC features and PLP feature extraction is frequently combined with the RASTA algorithm to produce RASTA-PLP features.

Both physiological and psychophysical data suggest that mammalian auditory systems include units in the brainstem that are sensitive to the specific modulation frequencies of amplitude-modulated signals, independent of carrier frequency [58]. Similarly, psychoacoustical findings also indicate that humans are sensitive to modulation frequency [114, 119], with the temporal modulation transfer functions indicating the greatest sensitivity to temporal modulations at approximately the same frequencies as in the physiological data, despite obvious species differences. This information has been used to implement features based on frequency components of the temporal envelopes of bandpass-filtered components of speech signals, which Kingsbury and others referred to as the modulation spectrum [64]. Typically, the modulation spectrum is obtained by passing the speech signal through bandpass peripheral auditory filters, computing the envelopes of the filter outputs, and passing these envelopes through a second set of parallel bandpass modulation filters with center frequencies between 2 and 16 Hz. As a result, the modulation spectrum is a joint function of the center frequencies of the initial peripheral auditory filters, which span the range of useful speech frequencies, and the center frequencies of the modulation filters. This is a useful representation, because speech signals typically exhibit temporal modulations with modulation frequencies in the range that is passed by this processing, while noise components often exhibit frequencies of amplitude modulation outside this range. Tchorz and Kollmeier [108], among other researchers, observed the greatest amount of temporal modulation at modulation frequencies of approximately 6 Hz, and that low-pass filtering the envelopes of the outputs of each channel generally reduced the variability introduced by background noise.

8.3.3.1 TempoRAl PatternS (TRAPS)

Hermansky and Sharma [49] developed the TRAPS representation, which operates on one-second segments of the log-spectral energies that emerge from each of 15 critical-band filters. In the original implementation, these outputs were classified directly by a multilayer perceptron (MLP). This work was extended by Zhu et al. [123], who developed HATS (for hidden activation TRAPS), which trains an additional MLP layer at the level of each critical-band filter to provide a set of basic functions optimized to maximize the discriminability of the data to be classified. TRAP-DCT features were proposed in [100], and are a variation of the previously proposed TRAP features, where a DCT was applied to 310 ms-long segments of critical spectral energies. The TRAP-DCT features reduce word error rates (WERs) for ASR tasks in noisy conditions [100].

8.3.3.2 Frequency-Domain Linear Prediction (FDLP)

Athineos and Ellis [4] developed FDLP, where the temporal envelopes of the outputs of critical-band filters are represented by linear prediction. Much as linear-predictive parameters computed from the time-domain signal within a short analysis window (e.g., 25 ms) represent the envelopes of the short-time spectrum within a slice of time, the FDLP parameters represent the Hilbert envelope of the temporal sequence within a slice of spectrum. This method was incorporated into a method called LP-TRAPs [5], in which the FDLP-derived Hilbert envelopes were used as input to MLPs that learned phonetically relevant transformations for later use in speech recognition. LP-TRAPS can be considered to be a parametric estimation approach to characterizing the trajectories of the temporal envelopes, while traditional TRAPS is nonparametric in nature. Traditional FDLP [5, 33] features approximate the temporal Hilbert envelope within spectral subbands by linear prediction on one-second-long cosine-transformed audio segments [33]. The derived set of temporal subband envelopes forms a two-dimensional representation that is convolved with an integration window of 25 ms before resampling at a frame rate of 100 Hz. Further, the spectral bands are integrated by using a mel filterbank, and cepstral coefficients are derived by applying a DCT. FDLP features demonstrate improved robustness against channel noise, additive noise, and room reverberation using a phoneme-recognition task with conversational telephone speech [33].

Mel filterbanks have served as the state-of-the-art spectral analysis filters for speech-processing tasks since their introduction. Recently, with more availability of computing resources, better-precision filterbanks, such as gammatone filterbanks, been used more frequently. The gammatone filterbanks address the limitations of the mel filterbanks, where the former use asymmetric filters to replace the computationally efficient triangular filters of the latter [39]. Gammatone filters are a linear approximation of the auditory filterbank found in the human ear. Gammatone filterbank (GFB) energies have been used for DNN acoustic-model training [81]. For GFB feature extraction, the power of the band-limited time signals within an analysis window of 26 ms is usually computed at a frame rate of 10 ms. The subband powers from 40 filters are then root compressed by using the 15th root.

8.3.3.3   Power-Normalized Cepstral Coefficients (PNCC)

The PNCC [62, 63] is a representative feature set that attempts to include many of the auditory processing attributes in a computationally efficient way. PNCC processing begins in traditional fashion with a short-time Fourier transform, with the outputs in each frame multiplied by gammatone frequency weighting along a power-function nonlinearity, and generation of cepstral-like coefficients using DCT and mean normalization. For the most part, noise and reverberation suppression is accomplished by a nonlinear series of operations that perform running noise suppression and temporal contrast enhancement, respectively, working in a medium-time context, with analysis intervals on the order of 50–150 ms. (The results of this longer-duration analysis are applied to signal representations extracted over traditional 20–35 ms analysis frames for speech recognition.) Multiple groups have found that PNCC processing provides effective noise robustness as well as suppression of reverberation effects, with minor modification, and the computation required is comparable to that used in MFCC and PLP feature extraction.

8.3.3.4   Modulation Spectrum Features

Modulation spectrum features incorporate low-pass filtering of critical bands with a cutoff frequency of 28 Hz and a subsequent AM bandpass filtering step [64]. The bandpass filter consists of a complex exponential function, which is windowed by a Hamming window and has its peak sensitivity at 4 Hz, matching the temporal characteristics of syllables. The filter emphasizes AM frequencies between approximately 0 and 8 Hz (i.e., the dominant AM range of speech) [36]. An approach to estimating the modulation spectrum of speech signals using the Hilbert envelopes in a nonparametric way was proposed in [112], where a modulation spectrum feature extracted from mel filterbanks was used in an ASR task. That work showed that the logarithm of a particular mel filterbank's Hilbert envelope over an analysis window of 100 ms produced a better AM estimate of the subband signals compared to shorter window lengths. Lower DCT coefficients (in the range 0–25 Hz) of the AM signal were used as the acoustic feature, which was named the fepstrum features. The fepstrum features' performance was evaluated on the Conversational Telephony Speech (CTS) recognition experiments on the Switchboard (SWB) corpus, where the results indicated that such features in combination with short-term features, such as MFCCs, provided up to 2.5% absolute improvement in phone recognition accuracy and up to 2.5–3.5% absolute word recognition accuracy improvement on the 1.5 h SWB test set [112].

8.3.3.5 Normalized Modulation Coefficient (NMC)

Studies [25, 38] have shown that AM of the speech signal plays an important role in speech perception and recognition. Hence, recent studies [92, 112] have treated the speech signal as a sum of amplitude-modulated narrowband signals. Demodulation of a narrowband signal into its AM and frequency modulation (FM) components can be performed through the use of the Discrete energy separation algorithm (DESA) [71], which uses the nonlinear Teager energy operator (TEO) to perform the demodulation operation. The TEO has been used in [57] to create mel-cepstral features that demonstrated robustness against car noise and improved ASR performance. The nonlinear DESA tracks the instantaneous AM energies quite reliably [71], which in turn provide better formant information [57] compared to conventional power-spectrum-based approaches. The NMC was proposed in [77] and uses the DESA algorithm to extract instantaneous AM estimates for generating acoustic features. The significance of DESA is twofold: (a) it does not impose a linear model to analyze speech, and (b) it tracks the frequency and amplitude variations at the sample level without imposing any stationary assumption as done by linear prediction or the Fourier transform. For DESA to give good AM/FM estimates, the input signal has to be sufficiently band-limited [92], for which a gammatone filterbank was used in NMC feature extraction.

The TEO used in DESA was first introduced in [109] as a nonlinear energy operator, Ψ, that tracks the instantaneous energy of a signal, where the signal's energy is defined to be a function of its amplitude and its frequency. Considering a discrete sinusoid $x[n]$, where A = const. amplitude, Ω = digital frequency, f = frequency of oscillation in hertz, f_s = sampling frequency in hertz, and θ = initial phase angle,

$$x[n] = A\cos[\Omega n + \theta]; \quad \Omega = 2\pi\left(\frac{f}{f_s}\right), \tag{8.1}$$

If $\Omega \leq \pi/4$ and is sufficiently small, then Ψ takes the form

$$\Psi\{x[n]\} = x^2[n] - x[n-1]x[n+1] \approx A^2\Omega^2, \tag{8.2}$$

where the maximum energy estimation error in Ψ will be 23% if $\Psi \leq \pi/4$.

DESA was formulated in [71], where Ψ was used to formulate a demodulation algorithm that can instantaneously separate the AM/FM components of a narrowband signal using the following sets of equations:

$$\Omega[n] \approx \cos^{-1}\left\{1 - \frac{\Psi(x[n]) + \Psi(x[N-1])}{4\Psi(x[n])}\right\}, \tag{8.3}$$

$$|a_i[n]| \approx \sqrt{\frac{\Psi\{x[n]\}}{1 - \cos(\Omega_i[n])^2}}. \tag{8.4}$$

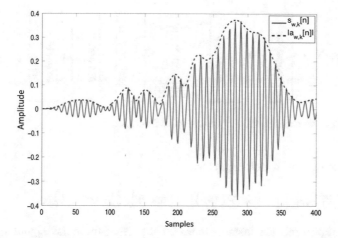

Fig. 8.1 A windowed narrowband speech signal and its corresponding AM signal from the modified DESA algorithm

Note that in (8.2), $x^2[n] - x[n-1]x[n+1]$ can be less than zero if $x^2[n] < x[n-1]x[n+1]$, while the right-hand side is strictly nonnegative, $A^2\Omega^2 \geq 0$; hence in [77] the TEO in (8.2) was modified to

$$\Psi\{x[n]\} = \left| x^2[n] - x[n-1]x[n+1] \right| \approx A^2\Omega^2, \qquad (8.5)$$

which tracks the magnitude of energy changes. Also, the AM/FM signals computed from (8.3) and (8.4) may contain discontinuities (that substantially increase their dynamic range). To prevent such discontinuities, the AM estimation equation (8.4) was modified for NMC feature extraction, as detailed in [77]. Figure 8.1 shows an overlay plot of the windowed narrowband time signal and its corresponding AM magnitude.

At this point the question remains, how robust is the TEO to different noisy conditions? To answer that, we need to revisit (8.5) and consider a noisy band-limited signal $s[n] = x[n] + v[n]$. The TEO, $\Psi s[n]$, is given as

$$\Psi\{s[n]\} = \Psi\{x[n]\} + \Psi\{v[n]\} + \widetilde{\Psi}\{x[n]v[n]\}, \qquad (8.6)$$

where $\widetilde{\Psi}\{x[n]v[n]\} = x[n]v[n] - (1/2)x[n-1]v[n+1] - (1/2)v[n-1]x[n+1]$ is the cross-TEO of $x[n]$ and $v[n]$. If we assume the subband noise $v[n]$ to be zero-mean and additive, then the expected value of the cross-term $\widetilde{\Psi}\{x[n]v[n]\}$ is zero, resulting in

$$E[\Psi\{s[n]\}] = E[\Psi\{x[n]\}] + E[\Psi\{v[n]\}]. \qquad (8.7)$$

If we assume that the noise is high-pass in every subband, then using low-pass filtering results in $E[\Psi\{v[n]\}] \ll E[\Psi\{x[n]\}]$, and Ω^2 is almost constant for

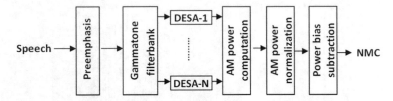

Fig. 8.2 Flow diagram of NMC feature extraction from speech

narrowband signals, which results in

$$E[\Psi\{s[n]\}] \approx E[\Psi\{x[n]\}], \text{ hence } E\left[A_s^2\right] \approx E\left[A_x^2\right], \tag{8.8}$$

where A_s represents the instantaneous amplitude of the noisy signal $s[n]$ and A_x represents the same for the clean signal $x[n]$. Thus (8.8) indicates how the estimated AM signals are robust to noise corruption.

The steps involved in obtaining the NMCC features are shown in Fig. 8.2. At the onset, the speech signal is preemphasized (using a preemphasis filter) and then analyzed using a 26 ms Hamming window with a 10 ms frame rate. The windowed speech signal $s^w[n]$ is passed through a gammatone filterbank having 40 channels between 200 and 7500 Hz (for a 16 kHz signal). The AM time signals $a_{k,j}[n]$ for the kth channel and jth frame are then obtained for each of the 40 channels using the modified DESA algorithm.

The normalized AM powers a then bias-subtracted using a similar approach, as specified in [63]. Figure 8.3 shows the spectrogram of a noise-corrupted signal, it's normalized AM power spectrum, and its corresponding bias-subtracted version. 15th-root power compression is performed on the bias-subtracted AM power spectrum and the resultant is typically used as the NMC feature set.

8.3.3.6 Modulation of Medium Duration Speech Amplitudes (MMeDuSA)

Note that, given (8.5) in Sect. 8.3.3.5, a simpler approach to estimating the instantaneous AM signal can be devised by assuming that the instantaneous FM signal will be approximately equal to the center frequency of the analysis gammatone filterbank when the subband signals are sufficiently band-limited, i.e.,

$$\Omega_i \approx f_c. \tag{8.9}$$

Given (8.9), the estimation of the instantaneous AM signal from (8.5) becomes very simple:

$$A_i \approx \sqrt{\frac{|x^2[n] - x[n-1]x[n+1]|}{\Omega_i^2}}. \tag{8.10}$$

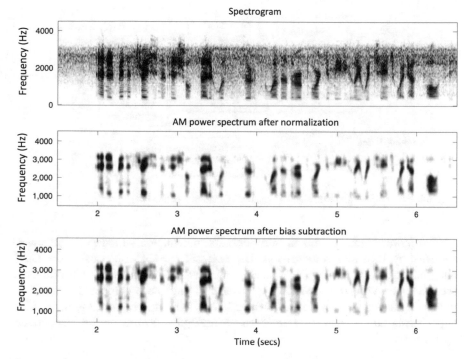

Fig. 8.3 Spectrogram of a noisy utterance corrupted with 15.6 dB nonstationary noise and it's normalized and bias-subtracted amplitude modulation spectrum

This simplification is essentially used in obtaining the MMeDuSA feature [79], as shown in Fig. 8.4. In the MMeDuSA pipeline the speech signal is first preemphasized and then analyzed using a Hamming window of 51 ms with a 10 ms frame rate. The windowed speech signal $s[n]$ is passed through a gammatone filterbank having 40 critical bands, with center frequencies spaced equally in the equivalent rectangular bandwidth (ERB) scale between 150 and 7400 Hz. For each of these 40 subband signals, their AM signals are computed using (8.10). The power of the estimated AM signals is computed followed by nonlinear dynamic range compression (using the 15th root). For a given analysis window, 40 power coefficients are obtained, and these are identified as the MMeDuSA1 features. Note that the feature operates in the medium duration as it uses an analysis window of size 52 ms compared to the traditionally used 10–25 ms windows. In parallel, each of the 40 estimated AM signals (as shown in Fig. 8.4) are bandpass filtered using the DCT, retaining information only within 5–350 Hz. These are the medium-duration modulations (represented as $a_{\mathrm{mod}_{k,j}}[n]$), which are summed across the frequency scale to obtain a medium-duration modulation summary

$$\overline{a_{\mathrm{mod}_j}} = \sum_{k=1}^{N} a_{\mathrm{mod}_{k,j}}[n]. \tag{8.11}$$

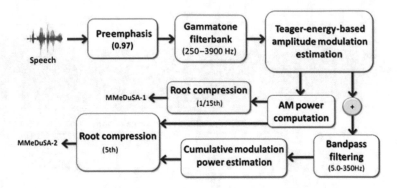

Fig. 8.4 MMeDuSA1 and MMeDuSA2 feature extraction pipeline [79]

The power signal of the medium-duration modulation summary is obtained, followed by 15th-root compression. The resultant is transformed using the DCT and the first n (typically 50%) coefficients are retained. These n coefficients are combined with the MMeDuSA1 feature to produce a combined feature set named MMeDuSA2. Typically, MMeDuSA2 is found to be more useful in ASR experiments and is usually termed the MMeDuSA feature unless categorized specifically.

8.3.3.7 Two Dimensional Modulation Extraction: Gabor Features

Most of the approaches discussed so far extracted modulation information across time only; extracting such information across time and frequency scales was performed in [73]. A 2D Gabor filter was used in [73] to extract specific modulation frequencies from the spectro-temporal information of speech. The design of the Gabor features is motivated by spectro-temporal receptive fields (STRFs), which provide an estimate of the stimulus that results in a high firing rate in isolated neurons [72]. It is observed that a significant proportion of the STRFs exhibit patterns that span durations of 200 ms, which is significantly longer than the traditionally used analysis durations in most speech features [73]. The Gabor/tandem posterior features use an MLP to predict the monophone class posteriors of each frame by using the Gabor features as input; the posteriors are then Karhunen–Loeve transformed to 22 dimensions and appended with standard 39-dimensional MFCCs to yield 64-dimensional features. In [15], a convolutional neural network called the Gabor convolutional neural network (GCNN) was proposed, which incorporated the Gabor functions into convolutional filter kernels. Features extracted using GCNNs showed significant performance improvement on noisy and channel-degraded speech over MFCC and other robust features such ETSI-AFE, PNCC, and Gabor-DNN posterior features.

8.3.3.8   Damped Oscillator Coefficient (DOC)

Studies have indicated that auditory hair cells exhibit damped oscillations in response to external stimuli [87] and such oscillations result in enhanced sensitivity and sharper frequency responses. To model such oscillations, a forced damped oscillator was proposed in [78] to generate acoustic features for ASR systems. The simplest oscillator is a simple harmonic oscillator, which is neither driven nor damped and is defined by the following equation:

$$m\frac{d^2x}{dt^2} + 2\zeta\omega_0 m\frac{dx}{dt} + \omega_0^2 mx = F_e(t), \tag{8.12}$$

where m is the mass of the oscillator, x is the position of the oscillator, ω_0 is the undamped angular frequency of the oscillator, and ζ is called the damping ratio. Assuming that the force can be represented as a sum of pulses, it can be shown that (8.12) can be written as

$$m\frac{d^2z(t)}{dt^2} + 2\zeta\omega_0 m\frac{dz(t)}{dt} + \omega_0^2 mz(t) = F_e e^{j\omega t}, \tag{8.13}$$

where $z(t) = x(t) + jy(t)$ and represents $\cos\omega t + j\sin\omega t = e^{j\omega t}$. Equation (8.13) suggests that there exists a solution of the form $z(t) = z_0 e^{\zeta t}$, where $d^2z(t)/dt^2 = \zeta^2 z_{(0)} e^{\zeta t}$ and $dz(t)/dt = \zeta z_0 e^{\zeta t}$.

If $z(t)$ is a complex exponential with the same frequency as the applied force, then the displacement $x(t)$ will also vary as a sine or cosine with a frequency ω. It can be shown [78] that the amplitude of oscillation in response to a force at frequency ω is given by

$$|z_0| = \frac{F_e/m}{\sqrt{(\omega_0^2 - \omega^2)^2 + (2\zeta\omega_0\omega)^2}}. \tag{8.14}$$

From (8.14), at resonance, i.e., $\omega_0 = \omega_t$, $|z_0|$ becomes

$$|z_0| = \frac{F_e}{2m\zeta\omega_0^2}, \tag{8.15}$$

indicating that the bank of oscillators behaves as a low-pass filter, where it uses lower gains for high-frequency bands and higher gains for the low-frequency bands. To counter this effect we selected m to be as follows:

$$m = \frac{1}{2\zeta\omega_0^2}. \tag{8.16}$$

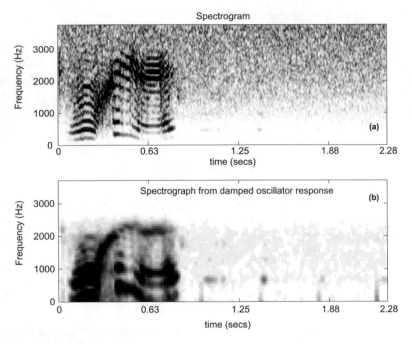

Fig. 8.5 (a) Spectrogram of signal corrupted with 3 dB noise and (b) spectral representation of the damped oscillator response after gammatone filtering

Note that ω_0 and ζ can be user defined, and for underdamped oscillation $\zeta < 1$. Modeling the damped-oscillator equation (8.12) in discrete time results in

$$x[n] = \frac{\left(2\zeta\Omega_0^2\right) F_e[n] + 2(1 + \zeta\Omega_0)x[n-1] - x[n-2]}{\left(1 + 2\zeta\Omega_0 + \Omega_0^2\right)}, \qquad (8.17)$$

where $\Omega_0 = \omega_0 T$ and $T = 1/f_s$.

The time response of the forced damped oscillators is obtained using (8.17) and their power over a Hamming analysis window of 25.6 ms is computed. Figure 8.5 shows the spectrogram of a speech signal and the damped oscillator response, where the oscillator model is found to successfully retain the harmonic structure of speech while suppressing the background noise.

The DOC feature extraction block diagram is shown in Fig. 8.6, where the damped oscillator response is computed using gammatone filterbank outputs as forcing functions. In DOC processing, the speech signal is preemphasized and then analyzed using a 25.6 ms Hamming window with a 10 ms frame rate. The windowed speech signal is passed through a gammatone filterbank having 40 channels with cutoff frequencies at 200–7000 Hz (for 16 kHz). The damped oscillator response is smoothed using a modulation filter with cutoff frequencies at 0.9 and 100 Hz, which helps to reduce the background subband noise. The powers of the resulting time

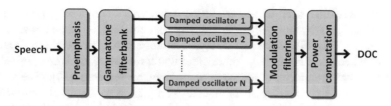

Fig. 8.6 Block diagram of DOC feature extraction

signals are computed, which are then root compressed (15th root) and the resulting 40-dimensional feature is used as the DOC features.

8.3.4   Current Trends

Recent advances in deep learning technology have redefined the common strategies used in acoustic modeling for ASR systems, where GMM-based models have been replaced by DNN-based models. DNNs [66, 85, 102] have demonstrated significant improvement in speech recognition performance compared to their GMM-HMM counterparts. Given the versatility of the DNN systems, [120] stated that speaker normalization techniques, such as VTLN [122], do not significantly improve speech recognition accuracy, as the DNN architecture's rich multiple projections through multiple hidden layers enable it to learn a speaker-invariant data representation. The current state-of-the-art architectures also deviate significantly from the traditional cepstral representation to simpler spectral representations, where MFCCs are usually replaced by mel-filterbank energy (MFB) features. While the basic assumptions in GMM-HMM architectures necessitated uncorrelated features due to their widely used diagonal covariance design (which in turn forced the observation to undergo a decorrelation step using the widely popular DCT), the current paradigm makes no such assumption. Rather, neural network architectures are known to benefit from cross-correlations [76] and hence demonstrate better performance by using spectral features rather than their cepstral versions [103]. Recent studies [23, 103] have demonstrated that DNNs work very well for noisy speech and improve performance significantly compared to GMM-HMM systems. CNNs [1, 50] have been found to perform as well as or sometimes better than fully connected DNN architectures [2]. CNNs are expected to be noise robust [1], especially in those cases where noise/distortion is localized in the spectrum. With CNNs, the localized convolution filters across frequency tend to normalize the spectral variations in speech arising from vocal tract length differences, enabling the CNNs to learn speaker-invariant data representations. Recent results [74, 80, 81] have also shown that CNNs are more robust to noise and channel degradations than DNNs.

In CNN/DNN-based ASR systems, speaker adaptation is usually done by using a generative framework that involves transforming features to a different space by using transforms such as feature space maximum likelihood linear regression (fMLLR) [32] or by applying a speaker-dependent bias by appending features like i-vectors [21, 98]. However, using i-vectors is often problematic, especially in mismatched conditions [90], where careful preprocessing, such as segmentation and additional architectural enhancements, may be required [34]. The i-vector framework was first developed for speaker verification as a way of summarizing the information in a variable-length utterance into a single fixed-length vector.

Much research has been performed on exploring and advancing feature-space adaptation methods, such as fMLLR approaches for GMM-HMM models. fMLLR applies a linear transform to the feature vectors for every frame, where the transform parameters are estimated by optimizing an auxiliary Q-function. DNN models are typically adapted by providing fMLLR transformed features as input. Use of fMLLR features has several advantages: firstly, it is efficient, as a few iterations of expectation maximization usually suffice; secondly, estimation of the fMLLR transform is quite robust even with very limited adaptation data; thirdly, it is quite versatile, as it can be applied to both supervised setting, where it is more robust to transcription errors than using a discriminative criterion, and to unsupervised setting, where reference transcription is not available.

Seide et al. [101] investigated the effectiveness of applying feature transforms developed for GMM-HMMs, including HLDA, VTLN, and fMLLR, to context-dependent deep neural network HMMs, or CD-DNN-HMMs. The authors observed that unsupervised speaker adaptation with discriminatively estimated fMLLR-like transforms works nearly as well as fMLLR for GMM-HMMs. Rath et al. [94] explored various methods of providing higher-dimensional features to DNNs, while still applying speaker adaptation with fMLLR of low dimensionality. The best-observed features consist of the baseline 40-dimensional speaker-adapted features that have been spliced again, followed by decorrelation and dimensionality reduction using another linear discriminant analysis (LDA). The authors believe that the whitening transform performed by LDA on the features will be favorable for DNN training, as the LDA would work as a preconditioner of the data, enabling setting higher learning rates, leading to faster learning (especially when pretraining is not used) [94]. Parthasarathi et al. [88] investigated fMLLR for DNN adaptation and proposed early fusion and late fusion to improve fMLLR performance, where early fusion with a bottleneck can act as a strong regularizer, and late fusion can provide significant robustness when fMLLR estimation is noisy.

In [43], the stacked bottleneck (SBN) neural network architecture was proposed to cope with limited data from a target domain, where the SBN net was used as a feature extractor. The SBN system was used to deal with unseen languages in [43] and, in [61], was extended to cope with unseen reverberation conditions. Unseen data conditions can significantly impact the performance of DNN ASR systems, and either supervised or unsupervised data adaptation is typically used to overcome such problems. In most such cases, labeled adaptation data is used to adapt the acoustic model (i.e., the DNN), with typically L2 [68] regularization employed. However, such approaches may veer the acoustic model away from the initial training acoustic conditions; hence, in [121], a KLD regularization was proposed for DNN model parameter adaptation, which differs from the typically used L2 regularization in the sense that it constrains the model parameters themselves rather than the output probabilities. Using such an approach, the model learns new acoustic conditions without digressing from what it had learned from the initial training data.

Recent results [95] have shown that a single DNN can be trained to learn both feature extraction and phonetic classification. Tüske et al. [111] proposed directly using the raw time signal as input to a DNN, and several others [12, 53, 97] have explored different ways to process the raw waveform and to train an acoustic model from it. In [97], using the raw signal resulted in better recognition performance than using conventional acoustic features. In a different study [12], conventional acoustic features were appended with DNN-generated features from the raw waveform, and the combination produced better performance than the conventional acoustic features did alone. While several research efforts have proposed different ways to learn data-driven feature extraction processes through DNN training, an open question remains about how robust such approaches are to unseen data conditions.

8.4 Case Studies

8.4.1 Speech Processing for Noise- and Channel-Degraded Audio

VAD is an essential stage of any ASR system. If a segment is not detected by VAD, it will not be processed by ASR, leading to word deletion errors. The performance of this stage greatly affects the quality of the final ASR hypothesis [35]. Robust features have been explored in recent years for the task of speech activity detection (i.e., VAD), in large part motivated by the challenges posed by noisy datasets [41]. Several robust features (such as PNCC, NMC, etc.) were explored for VAD in a DNN-based framework in [42] and it was observed that the fusion of all these features gave the best performance across different conditions. In [110], FDLP rate-scale features demonstrated significant robustness for the VAD task on noise and channel-degraded data.

DNN acoustic models using traditional MFB or MFCC features have been observed to suffer performance loss when the evaluation data is different from the training data [81]. Robust features are typically found to improve the performance of DNN/CNN acoustic models [15, 81]. In [74], baseline MFBs were compared with respect to MMeDuSA, NMC, and DOC features in a time–frequency CNN (TFCNN) acoustic-model-based Aurora-4 noisy word-recognition task [51], and the results demonstrated (see Table 8.1) a relative 5% reduction in WER compared to the baseline MFB features. Multiple robust features can be used in combination to provide a multiview representation of the acoustic signal, and these combinations typically improve recognition performance [75, 84].

Table 8.1 WER (averaged across all conditions) on multiconditioned training task of Aurora-4 (16 kHz) from using different features

Features	Avg. WER (%)
MFB	9.4
NMC	9.0
DOC	8.9
MMeDuSA	9.2

Table 8.2 Performance from different feature sets for Farsi KWS system from SRI's DARPA RATS submission

Features	$P(\text{fa})$	
	$P(\text{miss}) = 15\text{–}50\%$	$P(\text{miss}) = 15\%$
MFB	0.060	0.675
NMC	0.057	0.474
DOC	**0.054**	0.413
MMeDuSA	0.057	**0.389**

Robust features are found to be quite useful for performing keyword spotting (KWS) in mismatched training–testing conditions. Table 8.2 shows the performance of a CNN-based keyword-agnostic KWS system for Farsi datasets from the DARPA-RATS KWS evaluation conditions. Performance is given in terms of the average probability of false alarm [$P(\text{fa})$] for between 15% and 50% probability of a miss [$P(\text{miss})$]. As is evident from Table 8.2, the robust features demonstrated much better performance than the MFB features. Beyond the good individual performance of the robust features for KWS, the availability of multiple features enables creating systems that potentially capture complementary information, which in turn can be leveraged to provide even better results through system fusion [30].

Besides the good individual performance of the robust features in KWS, the availability of multiple features opens up the opportunity to create multiple systems that can potentially capture complementary information, which in turn can be leveraged to provide even better results through system fusion [30].

8.4.2 Speech Processing Under Reverberated Conditions

Robust acoustic features resistant to reverberation artifacts have shown significant promise in DNN acoustic models. Reverberation introduces mostly temporal distortion, where temporal smearing of information occurs whose duration is dependent on the impulse response of the room where the speech is recorded. The REVERB-2014 challenge [67] presented results from several research groups that used inverse filtering, nonnegative matrix factorization (NMF), modulation-based features, i-vectors, and several other methods to greatly improve the performance of DNN-based acoustic models under reverberated conditions [22, 117]. The results from REVERB-2014 indicate that sufficiently augmenting the training data with reverberation conditions similar to the evaluation conditions significantly improves

Table 8.3 WER from the ASpIRE dev set using GFB and DOC features, with different acoustic models

Features	Acoustic model	Avg. WER (%)
GFB	DNN	47.3
DOC	DNN	42.6
DOC	CNN	41.4
DOC	TFCNN	40.7

ASR performance (e.g., [22] showed an average relative reduction of 20% in WER through data augmentation).

The impact of training–testing data-condition mismatch was investigated in the ASpIRE [45] evaluation, where the training data consisted of the full Fisher training data [16], and the evaluation data contained reverberated speech recorded by far-field microphones. ASpIRE was a large-vocabulary continuous speech recognition (LVCSR) evaluation, where speech recognition robustness was investigated in a variety of acoustic environments and recording scenarios without having any knowledge about such conditions in the training and development data [45]. In ASpIRE, the participants were allowed to augment the training data by artificially introducing reverberation and/or noise into the training data. Data augmentation was found to be useful across all systems submitted to the challenge [54, 83, 90]. Speech enhancement using maximum kurtosis dereverberation reduced the WER by 2.3% absolute [24]. A TDNN using mel-cepstral features and i-vectors was presented in [90], where longer temporal information processing through the time-delay layer was found to be crucial for dealing with reverberation. A time–frequency CNN was presented in [83], which gave better performance than traditional CNN and DNNs, with the use of robust features found to be useful. Table 8.3 shows the results from baseline GFB features and DOC features, where DOC features, owing to their long temporal memory, were found to be robust against reverberation corruption.

In [54, 61], an autoencoder-based enhancement approach was proposed, where the role of the autoencoder was to denoise and dereverberate the degraded speech. In addition, FDLP and stacked-bottleneck features were also used in [54] along with DNN adaptation and data augmentation, which resulted in significant improvement compared to the baseline system.

In a recent study [52], robust features were used on top of multimicrophone beamforming-based dereverberation in the CHiME-3 challenge, where a significant reduction in error rate was observed compared to using mel-filterbank features. Beamforming techniques such as weighted delay-and-sum (WDAS) and minimum variance distortionless response (MVDR) are popular methods for leveraging multimicrophone data for coping with reverberation artifacts, and studies [22, 24] have shown impressive ASR performance under reverberated conditions when beamforming is used. The gain from robust features after beamforming in [52] was quite encouraging (see Table 8.4), as the results indicate that further performance improvement can be achieved from multimicrophone beamforming-based solutions when robust features are used.

Table 8.4 WER from DNN
acoustic models trained with
WDAS beamformed signals
using baseline and
noise-robust features for
CHiME-3 real evaluation data

Features	Real-test WER(%)
MFB	20.17
DOC	18.53
MMeDuSA	18.27
DOC+fMLLR	15.28
MMeDuSA+fMLLR	14.96

8.5   Conclusion

The use of robust features has helped in improving acoustic-model performance
under mismatched training–testing conditions across different flavors of deep
learning architectures. In recent speech recognition evaluations, it has been overwhelmingly witnessed that while DNN models produce state-of-the-art results
under matched training–testing conditions; they are susceptible to performance
degradation when the testing conditions are grossly mismatched to the training
conditions. Traditional approaches such as data augmentation and adaptation have
been found to be quite useful in data-mismatched cases, enabling the models to deal
with unseen data conditions. Robust features typically aim to create an invariant
representation of speech, such that data perturbation has minimal effect on its feature
space, hence providing a reliable feature representation to the acoustic models. The
use of robust features has been beneficial on top of data augmentation and adaptation
steps. The design of the signal-processing steps in acoustic feature engineering
has been largely motivated by signal theoretic approaches or speech perception
studies, where several different realizations of speech signal processing have been
explored and evaluated. Human auditory processing is a complex interaction of
several nonlinear processes, such as auditory attention, temporal filtering, masking,
etc., and on top of that it allows information to flow in both bottom-up and top-
down directions, providing human listeners with the capability to deal with varying
acoustic conditions and extremely quick adaptation skills. Researchers in auditory
neuroscience and psychoacoustics have been actively investigating the different
mechanisms of human auditory perception and their mutual interactions; such
observations may provide promising future directions for speech feature engineering
which can potentially lead to more versatile and robust acoustic features.

The surge in raw-signal processing in recent years has revolutionized the way
speech scientists and technologists think about speech systems. The current trend
replaces the signal-processing front end as an ad hoc step with an integrated
acoustic-modeling step where the neural networks are made to learn both signal
decomposition and phonetic discrimination all in one step using common objective
criteria. Raw-signal-based approaches are usually found to be data-hungry, where
several hundred (preferably a thousand or more) hours of training is necessary to
reliably learn the front-end transformation. The drawback of such systems is the
requirement on computational resources as the traditional acoustic models are no
longer using encoded/compressed feature forms, but are using information that is

five to ten times or more in size. Also, learning the front end in a data-driven way may result in overfitting the model to the training acoustic conditions; hence one may require a significant amount of diverse acoustic data to train acoustic models that can generalize well across unseen acoustic conditions. Given the recent impressive results from raw-signal-based systems, more and more researchers are investigating alternative models for raw-signal-based acoustic modeling, which provides optimism that some of the drawbacks of raw-signal-processing systems may be addressed in the near future, making such systems an integral part of our ASR systems.

References

1. Abdel-Hamid, O., Mohamed, A.R., Jiang, H., Penn, G.: Applying convolutional neural networks concepts to hybrid NN-HMM model for speech recognition. In: 2012 IEEE International Conference on Acoustics, Speech and Signal Processing (ICASSP), pp. 4277–4280. IEEE (2012)
2. Abdel-Hamid, O., Deng, L., Yu, D.: Exploring convolutional neural network structures and optimization techniques for speech recognition. In: Interspeech, pp. 3366–3370 (2013)
3. Atal, B.S., Hanauer, S.L.: Speech analysis and synthesis by linear prediction of the speech wave. J. Acoust. Soc. Am. **50**(2B), 637–655 (1971)
4. Athineos, M., Ellis, D.P.: Frequency domain linear prediction for temporal features. In: 2003 IEEE Workshop on Automatic Speech Recognition and Understanding, ASRU'30, pp. 261–266. IEEE (2003)
5. Athineos, M., Hermansky, H., Ellis, D.P.: LP-TRAP: linear predictive temporal patterns. Technical Report, IDIAP (2004)
6. Barker, J., Marxer, R., Vincent, E., Watanabe, S.: The third "chiME" speech separation and recognition challenge: dataset, task and baselines. In: 2015 IEEE Automatic Speech Recognition and Understanding Workshop (ASRU 2015) (2015)
7. Bartels, C., Wang, W., Mitra, V., Richey, C., Kathol, A., Vergyri, D., Bratt, H., Hung, C.: Toward human-assisted lexical unit discovery without text resources. In: SLT (2016)
8. Beh, J., Ko, H.: A novel spectral subtraction scheme for robust speech recognition: spectral subtraction using spectral harmonics of speech. In: 2003 IEEE International Conference on Acoustics, Speech, and Signal Processing, ICASSP'03, vol. 1, pp. I–648. IEEE (2003)
9. Bell, P., Gales, M., Hain, T., Kilgour, J., Lanchantin, P., Liu, X., McParland, A., Renals, S., Saz, O., Wester, M., et al.: The MGB challenge: evaluating multi-genre broadcast media recognition. In: 2015 Automatic Speech Recognition and Understanding Workshop (ASRU 2013) (2015)
10. Benesty, J., Makino, S.: Speech Enhancement. Springer Science & Business Media, New York (2005)
11. Bengio, Y.: Deep learning of representations for unsupervised and transfer learning. In: Unsupervised and Transfer Learning Challenges in Machine Learning, vol. 7, p. 19 (2012)
12. Bhargava, M., Rose, R.: Architectures for deep neural network based acoustic models defined over windowed speech waveforms. In: 16th Annual Conference of the International Speech Communication Association (2015)
13. Boll, S.F.: Suppression of acoustic noise in speech using spectral subtraction. IEEE Trans. Acoust. Speech Signal Process. **27**(2), 113–120 (1979)
14. Bregman, A.S.: Auditory Scene Analysis: The Perceptual Organization of Sound. MIT Press, Cambridge, MA (1994)

15. Chang, S.Y., Morgan, N.: Robust CNN-based speech recognition with Gabor filter kernels. In: Interspeech, pp. 905–909 (2014)
16. Cieri, C., Miller, D., Walker, K.: The fisher corpus: a resource for the next generations of speech-to-text. In: LREC, vol. 4, pp. 69–71 (2004)
17. Cohen, J.R.: Application of an auditory model to speech recognition. J. Acoust. Soc. Am. **85**(6), 2623–2629 (1989)
18. Cooke, M., Green, P., Josifovski, L., Vizinho, A.: Robust automatic speech recognition with missing and unreliable acoustic data. Speech Commun. **34**(3), 267–285 (2001)
19. Davis, S.B., Mermelstein, P.: Comparison of parametric representations for monosyllabic word recognition in continuously spoken sentences. IEEE Trans. Acoust. Speech Signal Process. **28**(4), 357–366 (1980)
20. Davis, K., Biddulph, R., Balashek, S.: Automatic recognition of spoken digits. J. Acoust. Soc. Am. **24**(6), 637–642 (1952)
21. Dehak, N., Kenny, P., Dehak, R., Dumouchel, P., Ouellet, P.: Front-end factor analysis for speaker verification. IEEE Trans. Audio Speech Lang. Process. **19**(4), 788–798 (2011)
22. Delcroix, M., Yoshioka, T., Ogawa, A., Kubo, Y., Fujimoto, M., Ito, N., Kinoshita, K., Espi, M., Hori, T., Nakatani, T., et al.: Linear prediction-based dereverberation with advanced speech enhancement and recognition technologies for the REVERB challenge. In: Proceedings of the REVERB Workshop (2014)
23. Deng, L., Hinton, G., Kingsbury, B.: New types of deep neural network learning for speech recognition and related applications: an overview. In: 2013 IEEE International Conference on Acoustics, Speech and Signal Processing (ICASSP), pp. 8599–8603. IEEE (2013)
24. Dennis, J., Dat, T.H.: Single and multi-channel approaches for distant speech recognition under noisy reverberant conditions: I2R's system description for the ASpIRE challenge. In: 2015 IEEE Workshop on Automatic Speech Recognition and Understanding (ASRU), pp. 518–524. IEEE (2015)
25. Drullman, R., Festen, J.M., Plomp, R.: Effect of reducing slow temporal modulations on speech reception. J. Acoust. Soc. Am. **95**(5), 2670–2680 (1994)
26. Elliott, T.M., Theunissen, F.E.: The modulation transfer function for speech intelligibility. PLoS Comput. Biol. **5**(3), e1000302 (2009)
27. ETSI: Speech processing, transmission and quality aspects (STQ); distributed speech recognition; front-end feature extraction algorithm; compression algorithms. ETSI ES 21 108, ver. 1.1.3 (2003)
28. ETSI: Speech processing, transmission and quality aspects (STQ); distributed speech recognition; advanced front-end feature extraction algorithm; compression algorithms. ETSI ES 202, 050, ver. 1.1.5 (2007)
29. Fine, S., Saon, G., Gopinath, R.A.: Digit recognition in noisy environments via a sequential GMM/SVM system. In: 2002 IEEE International Conference on Acoustics, Speech, and Signal Processing (ICASSP), vol. 1, pp. I–49. IEEE (2002)
30. Fiscus, J.G.: A post-processing system to yield reduced word error rates: recognizer output voting error reduction (ROVER). In: 1997 IEEE Workshop on Automatic Speech Recognition and Understanding, pp. 347–354. IEEE (1997)
31. Flynn, R., Jones, E.: Combined speech enhancement and auditory modelling for robust distributed speech recognition. Speech Commun. **50**(10), 797–809 (2008)
32. Gales, M.J., Woodland, P.C.: Mean and variance adaptation within the MLLR framework. Comput. Speech Lang. **10**(4), 249–264 (1996)
33. Ganapathy, S., Thomas, S., Hermansky, H.: Temporal envelope compensation for robust phoneme recognition using modulation spectrum. J. Acoust. Soc. Am. **128**(6), 3769–3780 (2010)
34. Garimella, S., Mandal, A., Strom, N., Hoffmeister, B., Matsoukas, S., Parthasarathi, S.H.K.: Robust i-vector based adaptation of DNN acoustic model for speech recognition. In: Interspeech (2015)
35. Gelly, G., Gauvain, J-L.: Minimum word error training of RNN-based voice activity detection. In: Interspeech, pp. 2650–2654 (2015)

36. Gemmeke, J.F., Virtanen, T.: Noise robust exemplar-based connected digit recognition. In: 2010 IEEE International Conference on Acoustics Speech and Signal Processing (ICASSP), pp. 4546–4549. IEEE (2010)
37. Ghitza, O.: Auditory nerve representation as a front-end for speech recognition in a noisy environment. Comput. Speech Lang. **1**(2), 109–130 (1986)
38. Ghitza, O.: On the upper cutoff frequency of the auditory critical-band envelope detectors in the context of speech perception. J. Acoust. Soc. Am. **110**(3), 1628–1640 (2001)
39. Gibson, J., Van Segbroeck, M., Narayanan, S.S.: Comparing time–frequency representations for directional derivative features. In: Interspeech, pp. 612–615 (2014)
40. Giegerich, H.J.: English Phonology: An Introduction. Cambridge University Press, Cambridge (1992)
41. Graciarena, M., Alwan, A., Ellis, D., Franco, H., Ferrer, L., Hansen, J.H., Janin, A., Lee, B.S., Lei, Y., Mitra, V., et al.: All for one: feature combination for highly channel-degraded speech activity detection. In: Interspeech, pp. 709–713 (2013)
42. Graciarena, M., Ferrer, L., Mitra, V.: The SRI system for the NIST open sad 2015 speech activity detection evaluation. In: Interspeech, pp. 3673–3677 (2016)
43. Grezl, F., Egorova, E., Karafiát, M.: Further investigation into multilingual training and adaptation of stacked bottle-neck neural network structure. In: 2014 IEEE Spoken Language Technology Workshop (SLT), pp. 48–53. IEEE (2014)
44. Gustafsson, H., Nordholm, S.E., Claesson, I.: Spectral subtraction using reduced delay convolution and adaptive averaging. IEEE Trans. Speech Audio Process. **9**(8), 799–807 (2001)
45. Harper, M.: The automatic speech recognition in reverberant environments (ASpIRE) challenge. In: ASRU (2015)
46. Harvilla, M.J., Stern, R.M.: Histogram-based subband powerwarping and spectral averaging for robust speech recognition under matched and multistyle training. In: 2012 IEEE International Conference on Acoustics, Speech and Signal Processing (ICASSP), pp. 4697–4700. IEEE (2012)
47. Hermansky, H.: Perceptual linear predictive (PLP) analysis of speech. J. Acoust. Soc. Am. **87**(4), 1738–1752 (1990)
48. Hermansky, H., Morgan, N.: RASTA processing of speech. IEEE Trans. Speech Audio Process. **2**(4), 578–589 (1994)
49. Hermansky, H., Sharma, S.: Temporal patterns (TRAPS) in ASR of noisy speech. In: 1999 IEEE International Conference on Acoustics, Speech, and Signal Processing, vol. 1, pp. 289–292. IEEE (1999)
50. Hinton, G., Deng, L., Yu, D., Dahl, G.E., Mohamed, A.R., Jaitly, N., Senior, A., Vanhoucke, V., Nguyen, P., Sainath, T.N., et al.: Deep neural networks for acoustic modeling in speech recognition: the shared views of four research groups. IEEE Signal Process. Mag. **29**(6), 82–97 (2012)
51. Hirsch, G.: Experimental framework for the performance evaluation of speech recognition front-ends on a large vocabulary task. ETSI STQ Aurora DSR Working Group (2002)
52. Hori, T., Chen, Z., Erdogan, H., Hershey, J.R., Roux, J., Mitra, V., Watanabe, S.: The MERL/SRI system for the 3rd CHiME challenge using beamforming, robust feature extraction, and advanced speech recognition. In: Proceedings of the IEEE ASRU (2015)
53. Hoshen, Y., Weiss, R.J., Wilson, K.W.: Speech acoustic modeling from raw multichannel waveforms. In: 2015 IEEE International Conference on Acoustics, Speech and Signal Processing (ICASSP), pp. 4624–4628. IEEE (2015)
54. Hsiao, R., Ma, J., Hartmann, W., Karafiat, M., Grézl, F., Burget, L., Szoke, I., Cernocky, J., Watanabe, S., Chen, Z., et al.: Robust speech recognition in unknown reverberant and noisy conditions. In: Proceedings of the IEEE Automatic Speech Recognition and Understanding Workshop (2015)
55. Itakura, F.: Minimum prediction residual principle applied to speech recognition. IEEE Trans. Acoust. Speech Signal Process. **23**(1), 67–72 (1975)

56. Itakura, F., Saito, S.: Statistical method for estimation of speech spectral density and formant frequencies. Electron. Commun. Jpn. **53**(1), 36 (1970)
57. Jabloun, F., Cetin, A.E., Erzin, E.: Teager energy based feature parameters for speech recognition in car noise. IEEE Signal Process. Lett. **6**(10), 259–261 (1999)
58. Joris, P., Schreiner, C., Rees, A.: Neural processing of amplitude-modulated sounds. Physiol. Rev. **84**(2), 541–577 (2004)
59. Juang, B.H., Rabiner, L.R.: Automatic speech recognition – a brief history of the technology development. In: Encyclopedia of Language and Linguistics. Elsevier, Amsterdam (2005)
60. Kanedera, N., Arai, T., Hermansky, H., Pavel, M.: On the importance of various modulation frequencies for speech recognition. In: 5th European Conference on Speech Communication and Technology (1997)
61. Karafiát, M., Grézl, F., Burget, L., Szöke, I., Černockỳ, J.: Three ways to adapt a CTS recognizer to unseen reverberated speech in BUT system for the ASpIRE challenge. In: 16th Annual Conference of the International Speech Communication Association (2015)
62. Kim, C., Stern, R.M.: Feature extraction for robust speech recognition based on maximizing the sharpness of the power distribution and on power flooring. In: ICASSP, pp. 4574–4577 (2010)
63. Kim, C., Stern, R.M.: Power-Cepstral Coefficients (PNCC) for Robust Speech Recognition. IEEE/ACM Trans. Audio, Speech, and Language Process. **24**(7), 1315–1329 (2016)
64. Kingsbury, B.E., Morgan, N., Greenberg, S.: Robust speech recognition using the modulation spectrogram. Speech Commun. **25**(1), 117–132 (1998)
65. Kingsbury, B., Saon, G., Mangu, L., Padmanabhan, M., Sarikaya, R.: Robust speech recognition in noisy environments: the 2001 IBM spine evaluation system. In: 2002 IEEE International Conference on Acoustics, Speech, and Signal Processing (ICASSP), vol. 1, pp. I–53. IEEE (2002)
66. Kingsbury, B., Sainath, T.N., Soltau, H.: Scalable minimum Bayes risk training of deep neural network acoustic models using distributed Hessian-free optimization. In: 13th Annual Conference of the International Speech Communication Association (2012)
67. Kinoshita, K., Delcroix, M., Yoshioka, T., Nakatani, T., Sehr, A., Kellermann, W., Maas, R.: The REVERB challenge: a common evaluation framework for dereverberation and recognition of reverberant speech. In: 2013 IEEE Workshop on Applications of Signal Processing to Audio and Acoustics (WASPAA), pp. 1–4. IEEE (2013)
68. Li, X., Bilmes, J.: Regularized adaptation of discriminative classifiers. In: 2006 IEEE International Conference on Acoustics, Speech and Signal Processing, ICASSP 2006, vol. 1, pp. I-237–I-240. IEEE (2006)
69. Lyon, R.F.: A computational model of filtering, detection, and compression in the cochlea. In: IEEE International Conference on Acoustics, Speech, and Signal Processing, ICASSP'82, vol. 7, pp. 1282–1285. IEEE (1982)
70. Makhoul, J., Cosell, L.: LPCW: an LPC vocoder with linear predictive spectral warping. In: IEEE International Conference on Acoustics, Speech, and Signal Processing, ICASSP'76, vol. 1, pp. 466–469. IEEE (1976)
71. Maragos, P., Kaiser, J.F., Quatieri, T.F.: Energy separation in signal modulations with application to speech analysis. IEEE Trans. Signal Process. **41**(10), 3024–3051 (1993)
72. Mesgarani, N., David, S., Shamma, S.: Representation of phonemes in primary auditory cortex: how the brain analyzes speech. In: 2007, IEEE International Conference on Acoustics, Speech and Signal Processing, ICASSP 2007, vol. 4, pp. IV–765. IEEE (2007)
73. Meyer, B.T., Ravuri, S.V., Schädler, M.R., Morgan, N.: Comparing different flavors of spectro-temporal features for ASR. In: Interspeech, pp. 1269–1272 (2011)
74. Mitra, V., Franco, H.: Time–frequency convolutional networks for robust speech recognition. In: 2015 IEEE Workshop on Automatic Speech Recognition and Understanding (ASRU), pp. 317–323. IEEE (2015)
75. Mitra, V., Franco, H.: Coping with unseen data conditions: investigating neural net architectures, robust features, and information fusion for robust speech recognition. In: Interspeech, pp. 3783–3787 (2016)

76. Mitra, V., Nam, H., Espy-Wilson, C.Y., Saltzman, E., Goldstein, L.: Retrieving tract variables from acoustics: a comparison of different machine learning strategies. IEEE J. Sel. Top. Signal. Process. **4**(6), 1027–1045 (2010)
77. Mitra, V., Franco, H., Graciarena, M., Mandal, A.: Normalized amplitude modulation features for large vocabulary noise-robust speech recognition. In: 2012 IEEE International Conference on Acoustics, Speech and Signal Processing (ICASSP), pp. 4117–4120. IEEE (2012)
78. Mitra, V., Franco, H., Graciarena, M.: Damped oscillator cepstral coefficients for robust speech recognition. In: Interspeech, pp. 886–890 (2013)
79. Mitra, V., Franco, H., Graciarena, M., Vergyri, D.: Medium-duration modulation cepstral feature for robust speech recognition. In: 2014 IEEE International Conference on Acoustics, Speech and Signal Processing (ICASSP), pp. 1749–1753. IEEE (2014)
80. Mitra, V., Wang, W., Franco, H.: Deep convolutional nets and robust features for reverberation-robust speech recognition. In: Spoken Language Technology Workshop (SLT), pp. 548–553. IEEE (2014)
81. Mitra, V., Wang, W., Franco, H., Lei, Y., Bartels, C., Graciarena, M.: Evaluating robust features on deep neural networks for speech recognition in noisy and channel mismatched conditions. In: Interspeech, pp. 895–899 (2014)
82. Mitra, V., Hout, J.V., McLaren, M., Wang, W., Graciarena, M., Vergyri, D., Franco, H.: Combating reverberation in large vocabulary continuous speech recognition. In: 16th Annual Conference of the International Speech Communication Association (2015)
83. Mitra, V., Van Hout, J., Wang, W., Graciarena, M., McLaren, M., Franco, H., Vergyri, D.: Improving robustness against reverberation for automatic speech recognition. In: 2015 IEEE Workshop on Automatic Speech Recognition and Understanding (ASRU), pp. 525–532. IEEE (2015)
84. Mitra, V., van Hout, J., Wang, W., Bartels, C., Franco, H., Vergyri, D., et al.: Fusion strategies for robust speech recognition and keyword spotting for channel- and noise-degraded speech. In: Interspeech, 2016 (2016)
85. Mohamed, A.R., Dahl, G.E., Hinton, G.: Acoustic modeling using deep belief networks. IEEE Trans. Audio Speech Lang. Process. **20**(1), 14–22 (2012)
86. Moore, B.: An Introduction to the Psychology of Hearing. Emerald Group Publishing Ltd., Bingley (1989)
87. Neiman, A.B., Dierkes, K., Lindner, B., Han, L., Shilnikov, A.L., et al.: Spontaneous voltage oscillations and response dynamics of a Hodgkin–Huxley type model of sensory hair cells. J. Math. Neurosci. **1**(11), 11 (2011)
88. Parthasarathi, S.H.K., Hoffmeister, B., Matsoukas, S., Mandal, A., Strom, N., Garimella, S.: fMLLR based feature-space speaker adaptation of DNN acoustic models. In: 16th Annual Conference of the International Speech Communication Association (2015)
89. Patterson, R.D., Robinson, K., Holdsworth, J., McKeown, D., Zhang, C., Allerhand, M.: Complex sounds and auditory images. Audit. Physiol. Percep. **83**, 429–446 (1992)
90. Peddinti, V., Chen, G., Manohar, V., Ko, T., Povey, D., Khudanpur, S.: JHU ASPIRE system: robust LVCSR with TDNNS, i-vector adaptation, and RNN-LMS. In: Proceedings of the IEEE Automatic Speech Recognition and Understanding Workshop (2015)
91. Peddinti, V., Povey, D., Khudanpur, S.: A time delay neural network architecture for efficient modeling of long temporal contexts. In: Interspeech (2015)
92. Potamianos, A., Maragos, P.: Time–frequency distributions for automatic speech recognition. IEEE Trans. Speech Audio Process. **9**(3), 196–200 (2001)
93. Rabiner, L.R., Levinson, S.E., Rosenberg, A.E., Wilpon, J.G.: Speaker-independent recognition of isolated words using clustering techniques. IEEE Trans. Acoust. Speech Signal Process. **27**(4), 336–349 (1979)
94. Rath, S.P., Povey, D., Veselỳ, K., Cernockỳ, J.: Improved feature processing for deep neural networks. In: Interspeech, pp. 109–113 (2013)
95. Sainath, T.N., Kingsbury, B., Mohamed, A.R., Ramabhadran, B.: Learning filter banks within a deep neural network framework. In: 2013 IEEE Workshop on Automatic Speech Recognition and Understanding (ASRU), pp. 297–302. IEEE (2013)

96. Sainath, T.N., Mohamed, A.R., Kingsbury, B., Ramabhadran, B.: Deep convolutional neural networks for LVCSR. In: 2013 IEEE International Conference on Acoustics, Speech and Signal Processing (ICASSP), pp. 8614–8618. IEEE (2013)

97. Sainath, T.N., Weiss, R.J., Senior, A., Wilson, K.W., Vinyals, O.: Learning the speech front-end with raw waveform CLDNNS. In: Proceedings of the Interspeech (2015)

98. Saon, G., Soltau, H., Nahamoo, D., Picheny, M.: Speaker adaptation of neural network acoustic models using i-vectors. In: ASRU, pp. 55–59 (2013)

99. Schroeder, M.R.: Recognition of complex acoustic signals. Life Sci. Res. Rep. 5(324), 130 (1977)

100. Schwarz, P.: Phoneme recognition based on long temporal context. Ph.D. thesis, Burno University of Technology (2009)

101. Seide, F., Li, G., Chen, X., Yu, D.: Feature engineering in context-dependent deep neural networks for conversational speech transcription. In: 2011 IEEE Workshop on Automatic Speech Recognition and Understanding (ASRU), pp. 24–29. IEEE (2011)

102. Seide, F., Li, G., Yu, D.: Conversational speech transcription using context-dependent deep neural networks. In: Interspeech, pp. 437–440 (2011)

103. Seltzer, M.L., Yu, D., Wang, Y.: An investigation of deep neural networks for noise robust speech recognition. In: 2013 IEEE International Conference on Acoustics, Speech and Signal Processing (ICASSP), pp. 7398–7402. IEEE (2013)

104. Seneff, S.: A joint synchrony/mean-rate model of auditory speech processing. In: Waibel, A., Lee, K.-F. (eds.) Readings in Speech Recognition, pp. 101–111. Morgan Kaufmann, Burlington, MA (1990)

105. Shao, Y., Srinivasan, S., Jin, Z., Wang, D.: A computational auditory scene analysis system for speech segregation and robust speech recognition. Comput. Speech Lang. 24(1), 77–93 (2010)

106. Srinivasan, S., Wang, D.: Transforming binary uncertainties for robust speech recognition. IEEE Trans. Audio Speech Lang. Process. 15(7), 2130–2140 (2007)

107. Stevens, S.S., Volkmann, J., Newman, E.B.: A scale for the measurement of the psychological magnitude pitch. J. Acoust. Soc. Am. 8(3), 185–190 (1937)

108. Tchorz, J., Kollmeier, B.: A model of auditory perception as front end for automatic speech recognition. J. Acoust. Soc. Am. 106(4), 2040–2050 (1999)

109. Teager, H.M.: Some observations on oral air flow during phonation. IEEE Trans. Acoust. Speech Signal Process. 28(5), 599–601 (1980)

110. Thomas, S., Saon, G., Van Segbroeck, M., Narayanan, S.S.: Improvements to the IBM speech activity detection system for the DARPA RATS program. In: 2015 IEEE International Conference on Acoustics, Speech and Signal Processing (ICASSP), pp. 4500–4504. IEEE (2015)

111. Tüske, Z., Golik, P., Schlüter, R., Ney, H.: Acoustic modeling with deep neural networks using raw time signal for LVCSR. In: Interspeech, pp. 890–894 (2014)

112. Tyagi, V.: Fepstrum features: design and application to conversational speech recognition. Technical Report, IBM Research Report (2011)

113. Van Hout, J.: Low complexity spectral imputation for noise robust speech recognition. M.S. thesis, UCLA (2012)

114. Viemeister, N.F.: Temporal modulation transfer functions based upon modulation thresholds. J. Acoust. Soc. Am. 66(5), 1364–1380 (1979)

115. Virag, N.: Single channel speech enhancement based on masking properties of the human auditory system. IEEE Trans. Speech Audio Process. 7(2), 126–137 (1999)

116. Wang, D.: On ideal binary mask as the computational goal of auditory scene analysis. In: Divenyi, P. (ed.) Speech Separation by Humans and Machines, pp. 181–197. Kluwer Academic, Dordrecht (2005)

117. Weninger, F., Watanabe, S., Le Roux, J., Hershey, J., Tachioka, Y., Geiger, J., Schuller, B., Rigoll, G.: The MERL/MELCO/TUM system for the REVERB challenge using deep recurrent neural network feature enhancement. In: Proceedings of the REVERB Workshop (2014)

118. Yoshioka, T., Ragni, A., Gales, M.J.: Investigation of unsupervised adaptation of DNN acoustic models with filter bank input. In: 2014 IEEE International Conference on Acoustics, Speech and Signal Processing (ICASSP), pp. 6344–6348. IEEE (2014)
119. Yost, W.A., Moore, M.: Temporal changes in a complex spectral profile. J. Acoust. Soc. Am. **81**(6), 1896–1905 (1987)
120. Yu, D., Seltzer, M.L., Li, J., Huang, J.T., Seide, F.: Feature learning in deep neural networks – studies on speech recognition tasks. arXiv:1301.3605 (2013, arXiv preprint)
121. Yu, D., Yao, K., Su, H., Li, G., Seide, F.: KL-divergence regularized deep neural network adaptation for improved large vocabulary speech recognition. In: 2013 IEEE International Conference on Acoustics, Speech and Signal Processing (ICASSP), pp. 7893–7897. IEEE (2013)
122. Zhan, P., Waibel, A.: Vocal tract length normalization for LVCSR. Technical Report, CMU-LTI-97-150, Carnegie Mellon University (1997)
123. Zhu, Q., Stolcke, A., Chen, B.Y., Morgan, N.: Incorporating tandem/HATS MLP features into SRIS conversational speech recognition system. In: Proceedings of the DARPA Rich Transcription Workshop (2004)

Chapter 9
Adaptation of Deep Neural Network Acoustic Models for Robust Automatic Speech Recognition

Khe Chai Sim, Yanmin Qian, Gautam Mantena, Lahiru Samarakoon, Souvik Kundu, and Tian Tan

Abstract Deep neural networks (DNNs) have been successfully applied to many pattern classification problems, including acoustic modelling for automatic speech recognition (ASR). However, DNN adaptation remains a challenging task. Many approaches have been proposed in recent years to improve the adaptability of DNNs to achieve robust ASR. This chapter will review the recent adaptation methods for DNNs, broadly categorising them into constrained adaptation, feature normalisation, feature augmentation and structured DNN parameterisation. Specifically, we will describe various methods of estimating reliable representations for feature augmentation, focusing primarily on comparing i-vectors and other bottleneck features. Moreover, we will also present an adaptable DNN layer parameterisation scheme based on a linear interpolation structure. The interpolation weights can be reliably adjusted to adapt the DNN to different conditions. This generic scheme subsumes many existing DNN adaptation methods, including speaker-code adaptation, learning hidden unit contribution factorised hidden layer and cluster adaptive training for DNNs.

9.1 Introduction

The deep neural network (DNN) is a powerful acoustic model that achieves superior performance in many automatic speech recognition (ASR) tasks compared to the traditional Gaussian-mixture-model (GMM)-based ASR systems [6, 21, 59]. However, DNN adaptation remains a challenging problem. Unlike the traditional

K.C. Sim (✉) • G. Mantena • L. Samarakoon • S. Kundu
National University of Singapore, Lower Kent Ridge Road, Singapore 119077, Singapore
e-mail: simkc@comp.nus.edu.sg

Y. Qian • T. Tan
Shanghai Jiao Tong University, Shanghai, China

© Springer International Publishing AG 2017
S. Watanabe et al. (eds.), *New Era for Robust Speech Recognition*,
DOI 10.1007/978-3-319-64680-0_9

Table 9.1 Comparison of DNN adaptation methods in terms of common strategies and methods

Method	Strategy		
	Test-time adaptation	Attribute-aware training	Adaptive training
Constrained adaptation	KL-div. regularization [90]	Multitask learning (MTL) [49, 50]	–
Feature normalisation	LIN [2]	CMLLR [58]	fDLR [58]
Feature augmentation	–	i-vector [57, 61], BSV [24, 32], NaT [60]	Speaker-code [1, 83]
Structured parameterisation	LHUC [67],LHN [14],LON [34]	FHL [54]	CAT [3, 9, 71, 72],FHL [54],SAT LHUC [68]

continuous density hidden Markov models (CDHMMs) [52], it is less obvious how one can systematically adapt a DNN in its generic multilayer architecture. This is largely due to the lack of interpretable structure in the model parameters. The parameters of a typical CDHMM system are given by the GMM parameters, defined for each hidden-Markov-model (HMM) state. It is therefore much easier to understand the meaning of the model parameters and establish meaningful relationships between them. For example, Gaussian distributions whose mean vectors are close to one another share similar acoustic attributes, such that shared transformations can be used to describe the speaker effect on the model parameters. This allows regression class trees to be constructed to dynamically control the complexity of the well-known maximum likelihood linear regression (MLLR) [33] adaptation method. The maximum a posteriori (MAP) [13] adaptation method is derived for the GMM distribution by defining appropriate prior distributions for the GMM parameters. Unfortunately, the DNN offers no such explicit definition of the model parameters, making it difficult to perform DNN adaptation. Many approaches have been proposed in recent years to improve the adaptability of DNNs to achieve robust ASR. In this book chapter, we will review these approaches, focusing primarily on speaker, noise and room adaptation for DNNs. Before we describe the various DNN adaptation methods, we will characterise these methods in terms of two criteria: *adaptation strategies* and *adaptation methods*, as summarised in Table 9.1.

9.1.1 DNN Adaptation Strategies

DNN adaptation methods can be broadly categorised into three common strategies: test-time adaptation, attribute-aware training and adaptive training. These strategies differ by how the speaker information is used (whether speaker labels are used only at test time or both during training and testing), and how speaker parameters are estimated (as part of a unified model or using a separate model).

9.1.1.1 Test-Time Adaptation

The test-time adaptation strategy does not involve any modification to the DNN model parameterisation and training procedures. All or some of the model parameters are updated during adaptation, for instance, by adapting the bias vectors in the hidden layers during test time [54]. Therefore, speaker labels are not needed during training. In some cases, additional parameters may also be introduced to perform adaptation at test time. For example, a condition-dependent transformation can be inserted into different parts of the DNN, to yield a linear input network (LIN) [2], linear hidden network (LHN) [14] and a linear output network (LON) [34]. Comparative studies of these methods can be found in [34, 36]. Similarly, the learning hidden unit contribution (LHUC) [67] introduces a scaling factor to each hidden unit to adapt the model at test time.

9.1.1.2 Attribute-Aware Training

The attribute-aware training strategy is used to describe the DNN adaptation methods that rely on incorporating attribute-specific information into the DNN by means of feature transformation or feature augmentation. The attribute-specific information is obtained using a separate system, such that the DNNs are trained to be *aware* of the presence of such information, but have no direct influence on how the information is estimated. For example, the speakers of [7] were estimated using a universal background model (UBM), which was separate from the DNN model. These speaker i-vectors are then appended to the acoustic features to yield the speaker-aware training method [57, 61]. Similarly, noise vectors obtained by averaging the acoustic feature frames from the head and tail of each utterance are used for noise-aware training (NaT) [60].

Attribute-aware training works based on the premise that the attribute-specific information can be reliably estimated during both training and deployment. This strategy requires a large variety of samples for different attributes in order to train the DNNs to have sufficient "awareness" of these attributes. This strategy may not generalise well if the attribute-specific information at test time differs considerably from that seen during training. As shown in [55], it is possible to further improve the performance by fine-tuning the attribute-specific information at test time.

9.1.1.3 Adaptive Training

The adaptive training strategy requires the DNN models to be parameterised by the global parameters as well as a relatively small number of attribute-specific parameters. These parameters can be estimated either jointly or in an interleaving manner. For example, the speaker-code adaptation method defines a set of speaker-dependent parameters (called a *speaker code*) and a set of global weights to derive a speaker-dependent bias in the DNN hidden layers.

Adaptive training typically yields a more accurate model compared to the attribute-aware training strategy. This is because the adaptive training strategy is able to learn a better *canonical* model, such that a more reliable estimation of the attribute-specific parameters can be achieved at deployment time.

9.1.2 Overview of DNN Adaptation Methods

Existing DNN adaptation methods can be broadly categorised into constrained adaptation, feature normalisation, feature augmentation and structured DNN parameterisation. In the following, we will provide an overview of these categories.

9.1.2.1 Constrained Adaptation

The constrained adaptation technique focuses on adding regularization to the adaptation criterion to avoid the overfitting problem due to the large number of parameters in a DNN. A straightforward method is to adapt only selected weights, such as weights with a maximum variance as computed on the adaptation data [66]. Doing adaptation with very small learning rate and early stopping can also be considered as regularization. In the Kullback–Leibler (KL)-divergence regularization adaptation [90], the intuition is that the outputs of the adapted and unadapted models should not deviate too much, so the KL divergence is added to the adaptation criteria as a regularization term. Multitask learning is another way to improve generalization by embedding several tasks into the DNN. This method can also be treated as soft constraints on parameters. For example, in [49, 50], training a DNN with speech enhancement and recognition jointly can improve the recognition performance; in [51, 86], embedding multiple factors into a DNN can further improve the accuracy in far-field speech recognition.

9.1.2.2 Feature Normalisation

The feature normalisation techniques, on the other hand, often treat the DNN as a black box and leverage independent feature-processing techniques to suppress the mismatch problem. This allows existing feature enhancement and normalisation techniques to be used for DNN adaptation. For example, a global constrained maximum likelihood linear regression (CMLLR) [11] transform estimated from a separate GMM/HMM system has been found to be very effective in reducing speaker variability in the acoustic features to improve the DNN-based

ASR performance [58]. Furthermore, feature-based discriminative linear regression (fDLR) [58, 87], which discriminatively estimates a CMLLR-like affine transformation, has also been successfully applied to unsupervised speaker adaptation of large-scale DNN systems. Feature-based vector Taylor series (VTS) [44], which use a Gaussian mixture model to perform a nonlinear mapping for estimating clean acoustic features from noisy acoustic features, has also been successfully applied to improve the noise robustness of DNN acoustic models [35]. Besides, with stereo data, advanced feature normalisation techniques based on deep learning, such as the denoising autoencoder [25, 40] and DNN-based speech enhancement [79, 80], have also been successfully applied to noisy speech recognition.

9.1.2.3   Feature Augmentation

The feature augmentation techniques incorporate a compact representation of speaker and noise information, such as speaker i-vectors [57, 61], bottleneck speaker vectors (BSVs) [24, 32] and noise vectors [60], to alleviate the mismatch problem. These techniques do not require explicit adaptation to be performed to the DNN, only the estimation of the speaker or noise representations.

9.1.2.4   Structured DNN Parameterisation

The final category of DNN adaptation methods imposes adaptable structures on the DNN hidden layers with a relatively small number of adaptation parameters associated with a speaker and/or noise type. We refer to these approaches as structured DNN parameterisation. Both the global and the adaptable parameters can be learned jointly during training, while only the adaptation parameters are updated during recognition. There are many existing DNN adaptation techniques that fall under this category, including LIN [2], LHN [14], LON [34], speaker-code adaptation [1, 83], LHUC [67], cluster adaptive training (CAT) for DNN [3, 9, 71, 72] and factorised hidden layer (FHL) [54].

9.1.3   Chapter Organisation

The remainder of the book chapter comprises two main parts. The first part will provide a detailed description of the DNN adaptation methods that belong to the family of feature augmentation. We will describe various methods of estimating reliable auxiliary information for feature augmentation, comparing i-vectors with other discriminative features that are extracted from a neural network, either trained separately from or jointly with the main recognition DNN. We will also introduce the multitask and joint-task learning methods for estimating the bottleneck features that provide a compact representation of multiple acoustic factors (such as speaker, noise and phones).

The second part will focus on structured DNN parameterisation. We will review and compare in detail the DNN adaptation methods that use special parameterisation structures to model the speaker-dependent DNNs with a compact representation to achieve robust estimations. We will describe the speaker-code adaptation method as a way of learning a structured form of speaker-dependent bias in the hidden layers and its close relationship to LHUC and the feature augmentation methods. We will also discuss how this method can be extended to model structured speaker-dependent affine transformations using an FHL structure or a CAT model.

9.2 Feature Augmentation

A popular approach to attribute-aware training is to explicitly provide attribute-specific information to the DNN and let the DNN learn the transformations necessary to compensate for the variability. In [89], it was shown that DNNs trained by such a method are able to generalise better to unseen conditions. If the goal is to compensate for the variability caused by the speaker, features that capture the speaker variability are explicitly provided to the DNN during training, and such an approach is referred to as speaker-aware training. For example, the activation of the first hidden layer for a typical DNN is given as follows:

$$h_t^1 = \text{sigmoid}\left(W^1 h_t^0 + b^1\right), \tag{9.1}$$

where the input to the DNN, h_t^0, is given by the acoustic features, o_t. Feature augmentation is achieved simply by extending the input to include an additional attribute-specific representation vector, v_t, such that $h_t^0 = \begin{bmatrix} o_t^\top & v_t^\top \end{bmatrix}^\top$. The corresponding transformation matrix can be expressed as $W^1 = \begin{bmatrix} A^1 & B^1 \end{bmatrix}$ to explicitly denote the terms associated with o_t and v_t, respectively, such that the new equation for the attribute-aware training is now given as follows [39, 89]:

$$h_t^1 = \text{sigmoid}\left(A^1 o_t + \underbrace{b^1 + B^1 v_t}_{\text{effective bias}}\right), \tag{9.2}$$

The result of augmenting the attribute-specific vector v_t is effectively to introduce an attribute-dependent bias, $b^1 + B^1 v_t$, to the first hidden layer. During training, B^1 learns to map v_t to an appropriate attribute-dependent bias in each hidden layer. This bias reduces the variability caused by nuisance attributes, such as speaker, noise and room conditions.

The attribute representation vector, v_t, can be estimated during the DNN training. In [1, 83], a speaker code was jointly learned along with the DNN parameters for each speaker. During decoding; speaker codes were learned for new speakers and were then used as the input for the DNN. The speaker features can also be obtained independently of the DNN training. In [19, 43, 57], i-vectors were used for speaker representation. In [24], bottleneck (BN) features were used. These BN features were obtained by training a separate bottleneck DNN to classify speakers. In [32], various learning techniques were explored to extract the BN features.

9.2.1   Speaker-Aware Training

There have been a number of successful attempts to apply feature augmentation for handling speaker variability for automatic speech recognition. The i-vector [8] has been found to be an effective speaker representation for speaker-aware training [19, 43, 57]. It is a popular technique used for speaker verification and recognition. Motivated by joint factor analysis (JFA) [28], the i-vector is a low-dimensional representation of the speaker characteristics. Typically, in JFA, speaker and noise subspaces are modelled separately. However, in the i-vector approach, all the variabilities are modelled together and the subspace is referred to as the total variability subspace. The speaker super-vector[1] μ_s is factorised into two parts:

$$\mu_s = \mu_b + T v_s, \tag{9.3}$$

where μ_b is the speaker- and channel-independent super-vector, which can be obtained from a universal background model. T is a total variability matrix and v_s is the i-vector. A maximum likelihood estimate is performed to obtain the parameters of (9.3). In [26], it was shown that the i-vector formulation is equivalent to CAT [12], where the columns of T are the cluster means and appropriate priors can be incorporated to enable a better estimation of the i-vector [27].

Alternatively, speaker representation vectors can also be obtained by training a separate bottleneck DNN to classify speaker classes. The BN features extracted using this DNN are then used to construct the speaker vectors. BN features have been used extensively in multilingual ASR and have been shown to improve the word error rate (WER) [16, 18, 75, 76]. The advantages of BN features are as follows [17]: (a) they are compressed features with lower dimension, and (b) the classification properties of the target class are reflected in the BN features.

In [24], bottleneck features were explored for speaker-aware training. In this work, the BN features were first extracted from the bottleneck layer per frame. A BSV were then computed per speaker by averaging the BN features obtained

---

[1] A speaker super-vector is a concatenation of the mean vectors of a Gaussian mixture model that represents the feature distribution for each speaker.

from all the features that belong to that speaker. The computation of the BSV is as follows:

$$\mathbf{v}_s = \frac{1}{T_s} \sum_{t \in T_s} \mathbf{b}_t, \tag{9.4}$$

where T_s is the total number of frames for a given speaker and $\mathbf{b}_t$ is the bottleneck feature obtained at a time t. Further, in order to capture phone-specific characteristics per speaker, the BN features can also be projected onto a super-vector space. These super-vectors are also referred to as bottleneck speaker super-vectors (BSSVs) [24], which are computed as follows:

$$\boldsymbol{\xi}_s = \left[\mathbf{v}_s(1)^\top, \, \mathbf{v}_s(2)^\top, \, \ldots, \, \mathbf{v}_s(c)^\top, \, \ldots, \, \mathbf{v}_s(C)^\top \right], \tag{9.5}$$

where $\mathbf{v}_s(c)$ is given as follows:

$$\mathbf{v}_s(c) = \frac{\sum_{t \in T_s} \gamma_c(t) \mathbf{b}_t}{\sum_{t \in T_s} \gamma_c(t)}. \tag{9.6}$$

$\gamma_c(t)$ is the probability of the speech frame at time t being associated with the phonetic class c. A disadvantage of obtaining a BSSV is that the dimensionality is large. To overcome this issue, in [32], two types of learning approaches, namely multitask and joint task learning, were explored. This will be described later in Sect. 9.2.4.

9.2.2 Noise-Aware Training

Similarly to speaker-aware training, NaT was proposed in [60] to handle noise variability for DNNs. In this work, a noise vector is estimated using the head and tail frames from each utterance and it is augmented with the regular acoustic features during DNN training. It was found to be more difficult for the DNN to learn how to handle noise variability from this kind of noise vector, and dropout [22] was used in combination with NaT to achieve performance improvement on the Aurora 4 dataset [23]. Likewise, it is also possible to train a bottleneck DNN with noise classes as the output targets to extract noise-dependent bottleneck features for NaT. This idea was explored in [32]. It was found to give some improvements on Aurora 4, but not as much as those achieved with speaker-aware training (SaT).

9.2.3   Room-Aware Training

DNN-based acoustic models have reduced the WER to a level that passed the threshold for adoption in many close-talk scenarios (e.g. voice search on a smartphone). However, ASR systems still perform poorly under the distant (far-field) talking condition [20], where the speech signals are captured by one or more microphones located further away from the speaker. Low signal strength is the main cause of the problem in this scenario, since it leads to low signal-to-noise ratio (SNR) and makes the system susceptible to reverberation and additive noise in the normal environment. Distant speech recognition is another main concerned scenario for robust speech recognition.

Many techniques [15, 69, 88] have been proposed to deal with the far-field speech recognition problem, and room-aware training is one effective approach to acoustic-model adaptation. Grouped as a feature augmentation adaptation, a representative room code is generated to encode room-related information. There is no standard method to encode the important features of a room; however, there are a number of metrics that describe various aspects of reverberation [46], and some of them, which have been verified effective for speech recognition, are listed below:

- $T60$: this is the room reverberation time, which reflects the time it takes the energy of an impulse to decay by 60 dB. The reverberation time, normally denoted as $T60$, can be calculated from a measured room impulse response by plotting its energy decay curve (EDC). There are several works in which it was proposed to estimate $T60$ [30], and a more delicate subband-based $T60$ estimated using non-negative matrix factorisation is referred to in [15].
- DDR: this is the direct-to-reverberant ratio, which is the ratio of the energy in the direct path to the energy from all reflected paths that cause the reverberation. $T60$ can describe some properties of reverberation, but it does not provide any indication of the amount of reverberant energy that is in the captured signal compared to the desired direct-path signal, so the DDR is designed to describe this knowledge. The detailed implementation is referred to in [15], and typically the DRR is lower as the user moves further away from the microphone.
- GCC: this is the generalized cross-correlation [29] between microphones in an array. The GCC encodes the time delay information between pairs of microphones and it is essential for determining the steering vector of the beamformer. Accordingly, the GCC containing time delay knowledge within different microphone pairs is believed to reflect some property of the room.
- $Distance$: the distance between the speaker and microphone will directly influence the reverberation strength and the final SNR. Most of the current methods assume the speaker–microphone distance to be unchanged throughout the recording; however, it is more likely be varied in real applications, considering that the speaker may walk around when speaking. In this case, if the distance can be estimated in advance, it should be helpful for acoustic modelling.

When these room-dependent codes are extracted, they can be utilized as the constant augmented auxiliary features to be fed into the DNN for room-aware training. Recently, some neural-network-based factor extractors were also proposed to represent the room information, such as the bottleneck feature from the distance-discriminant DNN [41] or the clean-feature-prediction DNN [50, 51], and they show a promising improvement in the reverberant scenario. More details can be found in [41, 50, 51].

9.2.4 Multiattribute-Aware Training

The feature augmentation approach makes it easy to incorporate multiple factors to achieve multiattribute-aware training. For example, two types of i-vectors can be estimated, one for the speakers and the other for the noise conditions. For each utterance, two i-vectors that correspond to the speaker and noise conditions of that utterance are appended to the standard acoustic features. In [26], a factorised i-vector was proposed to jointly learn the speaker and noise i-vectors by imposing orthogonality constraints to ensure that the speaker and noise subspaces are independent. This was found to achieve better performance compared to the simple case of appending speaker and noise i-vectors that have been trained separately.

Instead of extracting the representation vectors explicitly for each attribute, Kundu et al. [32] investigated the use of a bottleneck DNN to jointly extract a single bottleneck vector that encodes multiple attributes (speaker, noise and phones). The bottleneck DNN is trained to classify multiple attribute classes, using either an MTL approach or a joint-task learning (JTL) approach. For MTL, a single DNN is trained to predict the classes for multiple attributes (tasks). In this case, the output targets are simply the concatenation of the classes for all the attributes, as shown in Fig. 9.1. On the other hand, for JTL, the DNN is trained to predict the *cross product* of the classes from multiple attributes, as shown in Fig. 9.2. As shown in [32], both MTL and JTL achieved similar performance. However, the number of classes in the cross-product space increases dramatically with increasing number of attributes and the number of classes for each attribute. In general, it was found that the bottleneck vectors obtained from MTL and JTL performed better as compared to that of the conventional single-task learning (STL) set-up [32].

Table 9.2 shows the performance of the i-vectors and the BN vectors for multiattribute-aware training. The database used for the evaluation was Aurora 4 [23] and CMLLR-transformed features were used as the feature representation of the speech signal. In Table 9.2, the BN vectors were derived from a bottleneck DNN which was trained using a single task (also referred to as STL) and MTL and JTL. In all the experiments, the speaker (SP) was considered as the primary task. The auxiliary information used was noise (NS), context-independent (CI) phones, context-dependent (CD) phones and articulatory classes (AR) of speech. In [32], a more detailed description of the experimental set-up is provided. From Table 9.2, it

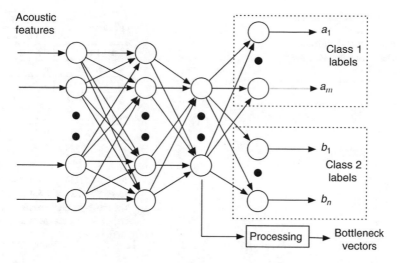

Fig. 9.1 A bottleneck DNN with output targets for MTL. a_i and b_i represent the output targets for different attribute classes

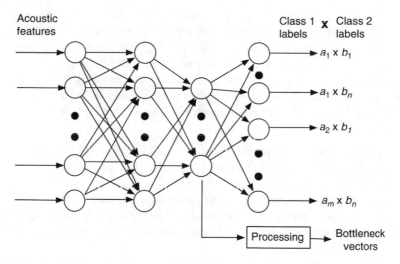

Fig. 9.2 A bottleneck DNN with output targets for JTL. a_i and b_i represent the output targets for different attribute classes

can be seen that (a) the BN vectors can be used to improve the performance of the ASR, (b) the best performance is achieved using the SP-AR BN vectors and (c) the noise (NS) attribute does not help as much as the other secondary attributes (CI, CD and AR).

Table 9.2 WER obtained using CMLLR in combination with i-vectors and other bottleneck features on Aurora 4

Features	WER (%)	
	MTL	JTL
CMLLR	9.6	
+ i-vectors	9.0	
+ SP BN	9.3	
+ SP-NS BN	9.2	9.2
+ SP-CI BN	8.9	9.0
+ SP-AR BN	8.9	**8.8**
+ SP-CD BN	8.9	–

Note that JTL was not used for the SP-CD BN vectors due to the large number of joint classes in the cross-product space. Number in bold indicates the best number in the table

9.2.5 Refinement of Augmented Features

The performance of the feature augmentation methods greatly depends on the quality of the representation vectors being appended to the acoustic features. If the testing conditions deviate greatly from the training conditions (e.g. speakers with different characteristics or unseen noise conditions), the DNN may not be able to take advantage of the augmented features. One way of addressing this problem is to refine the representation vectors to suppress the mismatch issue. In the following, we will describe several feature augmentation methods that include representation refinement:

- A simple refinement approach is to directly fine-tune the attribute-specific parameters at test time. In [55], it was found that performance improvement can be achieved by updating the i-vectors at test time.
- Instead of using the i-vectors as the speaker representation, the speaker code DNN adaptation method proposed in [1, 84] learns a *speaker code* for both the training and testing speakers by optimising the cross-entropy loss with respect to the DNN directly.
- Additional transformations can also be used to refine the i-vectors so that better representations can be obtained [42, 43].
- In the prediction–adaptation–correction recurrent neural network (PAC-RNN) [91], two additional DNNs are used, one for the prediction of the auxiliary information (used for feature augmentation) and the other to generate the correction for the prediction DNN, resulting in a feedback loop for adaptation.

9.3 Structured DNN Parameterisation

The DNN has a generic architecture and its parameters do not have a meaningful interpretation, which makes adaptation difficult as there are too many learnable parameters. The general goal of model-based DNN adaptation is to reliably estimate a condition-dependent affine transformation for each hidden layer, such that

$$h_t^l = \text{sigmoid}\left(W_s^l h_t^{l-1} + b_s^l\right), \tag{9.7}$$

where W_s^l and b_s^l are the condition-dependent weight matrix and bias vector, respectively. Without loss of generality, we will focus on speaker adaptation in this section, noting that the methods discussed here can also be extended to other conditions.

In order to perform model-based DNN adaptation reliably with a small amount of data, it is useful to introduce a systematic structure into the DNN so that a relatively small number of parameters can be adjusted to adapt the DNN. Many existing model-based DNN adaptation methods adopt a linear-basis interpolation structure to model the affine transformation in each hidden layer. The general form of such a kind of structured DNN parameterisation can be written as follows:

$$\mathbf{W}_s^l = \mathbf{W}^l + \sum_{i=1}^{n} \alpha_{s,i}^l \mathbf{W}_i^l, \tag{9.8}$$

$$\mathbf{b}_s^l = \mathbf{b}^l + \sum_{j=1}^{m} \beta_{s,j}^l \mathbf{b}_j^l = \mathbf{b}^l + \mathbf{B}^l \boldsymbol{\beta}_s^l, \tag{9.9}$$

where $\mathbf{W}^l, \mathbf{W}_i^l, \mathbf{W}_s^l \in \mathbb{R}^{|l| \times |l-1|}$, $\mathbf{b}^l, \mathbf{b}_j^l, \mathbf{b}_s^l \in \mathbb{R}^{|l| \times 1}$ and $m, n \ll l$, where $|l|$ is the size of layer l. $\mathbf{B}^l$ is a matrix whose columns are given by $\mathbf{b}_j^l$, and $\boldsymbol{\beta}_s^l$ is a column vector whose elements are given by $\beta_{s,j}^l$. In addition to the global transformation weight matrix $\mathbf{W}^l$ and the bias vector $\mathbf{b}_s^l$, the speaker-dependent (SD) transformation weight matrix is represented using a linear interpolation of the set of basis matrices, $\mathbf{W}_i^l$. Similarly, the SD bias vector is formed by linearly interpolating the basis vectors $\mathbf{b}_j^l$. $\alpha_{s,j}^l$ and $\beta_{s,j}^l$ are the interpolation weights that can be adjusted to adapt the DNN to different speakers. The number of bases can be adjusted to control the complexity of the adaptation depending on the amount of adaptation data available. Equation (9.8) refers to the CAT for DNNs formulation [3, 9, 71, 72], which is described later in Sect. 9.3.6.

9.3.1 Structured Bias Vectors

The feature augmentation approaches described in the previous section can also be viewed as a special form of structured parameterisation, applied only to the bias of the first hidden layer. Note the similarity between (9.9) and the effective bias term in (9.2). Moreover, the speaker code adaptation method [1, 62, 63] can also be viewed as an extension to the feature augmentation method. This method estimates a separate low-dimensional parametric space to model the speaker variability for the bias term using (9.9). The main differences are (1) the speaker codes are attached to several or all of the hidden layers, while feature augmentation methods apply to only the first hidden layer; and (2) the speaker codes are jointly estimated with the DNN parameters.

9.3.2 Structured Linear Transformation Adaptation

There are several DNN adaptation methods that introduce an additional linear transformation layer into the DNN, such as LIN [2, 34, 47, 73, 78], LHN [14] and LON [34]. In particular, for LIN and LHN, the additional linear transformation layer together with the succeeding hidden layer can be viewed as a single hidden layer with a structured parameterisation given below:

$$h_t^l = \text{sigmoid}\left(W^l \left(A_s^l h_t^{l-1} + k_s^l \right) + b^l \right). \tag{9.10}$$

Therefore, the resulting speaker-dependent transformation weight matrix and the bias vector are given by

$$\mathbf{W}_s^l = \mathbf{W}^l A_s^l \tag{9.11}$$

$$\mathbf{b}_s^l = \mathbf{W}^l k_s^l + b^l. \tag{9.12}$$

In [34], the transformation matrix A_s^l is represented as a linear-basis interpolation:

$$A_s^l = \sum_{i=1}^n \lambda_{s,i}^l \bar{A}_i^l, \tag{9.13}$$

where the basis matrices, $\bar{A}_i^l$, are estimated based on the principal components of the linear transforms estimated for the training speakers, and the speaker-dependent interpolation weights, $\lambda_{s,i}^l$, are estimated at test time. The effective transformation weight matrix becomes

$$\mathbf{W}_s^l = \sum_{i=1}^n \lambda_{s,i}^l \mathbf{W}^l \bar{A}_i^l. \tag{9.14}$$

Note the similarities in structure between (9.8) and (9.14), as well as (9.9) and (9.12). Although is less obvious, (9.11) can also be expressed in the same form as (9.8), by rewriting it as follows:

$$\mathbf{W}_s^l = \mathbf{W}^l \mathbf{A}_s^l \mathbf{I} = \sum_{i,j} \mathbf{A}_s^l(i,j) \mathbf{W}^l(i) \mathbf{I}(j)^\top, \qquad (9.15)$$

where $\mathbf{A}_s^l(i,j)$ is the (i,j)th element of $\mathbf{A}_s^l$, $\mathbf{W}^l(i)$ is the ith column of $\mathbf{W}^l$ and $\mathbf{I}(j)$ is the jth column of the identity matrix $\mathbf{I}$.

9.3.3 Learning Hidden Unit Contribution

LHUC [67, 68] is another DNN adaptation method that has been successfully applied to large-vocabulary continuous speech recognition. LHUC adapts a DNN by introducing a speaker-dependent scaling factor, $\alpha_{s,i}^i \in [0, 2]$, to modify the output activation of hidden unit i in layer l. In fact, this scaling factor can be viewed as an LHN with a diagonal transformation matrix without a bias. That is,

$$\mathbf{A}_s^l = \mathrm{diag}\left(\alpha_{s,i}^l\right), \qquad (9.16)$$

$$\mathbf{k}_s^l = \mathbf{0}. \qquad (9.17)$$

Therefore, LHUC can also be viewed as a structured paramaterised DNN with a linear-basis interpolation structure, whose interpolation weights are constrained to be in $[0, 2]$.

9.3.4 SVD-Based Structure

Low-rank representation of the transformation weight matrix, $\mathbf{W}^l$, has been used to achieved a compact model representation and improve computational efficiency [53, 82]. A low-rank representation of $\mathbf{W}^l$ can be approximated by a singular value decomposition (SVD)

$$\mathbf{W}_s^l \approx \mathbf{U}^l \Sigma^l \mathbf{V}^{l\top}, \qquad (9.18)$$

where $\mathbf{W}^l \in \mathbb{R}^{|l| \times |l-1|}$, $\mathbf{U}^l \in \mathbb{R}^{|l| \times k}$, $\Sigma^l \in \mathbb{R}^{k \times k}$, $\mathbf{V}^{l\top} \in \mathbb{R}^{k \times |l-1|}$, and $k \ll |l|, |l-1|$. Such a decomposition can be conveniently expressed as a linear interpolation of rank-1 matrices

$$\mathbf{W}_s^l \approx \mathbf{U}^l \Sigma_s^l \mathbf{V}^{l\top} = \sum_{i=1}^{n} \alpha_{s,i}^l \mathbf{u}_i^l \mathbf{v}_i^{l\top}, \qquad (9.19)$$

where $\Sigma_s^l = \mathrm{diag}(\alpha_s^l)$. Comparing this with (9.8), the basis matrices, $\mathbf{W}_i^l = \mathbf{u}_i^l \mathbf{v}_i^{lT}$, are rank-1 matrices. This low-rank representation provides a convenient structure to perform adaptation. This has been explored in [81], where the singular values are adapted in a supervised fashion. Furthermore, some approaches [31, 82, 85] also consider Σ_s^l as a full matrix during adaptation:

$$\mathbf{W}_s^l \approx \mathbf{U}^l \Sigma_s^l \mathbf{V}^{lT} = \sum_{i=1}^n \sum_{j=1}^n \alpha_{s,i,j}^l \mathbf{u}_i^l \mathbf{v}_j^{lT}. \tag{9.20}$$

9.3.5   Factorised Hidden Layer Adaptation

The FHL DNN adaptation method, as proposed in [54], represents the affine transformation of the hidden layers using the general-basis linear interpolation structure given (9.8) and (9.9), with the constraint that the basis matrices ($\mathbf{W}_i^l$) are rank-1. The resulting FHL formulation is given by

$$\mathbf{W}_s^l = \mathbf{W}^l + \mathbf{U}^l \Sigma_s^l \mathbf{V}^{lT}, \tag{9.21}$$

$$\mathbf{b}_s^l = \mathbf{W}^l k_s^l + \mathbf{b}^l. \tag{9.22}$$

FHL is different from the SVD-based structure previously described in two ways: (1) FHL represents both the transformation weight matrices and the bias vectors using linear interpolation; and (2) FHL has an additional full-rank term, $\mathbf{W}^l$. The rationale of having $\mathbf{W}^l$ is to ensure that the subspace needed to perform speaker adaptation is separated from the canonical transformation matrix needed by the DNN to perform classification.

Table 9.3 shows the effectiveness of various DNN structured parameterisation schemes for unsupervised speaker adaptation using three tasks, namely Aurora

Table 9.3 WER (%) for various structured parameterisation adaptation methods for Aurora 4 on LDA+STC features and for AMI IHM and SDM1 tasks on CMLLR features

Model	Refinement	Aurora 4 Eval	AMI Eval set	
			IHM	SDM1
Baseline	No	11.9	26.3	53.2
SaT (i-vector)	No	11.2	26.0	52.8
Speaker code	Yes	10.1	25.4	52.5
LHUC	Yes	10.0	24.9	52.6
4 FHLs	No	10.6	25.7	52.9
4 FHLs + diagonal adapt	Yes	9.0	**24.4**	**51.6**
4 FHLs + constrained full adapt	Yes	**8.4**	24.7	52.2

Numbers in bold indicate the best numbers for the respective test sets in the table

4, AMI IHM and AMI SDM1. All the DNN adaptation methods improved the performance over the baseline unadapted DNN system. In particular, the FHL systems performed better than the others. The *4 FHLs* refers to a system that applies FHL adaptation to the first 4 hidden layers. The i-vectors are used as the speaker parameters (interpolation weights). No refinement of speaker parameters is done at test time. For the *4 FHLs + diagonal adapt* system, the speaker parameters are adapted at test time, treating Σ_s^l as a diagonal matrix. For the *4 FHLs + constrained full adapt* system, the off-diagonal elements of Σ_s^l are shared across the 4 layers and refined at test time. Overall, the *4 FHLs + diagonal adapt* system achieved the best performance on the AMI evaluation sets, while the *4 FHLs + constrained full adapt* system performed best on the Aurora 4 test set. In general, the adaptation choice depends on the per-speaker footprint requirement, the availability of adaptation data and the quality of the adaptation hypotheses.

9.3.6 Cluster Adaptive Training for DNNs

CAT is an adaptation technique originally proposed for the GMM-HMM system [12]. Recently, CAT has also been applied to DNNs for speaker adaptation [3, 9, 71, 72]. The transformation weight matrix is represented by

$$\mathbf{W}_s^l = \mathbf{W}_{\text{nc}}^l + \sum_{i=1}^{n} \alpha_{s,i}^l \mathbf{W}_i^l. \tag{9.23}$$

The architecture of a CAT layer is depicted in Fig. 9.3, where the weight matrix basis of the layer l consists of the basis matrices $\mathbf{W}_i^l$, $1 \leq i \leq n$, and $\mathbf{W}_{\text{nc}}^l$, where $\mathbf{W}_{\text{nc}}^l$ is the weight matrix for the neutral cluster (the interpolation coefficient for the neutral cluster is always 1). $\alpha_{s,i}^l$ are the speaker-dependent interpolation weights, estimated for the training and testing speakers.

As a speaker-adaptive training technique, CAT for DNNs also has two sets of parameters, canonical model parameters and speaker-dependent parameters.

Fig. 9.3 The CAT layer architecture

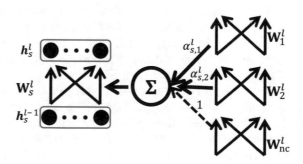

Table 9.4 WER (%) of
CAT-DNN trained on 310-h
Switchboard dataset

System	# cluster	WER	
		swb	fsh
SI	–	15.8	19.9
i-vector	100	14.8	18.3
spk-code	100	**14.3**	**17.5**
H1	2	15.2	18.8
	5	15.0	18.7
	10	14.6	17.8

'# Cluster' for 'i-vector' or 'spk-code' means the dimension of the speaker representation. Numbers in bold indicate the best numbers for the respective test sets in the table

- *Canonical parameters*: weight bases. Since the basis can be applied in multiple layers, the parameter sets of the canonical model in CAT for DNNs can be written as

$$M = \{\{\mathbf{M}^{l_1}, \ldots, \mathbf{M}^{l_L}\}, \{\mathbf{W}^{k_1}, \ldots, \mathbf{W}^{k_K}\}\}, \tag{9.24}$$

where $\mathbf{M}^l = \begin{bmatrix} \mathbf{W}^l_1 & \ldots & \mathbf{W}^l_n \end{bmatrix}$ is the weight matrix basis of layer l, L is the total number of CAT layers, $\mathbf{W}^k$ is the weight matrix of non-CAT layer k and K is the total number of non-CAT layers.
- *Speaker parameters*: speaker-dependent interpolation weight vectors

$$\boldsymbol{\alpha}^l_s = \begin{bmatrix} \alpha^l_{s,1} & \ldots & \alpha^l_{s,n} \end{bmatrix}^\top, \tag{9.25}$$

where $\boldsymbol{\alpha}^l_s$ denotes the speaker-dependent interpolation vector for layer l and speaker s.

The CAT parameters are optimized by using the back-propagation (BP) algorithm, where the detailed update formulas can be found in [72]. The interpolation vectors can be estimated from an i-vector [9] or trained jointly with the canonical model [3, 72]. More recently, in [4, 10], the interpolation weights were also predicted by a DNN using the i-vectors as input features.

Some results using CAT-DNN on the Switchboard task are illustrated in Table 9.4. H1 means applying CAT in the first layer. The results of the adaptation systems using i-vectors [57] and speaker codes [63] are also listed for comparison. Although the various models shown in the table do not have comparable numbers of free parameters, it has been demonstrated that increasing the number of layers or the number of neurons in each layer to a traditional DNN structure gives only small improvements [59]. Therefore, the performance improvements obtained by the CAT-DNN and other speaker adaptation methods are indeed the result of the new parameterisation structure. It is observed that, compared to the i-vector or speaker-code approaches using a 100-dim speaker-dependent representation, the

CAT-DNN achieves comparable performance using only ten clusters. More details can be found in [72].

9.4 Summary and Future Directions

The mismatch problem remains as a big challenge for deep-neural-network acoustic models, where the training and deployment conditions are vastly different. As with many machine learning techniques, the deep neural network learns to predict phone classes based on labelled training data. Speech recognition performance can be greatly affected when the distribution of the feature data deviates significantly from that of the training data. There are many factors that can contribute to the mismatch problem for speech recognition, including the speaker characteristics and speaking styles, changes due to the environment such as background noise and room conditions, and the recording conditions, such as the type of microphone, the transmission channels and the distance of the talker from the microphone.

Although there are other general approaches to improving the generalisation of DNNs to unseen data, such as multistyle training [37], data augmentation [5] and dropout [22, 60], it is still beneficial to incorporate DNN adaptation techniques to better address the mismatch problem. As reviewed in this chapter, many recent DNN adaptation techniques achieve promising improvements over state-of-the-art speech recognition systems. There is, however, room for improvement for many of these techniques. For example, it is unclear how to obtain reliable condition representation vectors for aware training methods. Although i-vectors [57, 61] and bottleneck speaker vectors [24, 32] have been found to yield improvements over the unadapted models, it has been shown in [55] that further refinements to the i-vectors at test time can achieve further performance gains. The problem lies in the fact that these vectors are extracted from and optimized for a separate system, so they are not always optimum for the DNNs. One attempt to address this problem is the PAC-RNN method [91], integrates the prediction of the condition vectors (auxiliary information), the adaptation of the RNN and the refinement of the predictor into a unified network.

Another aspect that has not been extensively explored is fast adaptation. Many of the existing research work for unsupervised speaker adaptation is based on speaker-level batch adaptation, where multiple utterances from each speaker are used for adaptation. When the amount of adaptation data is limited, it becomes more difficult to reliably adapt the DNNs, due to the tendency of overfitting to the adaptation data, the increased sensitivity to the errors in the supervision used to perform unsupervised adaptation and the amplification of other nuisance factors. Some recent work on fast DNN adaptation includes online i-vector estimation [48] and subspace LHUC [56].

One of the main difficulties in adapting the DNNs is the lack of interpretability of the DNN parameters. Unlike their generative-model counterparts, where the role and meaning of the model parameters are often well defined (such as the mean vector and covariance matrix of a GMM distribution), DNNs are typically used as a *black box*, making them less flexible for model adaptation. As discussed in this chapter, many model-based DNN adaptation methods use structured parameterisation to separate the *global* parameters that are responsible for phone classification and the *adaptable* parameters that can be adjusted to specific conditions. However, these adaptable parameters are often not directly interpretable. There are some methods that look at incorporating generative components into the DNN, allowing traditional adaptation techniques to be applied for adaptation. For example, Liu and Sim proposed using a temporally varying weight regression (TVWR) framework [38] to combine a DNN and GMM [64] to leverage the high-quality discriminative power of the DNN and the adaptability of the GMM. The work done by Variani et al. that incorporates a GMM layer into the DNN [74] can potentially be exploited to perform adaptation.

More recently, Nagamine et al. [45] and Sim [65] looked at analysing the activation pattern of the hidden units of the DNN to associate the roles of the hidden units with phone classes. Such information is useful for the interpretation of the DNN parameters and may be used to derive better DNN adaptation techniques. Furthermore, Tan et al. proposed stimulated deep learning [70] to explicitly constrain the DNN training process so that the hidden units of the network show interpretable activation patterns. These constraints were found to be effective as regularisation and potentially useful for adaptation [77].

References

1. Abdel-Hamid, O., Jiang, H.: Fast speaker adaptation of hybrid NN/HMM model for speech recognition based on discriminative learning of speaker code. In: Proceedings of IEEE International Conference on Acoustics, Speech and Signal Processing, ICASSP, pp. 7942–7946 (2013)
2. Abrash, V., Franco, H., Sankar, A., Cohen, M.: Connectionist speaker normalization and adaptation. In: Eurospeech, pp. 2183–2186. ISCA (1995)
3. Chunyang, W., Gales, M.J.: Multi-basis adaptive neural network for rapid adaptation in speech recognition. In: Proceedings of IEEE International Conference on Acoustics, Speech and Signal Processing, ICASSP, pp. 4315–4319. IEEE (2015)
4. Chunyang, W., Karanasou, P., Gales, M.J.: Combining i-vector representation and structured neural networks for rapid adaptation. In: Proceedings of IEEE International Conference on Acoustics, Speech and Signal Processing, ICASSP, pp. 5000–5004. IEEE (2016)
5. Cui, X., Goel, V., Kingsbury, B.: Data augmentation for deep neural network acoustic modeling. IEEE/ACM Trans. Audio Speech Lang. Process. **23**(9), 1469–1477 (2015)
6. Dahl, G., Yu, D., Deng, L., Acero, A.: Context-dependent pre-trained deep neural networks for large-vocabulary speech recognition. IEEE Trans. Audio Speech Lang. Process. **20**(1), 30–42 (2012)
7. Dehak, N., Dehak, R., Kenny, P., Brümmer, N., Ouellet, P., Dumouchel, P.: Support vector machines versus fast scoring in the low-dimensional total variability space for speaker verification. In: Proceedings of Interspeech, vol. 9, pp. 1559–1562 (2009)
8. Dehak, N., Kenny, P., Dehak, R., Dumouchel, P., Ouellet, P.: Front-end factor analysis for speaker verification. IEEE Trans. Audio Speech Lang. Process. **19**(4), 788–798 (2011)
9. Delcroix, M., Kinoshita, K., Hori, T., Nakatani, T.: Context adaptive deep neural networks for fast acoustic model adaptation. In: Proceedings of IEEE International Conference on Acoustics, Speech and Signal Processing, ICASSP, pp. 4535–4539. IEEE (2015)

10. Delcroix, M., Kinoshita, K., Chengzhu, Y., Atsunori, O.: Context adaptive deep neural networks for fast acoustic model adaptation in noise conditions. In: Proceedings of IEEE International Conference on Acoustics, Speech and Signal Processing, ICASSP, pp. 5270–5274. IEEE (2016)
11. Gales, M.J.F.: Maximum likelihood linear transformations for HMM-based speech recognition. Comput. Speech Lang. 12(2), 75–98 (1998)
12. Gales, M.J.: Cluster adaptive training of hidden Markov models. IEEE Trans. Speech Audio Process. 8(4), 417–428 (2000)
13. Gauvain, J.L., Lee, C.H.: Maximum a posteriori estimation for multivariate Gaussian mixture observations of Markov chains. IEEE Trans. Speech Audio Process. 2(2), 291–298 (1994)
14. Gemello, R., Mana, F., Scanzio, S., Laface, P., Mori, R.D.: Adaptation of hybrid ANN/HMM models using linear hidden transformations and conservative training. In: Proceedings of IEEE International Conference on Acoustics, Speech and Signal Processing, ICASSP, pp. 1189–1192. IEEE (2006)
15. Giri, R., Seltzer, M.L., Droppo, J., Yu, D.: Improving speech recognition in reverberation using a room-aware deep neural network and multi-task learning. In: Proceedings of IEEE International Conference on Acoustics, Speech and Signal Processing, ICASSP, pp. 5014–5018 (2015)
16. Grézl, F., Fousek, P.: Optimizing bottle-neck features for LVCSR. In: Proceedings of IEEE International Conference on Acoustics, Speech and Signal Processing, ICASSP, pp. 4729–4732 (2008)
17. Grézl, F., Karafiat, M., Kontar, S., Cernocky, J.: Probabilistic and bottle-neck features for LVCSR of meetings. In: Proceedings of IEEE International Conference on Acoustics, Speech and Signal Processing, ICASSP, vol. 4, pp. 757–760 (2007)
18. Grézl, F., Karafiát, M., Janda, M.: Study of probabilistic and bottle-neck features in multilingual environment. In: Proceedings of IEEE Automatic Speech Recognition and Understanding Workshop (ASRU), pp. 359–364 (2011)
19. Gupta, V., Kenny, P., Ouellet, P., Stafylakis, T.: I-vector-based speaker adaptation of deep neural networks for French broadcast audio transcription. In: Proceedings of IEEE International Conference on Acoustics, Speech and Signal Processing, ICASSP, pp. 6334–6338 (2014)
20. Hain, T., Burget, L., Dines, J., Garner, P.N., Grézl, F., Hannani, A.E., Huijbregts, M., Karafiat, M., Lincoln, M., Wan, V.: Transcribing meetings with the AMIDA systems. IEEE Trans. Audio Speech Lang. Process. 20(2), 486–498 (2012)
21. Hinton, G., Deng, L., Yu, D., Dahl, G., Mohamed, A., Jaitly, N., Senior, A., Vanhoucke, V., Nguyen, P., Sainath, T.N., Kingsbury, B.: Deep neural networks for acoustic modeling in speech recognition. IEEE Signal Process. Mag. 29, 82–97 (2012)
22. Hinton, G.E., Srivastava, N., Krizhevsky, A., Sutskever, I., Salakhutdinov, R.R.: Improving neural networks by preventing co-adaptation of feature detectors. arXiv:1207.0580 (2012, arXiv preprint)
23. Hirsch, G.: Experimental framework for the performance evaluation of speech recognition front-ends on a large vocabulary task, version 2.0. ETSI STQ-Aurora DSR Working Group (2002)
24. Huang, H., Sim, K.C.: An investigation of augmenting speaker representations to improve speaker normalisation for DNN-based speech recognition. In: Proceedings of IEEE International Conference on Acoustics, Speech and Signal Processing, ICASSP, pp. 4610–4613 (2015)
25. Ishii, T., Komiyama, H., Shinozaki, T., Horiuchi, Y., Kuroiwa, S.: Reverberant speech recognition based on denoising autoencoder. In: Proceedings of Interspeech, pp. 3512–3516 (2013)
26. Karanasou, P., Wang, Y., Gales, M.J.F., Woodland, P.C.: Adaptation of deep neural network acoustic models using factorised i-vectors. In: Proceedings of Interspeech, pp. 2180–2184 (2014)
27. Karanasou, P., Gales, M.J.F., Woodland, P.C.: I-vector estimation using informative priors for adaptation of deep neural networks. In: Interspeech, pp. 2872–2876 (2015)

28. Kenny, P., Ouellet, P., Dehak, N., Gupta, V., Dumouchel, P.: A study of interspeaker variability in speaker verification. IEEE Trans. Audio Speech Lang. Process. **16**(5), 980–988 (2008)
29. Knapp, C.H., Carter, G.C.: The generalized correlation method for estimation of time delay. IEEE Trans. Acoust. Speech Signal Process. **24**(4), 320–327 (1976)
30. Kumar, K., Singh, R., Raj, B., Stern, R.: Gammatone sub-band magnitude-domain dereverberation for ASR. In: Proceedings of IEEE International Conference on Acoustics, Speech and Signal Processing, ICASSP, pp. 4604–4607. IEEE (2011)
31. Kumar, K., Liu, C., Yao, K., Gong, Y.: Intermediate-layer DNN adaptation for offline and session-based iterative speaker adaptation. In: Proceedings of Interspeech. ISCA (2015)
32. Kundu, S., Mantena, G., Qian, Y., Tan, T., Delcroix, M., Sim, K.C.: Joint acoustic factor learning for robust deep neural network based automatic speech recognition. In: Proceedings of IEEE International Conference on Acoustics, Speech and Signal Processing, ICASSP (2016)
33. Leggetter, C.J., Woodland, P.C.: Maximum likelihood linear regression for speaker adaptation of continuous density hidden Markov models. Comput. Speech Lang. **9**(2), 171–185 (1995)
34. Li, B., Sim, K.: Comparison of discriminative input and output transformation for speaker adaptation in the hybrid NN/HMM systems. In: Proceedings of Interspeech, pp. 526–529. ISCA (2010)
35. Li, B., Sim, K.: Noise adaptive front-end normalization based on vector Taylor series for deep neural networks in robust speech recognition. In: Proceedings of IEEE International Conference on Acoustics, Speech and Signal Processing, ICASSP, pp. 7408–7412. IEEE (2013)
36. Liao, H.: Speaker adaptation of context dependent deep neural networks. In: Proceedings of IEEE International Conference on Acoustics, Speech and Signal Processing, ICASSP, pp. 7947–7951. IEEE (2013)
37. Lippman, R.P., Martin, E.A., Paul, D.B.: Multi-style training for robust isolated-word speech recognition. In: Proceedings of IEEE International Conference on Acoustics, Speech and Signal Processing, ICASSP, vol. 12, pp. 705–708. IEEE (1987)
38. Liu, S., Sim, K.C.: Temporally varying weight regression: a semi-parametric trajectory model for automatic speech recognition. IEEE/ACM Trans. Audio Speech Lang. Process. **22**(1), 151–160 (2014)
39. Liu, Y., Karanasou, P., Hain, T.: An investigation into speaker informed DNN front-end for LVCSR. In: Proceedings of IEEE International Conference on Acoustics, Speech and Signal Processing, ICASSP, pp. 4300–4304 (2015)
40. Lu, X., Tsao, Y., Matsuda, S., Hori, C.: Speech enhancement based on deep denoising autoencoder. In: Proceedings of Interspeech, pp. 436–440 (2013)
41. Miao, Y., Metze, F.: Distance-aware DNNS for robust speech recognition. In: Proceedings of Interspeech (2015)
42. Miao, Y., Jiang, L., Zhang, H., Metze, F.: Improvements to speaker adaptive training of deep neural networks. In: IEEE Spoken Language Technology Workshop (SLT), 2014, pp. 165–170. IEEE (2014)
43. Miao, Y., Zhang, H., Metze, F.: Towards speaker adaptive training of deep neural network acoustic models. In: Proceedings of Interspeech, pp. 2189–2193 (2014)
44. Moreno, P.J., Raj, B., Stern, R.M.: A vector Taylor series approach for environment-independent speech recognition. In: Proceedings of IEEE International Conference on Acoustics, Speech and Signal Processing, ICASSP, vol. 2, pp. 733–736. IEEE (1996)
45. Nagamine, T., Seltzer, M.L., Mesgarani, N.: Exploring how deep neural networks form phonemic categories. In: Proceedings of Interspeech (2015)
46. Naylor, P.A., Gaubitch, N.D.: Speech Dereverberation. Springer Science & Business Media, London (2010)
47. Neto, J., Almeida, L., Hochberg, M., Martins, C., Nunes, L., Renals, S., Robinson, T.: Speaker-adaptation for hybrid HMM-ANN continuous speech recognition system. In: Proceedings of Interspeech. ISCA (1995)
48. Peddinti, V., Chen, G., Povey, D., Khudanpur, S.: Reverberation robust acoustic modeling using i-vectors with time delay neural networks. In: Proceedings of Interspeech (2015)

49. Qian, Y., Yin, M., You, Y., Yu, K.: Multi-task joint-learning of deep neural networks for robust speech recognition. In: Proceedings of IEEE Automatic Speech Recognition and Understanding Workshop (ASRU), Scottsdale, AZ, pp. 310–316 (2015)
50. Qian, Y., Tan, T., Yu, D.: An investigation into using parallel data for far-field speech recognition. In: Proceedings of IEEE International Conference on Acoustics, Speech and Signal Processing, ICASSP, Shanghai, China, pp. 5725–5729 (2016)
51. Qian, Y., Tan, T., Yu, D., Zhang, Y.: Integrated adaptation with multi-factor joint-learning for far-field speech recognition. In: Proceedings of IEEE International Conference on Acoustics, Speech and Signal Processing, ICASSP, Shanghai, pp. 5770–5774 (2016)
52. Rabiner, L.R.: A tutorial on hidden Markov models and selected applications in speech recognition. Proc. IEEE **77**(2), 257–286 (1989)
53. Sainath, T.N., Kingsbury, B., Sindhwani, V., Arisoy, E., Ramabhadran, B.: Low-rank matrix factorization for deep neural network training with high-dimensional output targets. In: 2013 IEEE International Conference on Acoustics, Speech and Signal Processing (ICASSP), pp. 6655–6659. IEEE (2013)
54. Samarakoon, L., Sim, K.C.: Factorized hidden layer adaptation for deep neural network based acoustic modeling. IEEE/ACM Trans. Audio Speech Lang. Process. **24**(12), 2241–2250 (2016)
55. Samarakoon, L., Sim, K.C.: On combining i-vectors and discriminative adaptation methods for unsupervised speaker normalisation in DNN acoustic models. In: Proceedings of IEEE International Conference on Acoustics, Speech and Signal Processing, ICASSP (2016)
56. Samarakoon, L., Sim, K.C.: Subspace LHUC for fast adaptation of deep neural network acoustic models. In: Interspeech (2016)
57. Saon, G., Soltau, H., Nahamoo, D., Picheny, M.: Speaker adaptation of neural network acoustic models using i-vectors. In: Proceedings of IEEE Automatic Speech Recognition and Understanding Workshop (ASRU), pp. 55–59 (2013)
58. Seide, F., Li, G., Chen, X., Yu, D.: Feature engineering in context-dependent deep neural networks for conversational speech transcription. In: Proceedings of IEEE Automatic Speech Recognition and Understanding Workshop (ASRU), pp. 24–29. IEEE (2011)
59. Seide, F., Li, G., Yu, D.: Conversational speech transcription using context-dependent deep neural networks. In: Proceedings of Interspeech, pp. 437–440 (2011)
60. Seltzer, M.L., Yu, D., Wang, Y.: An investigation of deep neural networks for noise robust speech recognition. In: Proceedings of IEEE International Conference on Acoustics, Speech and Signal Processing, ICASSP, pp. 7398–7402 (2013)
61. Senior, A., Moreno, I.L.: Improving DNN speaker independence with i-vector inputs. In: Proceedings of IEEE International Conference on Acoustics, Speech and Signal Processing, ICASSP, pp. 225–229 (2014)
62. Shaofei, X., Abdel-Hamid, O., Hui, J., Lirong, D.: Direct adaptation of hybrid DNN/HMM model for fast speaker adaptation in LVCSR based on speaker code. In: Proceedings of IEEE International Conference on Acoustics, Speech and Signal Processing, ICASSP, pp. 6339–6343. IEEE (2014)
63. Shaofei, X., Abdel-Hamid, O., Hui, J., Lirong, D., Qingfeng, L.: Fast adaptation of deep neural network based on discriminant codes for speech recognition. IEEE/ACM Trans. Audio Speech Lang. Process. **22**(12), 1713–1725 (2014)
64. Shilin, L., Sim, K.C.: Joint adaptation and adaptive training of TVWR for robust automatic speech recognition. In: Proceedings of Interspeech (2014)
65. Sim, K.C.: On constructing and analysing an interpretable brain model for the DNN based on hidden activity patterns. In: Proceedings of Automatic Speech Recognition and Understanding (ASRU), pp. 22–29 (2015)
66. Stadermann, J., Rigoll, G.: Two-stage speaker adaptation of hybrid tied-posterior acoustic models. In: Proceedings of IEEE International Conference on Acoustics, Speech and Signal Processing, ICASSP, pp. 977–980 (2005)
67. Swietojanski, P., Renals, S.: Learning hidden unit contributions for unsupervised speaker adaptation of neural network acoustic models. In: Proceedings of IEEE Spoken Language Technology Workshop (SLT), pp. 171–176. IEEE (2014)

68. Swietojanski, P., Renals, S.: SAT-LHUC: speaker adaptive training for learning hidden unit contributions. In: Proceedings of IEEE International Conference on Acoustics, Speech and Signal Processing, ICASSP. IEEE (2016)
69. Swietojanski, P., Ghoshal, A., Renals, S.: Hybrid acoustic models for distant and multichannel large vocabulary speech recognition. In: Proceedings of IEEE Automatic Speech Recognition and Understanding Workshop (ASRU), pp. 285–290 (2013)
70. Tan, S., Sim, K.C., Gales, M.: Improving the interpretability of deep neural networks with stimulated learning. In: Proceedings of Automatic Speech Recognition and Understanding (ASRU), pp. 617–623 (2015)
71. Tan, T., Qian, Y., Yin, M., Zhuang, Y., Yu, K.: Cluster adaptive training for deep neural network. In: Proceedings of IEEE International Conference on Acoustics, Speech and Signal Processing, ICASSP, Brisbane, pp. 4325–4329 (2015)
72. Tan, T., Qian, Y., Yu, K.: Cluster adaptive training for deep neural network based acoustic model. IEEE/ACM Trans. Audio Speech Lang. Process. **24**(03), 459–468 (2016)
73. Trmal, J., Zelinka, J., Müller, L.: Adaptation of a feedforward artificial neural network using a linear transform. In: Sojka, P., et al. (eds.) Text, Speech and Dialogue, pp. 423–430. Springer, Berlin/Heidelberg (2010)
74. Variani, E., McDermott, E., Heigold, G.: A Gaussian mixture model layer jointly optimized with discriminative features within a deep neural network architecture. In: Proceedings of IEEE International Conference on Acoustics, Speech and Signal Processing, ICASSP, pp. 4270–4274. IEEE (2015)
75. Vesely, K., Karafiat, M., Grezl, F., Janda, M., Egorova, E.: The language-independent bottleneck features. In: Proceedings of IEEE Spoken Language Technology Workshop (SLT), pp. 336–341 (2012)
76. Vu, N.T., Metze, F., Schultz, T.: Multilingual bottle-neck features and its application for under-resourced languages. In: Proceedings of Workshop on Spoken Language Technologies for Under-Resourced Languages (SLTU), pp. 90–93 (2012)
77. Wu, C., Karanasou, P., Gales, M.J., Sim, K.C.: Stimulated deep neural network for speech recognition. In: Proceedings of Interspeech, pp. 400–404. ISCA (2016)
78. Xiao, Y., Zhang, Z., Cai, S., Pan, J., Yan, Y.: A initial attempt on task-specific adaptation for deep neural network-based large vocabulary continuous speech recognition. In: Proceedings of Interspeech. ISCA (2012)
79. Xu, Y., Du, J., Dai, L.R., Lee, C.H.: An experimental study on speech enhancement based on deep neural networks. IEEE Signal Process. Lett. **21**(1), 65–68 (2014)
80. Xu, Y., Du, J., Dai, L.R., Lee, C.H.: A regression approach to speech enhancement based on deep neural networks. IEEE/ACM Trans. Audio Speech Lang. Process. **23**(1), 7–19 (2015)
81. Xue, J., Li, J., Gong, Y.: Restructuring of deep neural network acoustic models with singular value decomposition. In: Proceedings of Interspeech, pp. 2365–2369. ISCA (2013)
82. Xue, J., Li, J., Yu, D., Seltzer, M., Gong, Y.: Singular value decomposition based low-footprint speaker adaptation and personalization for deep neural network. In: Proceedings of IEEE International Conference on Acoustics, Speech and Signal Processing ICASSP, pp. 6359–6363. IEEE (2014)
83. Xue, S., Abdel-Hamid, O., Jiang, H., Dai, L.: Direct adaptation of hybrid DNN/HMM model for fast speaker adaptation in LVCSR based on speaker code. In: Proceedings of IEEE International Conference on Acoustics, Speech and Signal Processing, ICASSP, pp. 6339–6343 (2014)
84. Xue, S., Abdel-Hamid, O., Jiang, H., Dai, L., Liu, Q.: Fast adaptation of deep neural network based on discriminant codes for speech recognition. IEEE/ACM Trans. Audio Speech Lang. Process. **22**(12), 1713–1725 (2014)
85. Xue, S., Jiang, H., Dai, L.: Speaker adaptation of hybrid NN/HMM model for speech recognition based on singular value decomposition. In: ISCSLP, pp. 1–5. IEEE (2014)
86. Yanmin Qian, T.T., Yu, D.: Neural network based multi-factor aware joint training for robust speech recognition. IEEE/ACM Trans. Audio Speech Lang. Process. **24**(12), 2231–2240 (2016)

87. Yao, K., Yu, D., Seide, F., Su, H., Deng, L., Gong, Y.: Adaptation of context-dependent deep neural networks for automatic speech recognition. In: IEEE Spoken Language Technology Workshop (SLT), 2012, pp. 366–369. IEEE (2012)
88. Yoshioka, T., Sehr, A., Delcroix, M., Kinoshita, K., Maas, R., Nakatani, T., Kellermann, W.: Making machines understand us in reverberant rooms: robustness against reverberation for automatic speech recognition. IEEE Signal Process. Mag. **29**(6), 114–126 (2012)
89. Yu, D., Deng, L.: Automatic Speech Recognition: A Deep Learning Approach. Springer, London (2014)
90. Yu, D., Yao, K., Su, H., Li, G., Seide, F.: KL-divergence regularized deep neural network adaptation for improved large vocabulary speech recognition. In: Proceedings of IEEE International Conference on Acoustics, Speech and Signal Processing, ICASSP, pp. 7893–7897. IEEE (2013)
91. Zhang, Y., Yu, D., Seltzer, M.L., Droppo, J.: Speech recognition with prediction–adaptation–correction recurrent neural networks. In: Proceedings of IEEE International Conference on Acoustics, Speech and Signal Processing, ICASSP, pp. 5004–5008 (2015)

Chapter 10
Training Data Augmentation and Data Selection

Martin Karafiát, Karel Veselý, Kateřina Žmolíková, Marc Delcroix,
Shinji Watanabe, Lukáš Burget, Jan "Honza" Černocký, and Igor Szőke

Abstract Data augmentation is a simple and efficient technique to improve the robustness of a speech recognizer when deployed in mismatched training-test conditions. Our work, conducted during the JSALT 2015 workshop, aimed at the development of: (1) Data augmentation strategies including noising and reverberation. They were tested in combination with two approaches to signal enhancement: a carefully engineered WPE dereverberation and a learned DNN-based denoising autoencoder. (2) Proposing a novel technique for extracting an informative vector from a Sequence Summarizing Neural Network (SSNN). Similarly to i-vector extractor, the SSNN produces a "summary vector", representing an acoustic summary of an utterance. Such vector can be used directly for adaptation, but the main usage matching the aim of this chapter is for selection of augmented training data. All techniques were tested on the AMI training set and CHiME3 test set.

10.1 Introduction

Training (or "source") versus evaluation ("target") data match or mismatch is a well-known problem in statistical machine learning. It was shown [1] that an automatic speech recognizer trained on clean data performs accurately on clean test data but poorly on noisy evaluation data. But it holds also vice versa—a recognizer trained on noisy data performs accurately on noisy but poorly on clean evaluation data.

M. Karafiát (✉) • K. Veselý • K. Žmolíková • L. Burget • J. "Honza" Černocký • I. Szőke
Brno University of Technology, Speech@FIT and IT4I Center of Excellence, Brno,
Czech Republic
e-mail: karafiat@fit.vutbr.cz; iveselyk@fit.vutbr.cz; izmolikova@fit.vutbr.cz; burget@fit.vutbr.cz; cernocky@fit.vutbr.cz; szoke@fit.vutbr.cz

M. Delcroix
NTT Corporation, 2-4, Hikaridai, Seika-cho, Kyoto, Japan

S. Watanabe
Mitsubishi Electric Research Laboratories (MERL), Cambridge, MA, USA

© Springer International Publishing AG 2017   245
S. Watanabe et al. (eds.), *New Era for Robust Speech Recognition*,
DOI 10.1007/978-3-319-64680-0_10

Unfortunately, "clean" and "noisy" are only very broad categories; the source vs. target data mismatch can have many forms and can depend on the speakers, acoustic conditions, and many other factors.

Typical solutions for dealing with a data mismatch include *speech enhancement*, where we modify the (possibly noisy) target data to fit a system trained on clean source data [13, 33], and *model adaptation*, trying to adjust/adapt a model to deal with the mismatch condition [10, 15, 30]. A third technique, investigated in this chapter, is *data augmentation*. Here, we are trying to change the source data in order to obtain data with similar characteristics to the target data. Such generated data (generally a greater amount than the original one, as we are trying to cover a wide variety of target conditions) is called "augmented."

Note that two terms are generally used in the literature: *data augmentation* usually means filling sparse target data by borrowing from rich source data, while *data perturbation* means adding more variation into the target data by using the target data only. In our approach to data augmentation, we usually combine both, i.e., we borrow from source data *and* we modify it to fit the target data characteristics.

10.1.1 Data Augmentation in the Literature

One of the first attempts to augment source data to fit the target data was made by Bellegarda et al. [4]. In experiments conducted on the Wall Street Journal corpus, the goal was to populate the target speaker feature space with transformed data from source speakers. A feature rotation ("metamorphing") algorithm [3] was used: both source and target speaker features are first transformed onto a unit sphere by phoneme-dependent normalization, then a transformation is estimated to map the source features to target ones. For a target speaker, the closest source speakers can be found using a distance metric among phoneme clusters on the sphere. When these source speakers are found, one can remap their features to the target speaker space, thus increasing the data for speaker-dependent model training. The conclusion was that 100 target speaker sentences augmented by 1500 source speakers' sentences led to a speaker-dependent model with the same accuracy as a model trained on 600 target speaker sentences.

More recent approaches can be split into categories depending on the level on which the data augmentation is done:

On the *audio level*, the goal is to perturb the audio to minimize source/target data mismatch. The original voice is not modified in the sense of generating an unseen speaker; the augmentation is done to neglect different acoustic environments by using artificial reverberation, noising, or some other perturbation of the source data. In this scenario, we typically have enough source speakers, but a nonmatching acoustic environment (for example a quiet place versus a crowded public place). The usual procedure includes adding artificial [17] or real [18] noise. Reverberation can also be simulated using real or artificial room impulse responses [18], or both methods can be combined.

The audio itself can also be modified. The approaches covered in [20] include upsampling and downsampling the audio with an appropriate change of reference labels timing, or changing the pitch or tempo of the raw audio using an appropriate audio editor. The results of [20] show that reverberation/noising and resampling/pitch modifications are not completely complementary.

Finally, new speech data can be artificially generated using either statistical or concatenative text-to-speech (TTS) synthesis. This approach may not generate a large variety of new speakers but can still generate unseen sentences to augment training data for less frequent phoneme contexts. The latter technique, where speaker and prosody parameters are generated using hidden Markov models [27], is used more often. For use with automatic speech recognition (ASR) training, one can skip audio generation and use statistical speech synthesis to generate directly perceptual linear prediction (PLP) or mel-frequency cepstral coefficient (MFCC) features [22, 34].

On the *feature level*, the augmentation is done on the level of extracted features: low-level (Mel-banks, PLPs) or high-level (neural network bottle-neck features). A typical example is vocal tract length perturbation (VTLP) [14], modifying the speaker's vocal tract shape by spectral shifts. This is an "inverse" technique to the commonly used vocal tract length normalization (VTLN), where the goal is to normalize different speakers to a generic one. In VTLP, new data is generated by altering the true VTLN factor; for example, the authors of [14] generated five versions of data and tested them on TIMIT phoneme recognition. A nice improvement was observed; moreover, they found a positive effect of perturbing the test data with several warp factors and then averaging the acoustic-model (deep-neural network, DNN) scores before the decoding.

Stochastic feature mapping (SFM) [5] falls under feature-level data augmentation too. Here, feature transformations are used to create artificial speakers. This approach is partly complementary to VTLP but its main advantage is that features can be easily generated on the fly.

10.1.2 Complementary Approaches

In a broader sense [23], data augmentation can be seen also from a different perspective: We can "fill" the sparsity of the actual training data with either untranscribed data from the target language, synthesized data from the target language, or other language data.

Untranscribed data can be used in the case where we have a few transcribed resources but a vast amount of untranscribed data (for example from Internet sources). In so-called self-training [39] (a variation of unsupervised or semisupervised training), a speech recognition system is first bootstrapped using the available little amount of transcribed data, then used to label the untranscribed data. A confidence measure is used to select the reliably transcribed segments, which are then added to the training set, and the system is retrained. The process can be done iteratively.

Data from *other languages* can be used, which has the great advantage of using real, not artificially prepared, data. However, it is more complicated because of language mismatch, which can be partially overcome by using universal phone sets, phone-to-phone mappings, or hidden-layer unit-to-target mappings in multiple-layer perceptrons [9]. Note, however that multilingual training is still a very active research topic, which is being investigated in several projects, such as the U.S. IARPA-sponsored Babel.[1]

The experimental part of this chapter concentrates only on the signal-processing approaches described in Sect. 10.1.1.

10.2   Data Augmentation in Mismatched Environments

10.2.1   Data Generation

In this section, we describe the noising and reverberation techniques used and the strategies to construct the training dataset.

- *Noise.* Our training data was processed by artificially adding two types of noises *real background noises* were downloaded from various sources, for example Freesound,[2] and "babbling noises" were created by merging speech from random speakers.
- *Reverberation.* We generated artificial room impulse responses (RIRs) using a "room impulse response generator" tool from E. Habets.[3] The tool can model the size of the room (three dimensions), the reflectivity of each wall, the type of microphone, the position of the source and the microphone, the orientation of microphone toward the audio source, and the number of bounces (reflections) of the signal. In each room, we created a pair of RIRs: one was used to reverberate (by convolution with the RIR) the speech signal and the other was used to reverberate the noise. Both signals were then mixed into a single recording. Each pair of RIRs differed only by the coordinates of the audio sources (speech/noise). We randomly set all parameters of the room for each room model.
- *Data augmentation strategies.* For each original clean utterance (independent headset microphones - IHM) from the AMI corpus, a corrupted version of the utterance was created by randomly choosing one of the following four speech corruption methods:

---

[1] https://www.iarpa.gov/index.php/research-programs/babel.

[2] www.freesound.org.

[3] https://github.com/ehabets/RIR-Generator.

1. One of the *real background noises* (see above) was randomly selected and added to the speech signal at a signal-noise-ratio (SNR) randomly selected from values: of −5, 0, 5, 10, and 15 dB.
2. One of the RIRs was randomly selected and used to reverberate the speech signal. In this case, no noise was added. Note that when adding reverberation, we compensated for the incurred delay to match the timing with the original signal.
3. The third option is a combination of the previous two. A random stationary noise and random reverberation were added. Speech and noise were reverberated by two different RIRs as described above. The two signals were then mixed at a randomly selected SNR level from the same range as before.
4. The same as the previous option, but babbling noise was used instead of stationary.

10.2.2   Speech Enhancement

In addition to the above data augmentation, we investigated two front-end approaches to handle source and target data mismatch caused by noise and reverberation. Here, we compare two main approaches: denoising by a neural-network(NN)-based autoencoder and signal-processing enhancement using the weighted prediction error (WPE).

10.2.2.1   WPE-Based Dereverberation

Reverberation is responsible for nonstationary distortions that are correlated with the speech signal and, consequently, it cannot be suppressed using the conventional noise reduction approaches. Therefore, we used the WPE dereverberation method [35, 36], which was shown to greatly improve ASR in reverberant conditions for several tasks [7, 37]. WPE is discussed in detail in Chap. 2.

WPE is based on long-term linear prediction (LP), but introduces modifications to conventional LP to make it effective for dereverberation. It is well known that multichannel LP can be used for channel equalization [11]; however, using conventional LP for speech signals causes excessive degradation because LP equalizes not only the room impulse responses but also the (useful) speech production process. To address this issue, WPE modifies the conventional LP algorithm in two ways: by modeling speech with a short-term Gaussian distribution with a time-varying variance [21], and by introducing a short time delay in the LP filters that prevents the equalization of the speech production [19].

WPE has a number of characteristics that make it particularly suitable for distant speech recognition: it is based on linear filtering, which ensures a low level of distortion in the processed speech. WPE can be formulated for single-channel or multichannel cases. It has also been shown to be relatively robust to ambient noise.

Note that the WPE algorithm does not require a pretrained model of speech and operates in a per-utterance manner.

More details about this technique can be found in Chap. 2.

10.2.2.2 Denoising Autoencoder

An artificial neural network was also employed as a denoising autoencoder to enhance (denoise and dereverberate) the speech signal. It was trained on the artificially created parallel clean–noisy AMI corpora. The reverberated and noised data described above (option three) was used for this purpose.

The input of the NN consisted of 257-dimensional vectors of log spectra stacked over 31 frames (i.e., a 7967-dimensional vector). The desired output was a 257-dimensional vector (again a log spectrum) corresponding to the clean version of the central input frame. A standard feedforward architecture was used: 7967 inputs, 3 hidden layers each with 1500 neurons, 257 outputs, and tanh nonlinearities in the hidden layers. The NN was initialized in such a way that it (approximately) replicated its input to the output and it was trained using the conventional stochastic gradient descent to minimize the mean squared error (MSE) objective.

We have experimented with different strategies for normalizing the NN input and output. To achieve a good performance, utterance-level mean and variance normalization was applied to both the NN input and the desired NN output. To synthesize the cleaned-up speech log spectrum, the NN output was denormalized based on the global mean and variance of clean speech.

10.2.3 Results with Speech Enhancement on Test Data

In the following experiments, we compare the effectiveness of the two speech enhancement techniques described in the previous sections: the denoising autoencoder and WPE [7] dereverberation. Table 10.1 shows results on the noisy CHiME-3 data obtained with the baseline DNN-based ASR system trained on clean independent-headset-microphone (IHM) AMI data. The table compares results obtained on the original unprocessed noisy test data and those obtained on data enhanced with the two techniques. The results are also compared for systems trained

Table 10.1 Performance of speech enhancement techniques on CHiME-3

Test data enhancement	XE (%WER)	sMBR (%WER)
None	48.86	46.99
WPE	45.36	43.63
Autoencoder	**30.58**	**30.59**

WER word error rate

The bold numbers indicate the best values in the table that help the orientation

Table 10.2 Results on CHiME-3 with different data augmentation variants

DNN training data		Test data	XE	sMBR
Noise type	Reverb	enhancement	(%WER)	(%WER)
None	None	None	48.86	46.99
Babble	Artificial	None	26.41	–
Stationary	Artificial	None	25.8	–
Stationary	None	None	24.26	20.47
Stationary	None	WPE	**22.72**	**19.28**

The bold numbers indicate the best values in the table that help the orientation

using a frame-by-frame cross-entropy (XE) objective and systems further retrained using state-level minimum Bayes risk (sMBR) discriminative training [28].

A relatively small improvement was obtained with WPE. This can be easily explained, as WPE aims to only dereverberate the signals, and relatively low reverberation is present in the CHiME-3 data due to the small distance between the speaker and the microphones. This trend might, however, differ for other test data. On the contrary, the denoising autoencoder provides significant gains, as it was trained to reduce both noise and reverberation.

No improvement was obtained from sMBR training. A possible reason is that new types of errors caused by the presence of noise in the data were not seen during the training, where only the IHM data were used. Note that much larger relative gains from sMBR will be reported in the following sections, where the training data was augmented with artificially corrupted speech data (see, e.g. Table 10.2).

10.2.4   Results with Training Data Augmentation

Table 10.2 presents results obtained when adding different types of noise to the training data (babbling vs. stationary noise). We also tested whether it helps to add reverberation to the training data or if it is sufficient to corrupt the training data only with additive noise. Interestingly, the best results were obtained with no reverberation added to the training data, but with the test data enhanced using the WPE dereverberation technique. WPE enhancement brings almost 1.5% absolute improvement. This indicates that signal-level dereverberation is more effective than training the acoustic model on reverberated speech.

Table 10.2 also shows a nice improvement (over ~3% absolute) from retraining DNNs using sMBR sequence training. Comparing the results in Tables 10.1 and 10.2, we observe that better performance could be achieved by training a recognizer on artificially corrupted data than using a denoising autoencoder trained on the same data to reduce noise.

Table 10.3 shows the results obtained for the different training datasets on the REVERB dev set. As for the CHiME-3 experiment, we observe that adding noise to the training data greatly improved performance. Not surprisingly, in the case of the REVERB data, adding reverberation to the training data also significantly improved

Table 10.3 Results on REVERB dev set with different data augmentation variants (sMBR models)

DNN training data			dev	
Noise type	Reverb	Test data enhancement	Near (%WER)	Far (%WER)
None	None	None	92.48	90.89
Stationary	None	None	52.49	49.72
Stationary	Artificial	None	40.33	37.93
None	None	WPE	56.19	49.28
Stationary	None	WPE	19.75	22.52
Stationary	Artificial	WPE	**19.75**	**20.77**

The bold numbers indicate the best values in the table that help the orientation

performance. However, the improvement brought by the reverberant training data is significantly reduced when using a dereverberation front end.

On the CHiME-3 test, we found no effect from adding reverberation into the training data for acoustic-model training. On the contrary, the WPE dereverberation technique was found effective. We also showed that greater performance improvement could be achieved by retraining the acoustic model on artificially corrupted speech than using a denoising autoencoder trained on the same training data to remove noise. On the REVERB test, adding reverberation was found beneficial, but the gains become smaller when using the WPE dereverberation front end. On the other hand, the impact of adding noisy training data remains. In future, we would like to verify our findings on other databases.

Finally, sMBR was found effective when the system was trained on noised data, in comparison with a system based on an enhancement autoencoder, where no gain from discriminative training was observed.

Note that the effect of data augmentation has also been investigated for the REVERB task in [8], with the significant difference that they used (instead of the AMI corpus) the REVERB training data, which they augmented with noise and various SNRs. In addition, in CHiME-3 and AMI, data augmentation has also been performed by treating each microphone signal recording as an independent training sample. This simple approach was also found to improve performance [26, 38].

10.3 Data Selection

10.3.1 Introduction

Using the methods above, we can generate large amounts of augmented data, from which it is possible to choose the training set. Different approaches to data selection have been explored in the past. Many existing methods are based on inspecting the frequency of speech units to evaluate the benefit of adding an utterance to the training. In [32], the data selection strategy aimed to choose such a subset which has a uniform distribution over phonemes or words. Similarly, the selection method in [31] was guided by the term frequency and inverse document frequency (TF-IDF) of triphones. However, these methods are not very suitable for the case of data created by artificial noising as they do not explore the acoustic diversity.

A different approach was used in [2], where the aim was to select a subset of the data with the most similar acoustic characteristics to the target domain. To achieve this, the vector of the posterior probabilities of components in Gaussian mixture models was used to represent each utterance. The idea of using a fixed-length vector for characterization of utterances was also exploited in [25], where the selection was based on the distributions of i-vectors.

Here we investigate use of a fixed-length "summary vector" representing the acoustic conditions to select training utterances that are the most similar to the test conditions. Unlike the existing data selection approaches, the proposed summary vector extraction exploits a neural network framework. A special NN is used to compensate for the mismatch between clean and noisy conditions. This is realized by appending a compensation network to an NN trained on clean speech. The compensation network is trained to perform utterance-level bias compensation. Consequently, the output of the compensation network summarizes the information about the noise conditions of an entire utterance and can thus be used to select useful training data.

The extracted "summary vector" has the desired property of discriminating specific noise types, which can be proved by visualizing the vectors. Moreover, we have also confirmed experimentally that the proposed "summary vector" can be used to select training data and that it outperforms random training data selection and i-vector-based data selection.

10.3.2 Sequence-Summarizing Neural Network

We describe each utterance using a fixed-length vector summarizing the acoustic conditions of the utterance. In other words, a vector is extracted from each utterance that is in some sense similar to i-vectors known from speaker recognition [6]. However, instead of relying on a conventional i-vector extraction, we train a special neural network able to "summarize each utterance."

To extract summary vectors, we train a composite architecture combining two neural networks as sketched in Fig. 10.1 (a similar architecture was previously used for speaker adaptation [29]). It consists of the main NN and the sequence-summarizing NNs (SSNN), both sharing the same input features. To train this scheme for the extraction of summary vectors, we proceed as follows:

1. First, the main network (DNN($\cdot$), upper part of Fig. 10.1) is trained on clean data $\mathscr{X}^{\text{clean}}$ as a standard DNN classifier with triphone state targets $\mathscr{Y}$ and a cross-entropy criterion (XE[$\cdot$]):

$$\hat{\Theta}^{\text{clean}} = \arg \max_{\Theta} \text{XE}\left[\mathscr{Y}, \text{DNN}\left(\mathscr{X}^{\text{clean}}; \Theta\right)\right]. \tag{10.1}$$

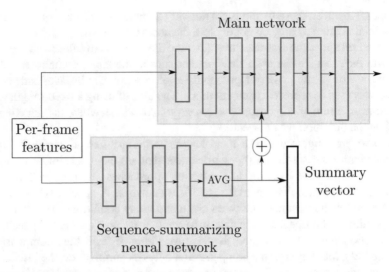

Fig. 10.1 Training of acoustic-condition estimator

The estimated parameters of the main network $\hat{\Theta}^{\text{clean}}$ then stay fixed for the rest of the training.

2. The sequence-summarizing NN is added to the scheme (SSNN($\cdot$), lower part of Fig. 10.1), which also receives frame-by-frame speech features (the same as in the main network) as its input. The last layer of the SSNN involves averaging (global pooling over the frames) and produces one fixed-length summary vector for each utterance, which is then added to the hidden-layer activations of the main network. The whole architecture is now trained on noisy data $\mathscr{X}^{\text{noisy}}$ with the same objective as used in the first step:

$$\arg \max_{\Phi} \text{XE} \left[\mathscr{Y}, \text{DNN} \left(\mathscr{X}^{\text{noisy}}, \overline{\text{SSNN}(\mathscr{X}^{\text{noisy}}; \Phi)}; \hat{\Theta}^{\text{clean}} \right) \right]. \quad (10.2)$$

Note that only the parameters of the sequence summarizing neural network Φ are trained at this point.

The idea of the training procedure is that the SSNN should learn to compensate for the mismatch caused by presenting noisy data $\mathscr{X}^{\text{noisy}}$ to the main network, which was previously trained only on clean data $\mathscr{X}^{\text{clean}}$. Thus, the summary vector extracted by the SSNN should contain important information about the acoustic conditions to characterize the noise component of an utterance. In the final application (data selection), we discard the main network and only use the SSNN to extract the summary vectors.

10.3.3   Configuration of the Neural Network

To find the optimal configuration of the summarizing neural network, we performed a set of experiments varying the hidden layer where the summary vector is added (connection layer), the size of this layer (thus the size of the extracted vector), and the amount of data used to train the network. Although the whole composite network as seen in Fig. 10.1 was not intended to be used for decoding in the final application, for these experiments, we used it directly to test the noisy data. This scenario, channel adaptation, allowed us to find the best configuration without the final time-consuming procedure—data selection and system rebuild.

First, the optimal connection layer was evaluated. In these experiments, the sizes of all hidden layers in both DMM and SNN were 2048 and the amount of noised training data was equal to the original clean set. The results on CHiME-3 are shown in Table 10.4. It seems that adding the summary vector to the second hidden layer is the most effective for adapting the clean DNN to the noisy data. Moreover, the results present over 8% absolute WER reduction by adding a summary-vector extractor to the clean DNN. It is a nice improvement, taking into account that the extractor performs just a simple per-utterance bias compensation in the hidden layer of the clean DNN classifier.

The second hidden layer was taken as the connection layer and the optimal size of the summary vector was evaluated. To be able to train summary-vector extractors of different sizes, we had to retrain the original DNN classifier with various sizes of the second hidden layer on the clean data. Table 10.5 shows the WER reduction as a function of dimensionality of the summary vector. It degrades with decreasing dimensionality; therefore we decided to keep its original dimensionality of 2048.

Finally, the effect of adding data for the training of the summarizing network was evaluated. We generated several random selections of noised data and trained

Table 10.4 Optimal connection layer for training of summary-vector extractor (CHiME-3)

Connection layer	XE (%WER)
None	47.72
1	39.89
2	**39.32**
3	40.09
4	41.08
5	40.70

The bold numbers indicate the best values in the table that help the orientation

Table 10.5 Dimensionality of summary-vector extractor (CHiME-3 %WER)

Size of second layer	256	512	1024	2048
Clean DNN	49.03	48.41	47.83	47.72
Joint NN	49.39	42.70	40.15	39.32
Abs. improvement	−0.36	5.71	7.68	**8.40**

The bold numbers indicate the best values in the table that help the orientation

Table 10.6 Data sizes for summary-vector extractor training (CHiME-3 %WER)

Summary-vector dimensionality	Data size		
	1×train	2×train	3×train
1024	40.15	38.99	37.30
2048	39.32	40.24	**37.15**

The bold numbers indicate the best values in the table that help the orientation

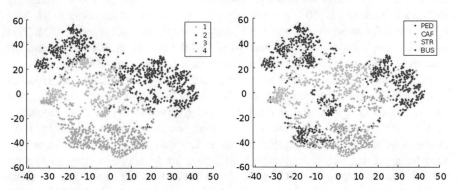

Fig. 10.2 t-SNE plots of summary vectors estimated from CHiME-3 data. The *colors* correspond to the clusters obtained by k-means (*left*) and the actual noise conditions (*right*)

the SSNN on them. Table 10.6 shows the positive effect of a sufficient amount of training data (3× original clean set).

10.3.4 Properties of the Extracted Vectors

To see whether the method generates vectors reflecting the noise conditions in the data, we extracted the vectors for CHiME-3 utterances and observed their properties. The CHiME-3 test set contains four different recording environments—bus (BUS), cafe (CAF), street (STR), and pedestrian area (PED). We performed clustering of the extracted vectors into four clusters using k-means and compared the obtained clusters to the real environments in the data. Figure 10.2 shows two plots created by t-SNE [12]—the right one shows the four real environments in the data and the left one the clusters created by k-means. Although the clusters were created by an unsupervised technique, there are clear similarities with the real ones.

It is also worth comparing the newly proposed summary vector with i-vectors [6] as i-vectors are also known to capture information about the channel. Note that i-vectors were also recently used for adapting DNNs in speech recognition tasks [16, 24]. Figure 10.3 shows i-vectors and summary vectors extracted from CHiME-3 projected onto the first two liner discriminant analysis (LDA) bases. The recording environment labels were used as the classes for LDA. It seems that the environments

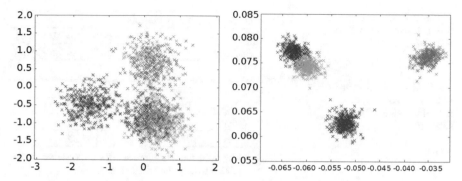

Fig. 10.3 Plot of the first and second LDA bases on CHiME-3 data for i-vectors (*left*) and summary vectors (*right*)

are better separated in the summary-vector space than in the i-vector space.[4] This shows that summary vectors contain information suitable for CHiME-3 recording environment clustering, even though the extractor was trained on different data (a corrupted AMI corpus).

10.3.5   Results with Data Selection

To perform data selection, we extract summary vectors for each generated training utterance and each test utterance. We select a subset of the generated training data by selecting the utterances that are the closest to the test set conditions. For this, we compute the mean over all summary vectors of the test set and measure its distance to the summary vector of each utterance of the training data. Only the closest utterances are kept in the training subset. By this, we aim to select that type of added noise which matches best the noise in the test data. The amount of selected data is equal to the size of the clean training set.

As few different types of noise are present in the test data, computing the mean of summary vectors from the whole test set may not be the best way to represent it. Therefore, further experiments were performed by clustering the summary vectors of the test set and computing the means of these clusters. The training utterances were then selected to have the shortest distance to the summary vector centroid of one of the clusters.

For measuring the distance between vectors, we used the cosine and Euclidean distances. Table 10.7 shows results obtained using these two measures and different numbers of clusters of test data. The results indicate that using the cosine distance is more effective. The best results are obtained using four clusters of test data, which corresponds to the fact that there are four real recording environments in the test set.

---

[4]Brno University of Technology open i-vector extractor (see http://voicebiometry.org) was used for these experiments.

Table 10.7 Comparison of different selection methods (CHiME-3 %WER)

Distance measure/# clusters	1	4	10
Cosine	25.09	24.72	24.98
Euclidean	26.75	26.55	26.59

Table 10.8 Comparison of random vs. automatic selection results (CHiME-3 %WER)

Dataset	Selection		
	Random	i-Vector	Summary vector
dev	25.8	25.61	**24.72**
eval	45.58	44.02	**43.23**

The bold numbers indicate the best values in the table that help the orientation

Table 10.8 shows the best result obtained with the summary-vector data selection compared to random data selection and selection using i-vectors. About 1% absolute improvement on the dev set and 2% on the eval set was obtained with the proposed data selection method compared to random data selection, showing the effectiveness of the proposed data selection method.

10.4 Conclusions

We have shown that, despite its simplicity, data augmentation is an effective technique to improve the robustness of a speech recognizer when deployed in mismatched training–test conditions. Noising of the data was found to be more effective than NN-based denoising strategies. We have also proposed a new promising approach for selecting data within the augmented set, based on a summarizing neural network that is able to generate one fixed-dimensional vector per utterance. On the CHiME-3 test set, we observed 1% absolute improvement over random data selection and the technique also compared favorably to data selection based on i-vectors.

Acknowledgements Besides the funding for the JSALT 2015 workshop, BUT researchers were supported by the Czech Ministry of Interior project no. VI20152020025, "DRAPAK," and by the Czech Ministry of Education, Youth, and Sports from the National Program of Sustainability (NPU II) project "IT4 Innovations Excellence in Science—LQ1602."

References

1. Ager, M., Cvetkovic, Z., Sollich, P., Bin, Y.: Towards robust phoneme classification: augmentation of PLP models with acoustic waveforms. In: 16th European Signal Processing Conference, 2008, pp. 1–5 (2008)
2. Beaufays, F., Vanhoucke, V., Strope, B.: Unsupervised discovery and training of maximally dissimilar cluster models. In: Proceedings of Interspeech (2010)

3. Bellegarda, J.R., de Souza, P.V., Nadas, A., Nahamoo, D., Picheny, M.A., Bahl, L.R.: The metamorphic algorithm: a speaker mapping approach to data augmentation. IEEE Trans. Speech Audio Process. **2**(3), 413–420 (1994). doi:10.1109/89.294355

4. Bellegarda, J., de Souza, P., Nahamoo, D., Padmanabhan, M., Picheny, M., Bahl, L.: Experiments using data augmentation for speaker adaptation. In: International Conference on Acoustics, Speech, and Signal Processing, 1995, ICASSP-95, vol. 1, pp. 692–695 (1995). doi:10.1109/ICASSP.1995.479788

5. Cui, X., Goel, V., Kingsbury, B.: Data augmentation for deep neural network acoustic modeling. IEEE/ACM Trans. Audio Speech Lang. Process. **23**(9), 1469–1477 (2015). doi:10.1109/TASLP.2015.2438544

6. Dehak, N., Kenny, P., Dehak, R., Dumouchel, P., Ouellet, P.: Front-end factor analysis for speaker verification. IEEE Trans. Audio Speech Lang. Process. **19**(4), 788–798 (2011). doi:10.1109/TASL.2010.2064307. http://dx.doi.org/10.1109/TASL.2010.2064307

7. Delcroix, M., Yoshioka, T., Ogawa, A., Kubo, Y., Fujimoto, M., Ito, N., Kinoshita, K., Espi, M., Hori, T., Nakatani, T., Nakamura, A.: Linear prediction-based dereverberation with advanced speech enhancement and recognition technologies for the REVERB challenge. In: Proceedings of REVERB'14 (2014)

8. Delcroix, M., Yoshioka, T., Ogawa, A., Kubo, Y., Fujimoto, M., Ito, N., Kinoshita, K., Espi, M., Araki, S., Hori, T., Nakatani, T.: Strategies for distant speech recognition in reverberant environments. EURASIP J. Adv. Signal Process. **2015**, Article ID 60, 15 pp. (2015)

9. Egorova, E., Veselý, K., Karafiát, M., Janda, M., Černocký, J.: Manual and semi-automatic approaches to building a multilingual phoneme set. In: Proceedings of ICASSP 2013, pp. 7324–7328. IEEE Signal Processing Society, Piscataway (2013). http://www.fit.vutbr.cz/research/view_pub.php?id=10323

10. Gales, M.J.F., College, C.: Model-Based Techniques for Noise Robust Speech Recognition. University of Cambridge, Cambridge (1995)

11. Haykin, S.: Adaptive Filter Theory, 3rd edn. Prentice-Hall, Upper Saddle River, NJ (1996)

12. Hinton, G., Bengio, Y.: Visualizing data using t-SNE. In: Cost-Sensitive Machine Learning for Information Retrieval 33 (2008)

13. Hu, Y., Loizou, P.C.: Subjective comparison of speech enhancement algorithms. In: Proceedings of IEEE International Conference on Speech and Signal Processing, pp. 153–156 (2006)

14. Jaitly, N., Hinton, G.E.: Vocal tract length perturbation (VTLP) improves speech recognition. In: Proceedings of the 30th International Conference on Machine Learning, Atlanta, GA (2013)

15. Kalinli, O., Seltzer, M.L., Acero, A.: Noise adaptive training using a vector Taylor series approach for noise robust automatic speech recognition. In: Proceedings of the 2009 IEEE International Conference on Acoustics, Speech and Signal Processing, ICASSP'09, pp. 3825–3828. IEEE Computer Society, Washington (2009) doi:10.1109/ICASSP.2009.4960461. http://dx.doi.org/10.1109/ICASSP.2009.4960461

16. Karafiát, M., Burget, L., Matějka, P., Glembek, O., Černocký, J.: iVector-based discriminative adaptation for automatic speech recognition. In: Proceedings of ASRU 2011, pp. 152–157. IEEE Signal Processing Society, Piscataway (2011). http://www.fit.vutbr.cz/research/view_pub.php?id=9762

17. Karafiát, M., Veselý, K., Szőke, I., Burget, L., Grézl, F., Hannemann, M., Černocký, J.: BUT ASR system for BABEL surprise evaluation 2014. In: Proceedings of 2014 Spoken Language Technology Workshop, pp. 501–506. IEEE Signal Processing Society, Piscataway (2014). http://www.fit.vutbr.cz/research/view_pub.php?id=10799

18. Karafiát, M., Grézl, F., Burget, L., Szőke, I., Černocký, J.: Three ways to adapt a CTS recognizer to unseen reverberated speech in BUT system for the ASpIRE challenge. In: Proceedings of Interspeech 2015, pp. 2454–2458. International Speech Communication Association, Grenoble (2015). http://www.fit.vutbr.cz/research/view_pub.php?id=10972

19. Kinoshita, K., Delcroix, M., Nakatani, T., Miyoshi, M.: Suppression of late reverberation effect on speech signal using long-term multiple-step linear prediction. IEEE Trans. Audio Speech Lang. Process. **17**(4), 534–545 (2009)

20. Ko, T., Peddinti, V., Povey, D., Khudanpur, S.: Audio augmentation for speech recognition. In: INTERSPEECH, pp. 3586–3589. ISCA, Grenoble (2015)

21. Nakatani, T., Yoshioka, T., Kinoshita, K., Miyoshi, M., Juang, B.H.: Blind speech dereverberation with multi-channel linear prediction based on short time Fourier transform representation. In: Proceedings of ICASSP'08, pp. 85–88 (2008)
22. Ogata, K., Tachibana, M., Yamagishi, J., Kobayashi, T.: Acoustic model training based on linear transformation and MAP modification for HSMM-based speech synthesis. In: INTERSPEECH, pp. 1328–1331 (2006)
23. Ragni, A., Knill, K.M., Rath, S.P., Gales, M.J.F.: Data augmentation for low resource languages. In: INTERSPEECH 2014, 15th Annual Conference of the International Speech Communication Association, Singapore, September 14–18, 2014, pp. 810–814 (2014)
24. Saon, G., Soltau, H., Nahamoo, D., Picheny, M.: Speaker adaptation of neural network acoustic models using i-vectors. In: 2013 IEEE Workshop on Automatic Speech Recognition and Understanding (ASRU), pp. 55–59. IEEE, New York (2013)
25. Siohan, O., Bacchiani, M.: iVector-based acoustic data selection. In: Proceedings of INTERSPEECH, pp. 657–661 (2013)
26. Swietojanski, P., Ghoshal, A., Renals, S.: Hybrid acoustic models for distant and multichannel large vocabulary speech recognition. In: 2013 IEEE Workshop on Automatic Speech Recognition and Understanding (ASRU). IEEE, New York (2013)
27. Tokuda, K., Zen, H., Black, A.: An HMM-based approach to multilingual speech synthesis. In: Text to Speech Synthesis: New Paradigms and Advances, pp. 135–153. Prentice Hall, Upper Saddle River (2004)
28. Veselý, K., Ghoshal, A., Burget, L., Povey, D.: Sequence-discriminative training of deep neural networks. In: Proceedings of INTERSPEECH 2013, pp. 2345–2349. International Speech Communication Association, Grenoble (2013). http://www.fit.vutbr.cz/research/view_pub.php?id=10422
29. Veselý, K., Watanabe, S., Žmolíková, K., Karafiát, M., Burget, L., Černocký, J.: Sequence summarizing neural network for speaker adaptation. In: Proceedings of ICASSP (2016)
30. Wang, Y., Gales, M.J.F.: Speaker and noise factorization for robust speech recognition. IEEE Trans. Audio Speech Lang. Process. 20(7), 2149–2158 (2012). http://dblp.uni-trier.de/db/journals/taslp/taslp20.html#WangG12
31. Wei, K., Liu, Y., Kirchhoff, K., Bartels, C., Bilmes, J.: Submodular subset selection for large-scale speech training data. In: Proceedings of ICASSP, pp. 3311–3315 (2014)
32. Wu, Y., Zhang, R., Rudnicky, A.: Data selection for speech recognition. In: Proceedings of ASRU, pp. 562–565 (2007)
33. Xu, Y., Du, J., Dai, L.R., Lee, C.H.: An experimental study on speech enhancement based on deep neural networks. IEEE Signal Process Lett. 21(1), 65–68 (2014)
34. Yoshimura, T., Masuko, T., Tokuda, K., Kobayashi, T., Kitamura, T.: Speaker interpolation in HMM-based speech synthesis system. In: Eurospeech, pp. 2523–2526 (1997)
35. Yoshioka, T., Nakatani, T.: Generalization of multi-channel linear prediction methods for blind MIMO impulse response shortening. IEEE Trans. Audio Speech Lang. Process. 20(10), 2707–2720 (2012)
36. Yoshioka, T., Nakatani, T., Miyoshi, M., Okuno, H.G.: Blind separation and dereverberation of speech mixtures by joint optimization. IEEE Trans. Audio Speech Lang. Process. 19(1), 69–84 (2011)
37. Yoshioka, T., Chen, X., Gales, M.J.F.: Impact of single-microphone dereverberation on DNN-based meeting transcription systems. In: Proceedings of ICASSP'14 (2014)
38. Yoshioka, T., Ito, N., Delcroix, M., Ogawa, A., Kinoshita, K., Fujimoto, M., Yu, C., Fabian, W.J., Espi, M., Higuchi, T., Araki, S., Nakatani, T.: The NTT CHiME-3 system: advances in speech enhancement and recognition for mobile multi-microphone devices. In: Proceedings of ASRU'15, pp. 436–443 (2015)
39. Zavaliagkos, G., Siu, M.-H., Colthurst, T., Billa, J.: Using untranscribed training data to improve performance. In: The 5th International Conference on Spoken Language Processing, Incorporating the 7th Australian International Speech Science and Technology Conference, Sydney, Australia, 30 November–4 December 1998 (1998)

Chapter 11
Advanced Recurrent Neural Networks for Automatic Speech Recognition

Yu Zhang, Dong Yu, and Guoguo Chen

Abstract A recurrent neural network (RNN) is a class of neural network models in which connections between its neurons form a directed cycle. This creates an internal state of the network which allows it to exhibit dynamic temporal behavior. In this chapter, we describe several advanced RNN models for distant speech recognition (DSR). The first set of models are extensions of the prediction-adaptation-correction RNNs (PAC-RNNs). These models were inspired by the widely observed behavior of prediction, adaptation, and correction in human speech recognition. The second set of models, include highway long short-term memory (LSTM) RNNs, latency-controlled bidirectional LSTM RNNs, Grid LSTM RNNs, and Residual LSTM RNNs, are all extensions of deep LSTM RNNs. These models are so built that their optimization can be more effective than the basic deep LSTM RNNs. We evaluate and compare these advanced RNN models on DSR tasks using the AMI corpus.

11.1 Introduction

Deep-neural-network (DNN)-based acoustic models (AMs) greatly improved automatic speech recognition (ASR) accuracy on many tasks [7, 15, 24, 25]. Further improvements were reported by using more advanced models, e.g., convolutional neural networks (CNNs) [1, 3, 28, 32] and recurrent neural networks (RNNs) such as long short-term memory (LSTM) networks [10, 11, 22].

Although these new techniques help to decrease the word error rate (WER) on distant speech recognition (DSR) [27] tasks, DSR remains a challenging problem

---

Y. Zhang (✉)
Massachusetts Institute of Technology, Cambridge, MA, USA
e-mail: yzhang87@csail.mit.edu

D. Yu
Tencent AI Lab, Seattle, WA, USA
e-mail: dyu@tencent.com

G. Chen
Johns Hopkins University, Baltimore, MD, USA

© Springer International Publishing AG 2017   261
S. Watanabe et al. (eds.), *New Era for Robust Speech Recognition*,
DOI 10.1007/978-3-319-64680-0_11

due to reverberation and overlapping acoustic signals, even with sophisticated front-end processing techniques [12, 20, 26] and multipass decoding schemes.

In this chapter, we explore several advanced back-end techniques for DSR. These techniques are all built upon recurrent neural networks.

The first set of models are extensions of the prediction–adaptation–correction RNN (PAC-RNN) proposed in [33]. These models were inspired by the widely observed behavior of prediction, adaptation, and correction in human speech recognition. In this chapter, we extend the PAC-RNN by introducing more advanced prediction components and evaluate them on large-vocabulary continuous speech recognition (LVCSR) tasks using Babel and the AMI corpus.

The second set of models are extensions of deep LSTM (DLSTM) RNNs. DLSTM RNNs help improve generalization and often outperform single-layer LSTM RNNs [22]. However, they are harder to train and slower to converge when the model becomes deeper. In this chapter, we extend DLSTM RNNs in several directions. First, we introduce gated direct connections, called highway connections, between memory cells of adjacent layers to form highway LSTM (HLSTM) RNNs [34]. The highway connections provide a path for information to flow between layers more directly without decay. They alleviate the vanishing-gradient problem and enable DLSTM RNNs to go deeper, especially when dropout is exploited to control the highway connections. HLSTM RNNs can be easily extended from unidirectional to bidirectional. To speed up the training of and reduce the latency in bidirectional HLSTM (BHLSTM) RNNs, we further introduce the latency-controlled BHLSTM RNNs, in which the whole past history and a window of the future context of the utterance is exploited.

We further present two other extensions of the DLSTM RNNs. The grid LSTM (GLSTM) [19] uses separate LSTM blocks along the time and depth axes to improve modeling power. The residual LSTM (RLSTM) RNNs inspired by linearly augmented DNNs [9] and residual CNNs [13] contain direct links between the lower-layer outputs and the higher-layer inputs in DLSTM RNNs. Both GLSTM and RLSTM RNNs enable us to train deeper models and achieve better accuracy.

11.2 Basic Deep Long Short-Term Memory RNNs

In this section, we review the basic single-layer LSTM RNNs, and their deep version.

11.2.1 Long Short-Term Memory RNNs

The LSTM RNN was initially proposed in [17] to solve the gradient-diminishing and explosion problem that often happens when training RNNs. It introduces a linear dependency between c_t, the memory cell state at time t, and c_{t-1}, the same cell's state at $t - 1$. Nonlinear gates are introduced to control the information flow. The operation of the network follows the equations

$$\mathbf{i}_t = \sigma(\mathbf{W}_{xi}\mathbf{x}_t + \mathbf{W}_{mi}\mathbf{m}_{t-1} + \mathbf{W}_{ci}\mathbf{c}_{t-1} + \mathbf{b}_i), \tag{11.1}$$

$$\mathbf{f}_t = \sigma(\mathbf{W}_{xf}\mathbf{x}_t + \mathbf{W}_{mf}\mathbf{m}_{t-1} + \mathbf{W}_{cf}\mathbf{c}_{t-1} + \mathbf{b}_f), \tag{11.2}$$

$$\mathbf{c}_t = \mathbf{f}_t \odot \mathbf{c}_{t-1} + \mathbf{i}_t \odot \tanh(\mathbf{W}_{xc}\mathbf{x}_t + \mathbf{W}_{mc}\mathbf{m}_{t-1} + \mathbf{b}_c), \tag{11.3}$$

$$\mathbf{o}_t = \sigma(\mathbf{W}_{xo}\mathbf{x}_t + \mathbf{W}_{mo}\mathbf{m}_{t-1} + \mathbf{W}_{co}\mathbf{c}_t + \mathbf{b}_o), \tag{11.4}$$

$$\mathbf{m}_t = \mathbf{o}_t \odot \tanh(\mathbf{c}_t) \tag{11.5}$$

iteratively from $t = 1$ to $t = T$, where $\sigma()$ is the logistic sigmoid function, and $\mathbf{i}_t, \mathbf{f}_t, \mathbf{o}_t, \mathbf{c}_t$ and $\mathbf{m}_t$ are vectors to represent values of the input gate, forget gate, output gate, cell activation, and cell output activation respectively, at time t. $\odot$ denotes the elementwise product of vectors. $\mathbf{W}_*$ are the weight matrices connecting different gates, and $\mathbf{b}_*$ are the corresponding bias vectors. All these matrices are full except the matrices $\mathbf{W}_{ci}$, $\mathbf{W}_{cf}$, and $\mathbf{W}_{co}$ that connect the cell to gates, which are diagonal.

11.2.2 Deep LSTM RNNs

Deep LSTM RNNs are formed by stacking multiple layers of LSTM cells. Specifically, the output of the lower-layer LSTM cells $\mathbf{y}_t^l$ is fed to the upper layer as an input $\mathbf{x}_t^{l+1}$. Although each LSTM layer is deep in time, since it can be unrolled in time to become a feedforward neural network in which each layer shares the same weights, deep LSTM RNNs still outperform single-layer LSTM RNNs significantly. It is conjectured [22] that DLSTM RNNs can make better use of parameters by distributing them over the space of multiple layers. Note that in the conventional DLSTM RNNs there is no direct interaction between cells in different layers.

11.3 Prediction–Adaptation–Correction Recurrent Neural Networks

In this section, we present the PAC-RNN, which combines the abilities of prediction, adaptation, and correction in the same model, which is often observed in human speech recognition.

The PAC-RNN, illustrated in Fig. 11.1, was originally proposed by Zhang et al. in [33]. It has two main components: a *correction* DNN and a *prediction* DNN. The correction DNN estimates the state posterior probability $p^{\text{corr}}(s_t|\mathbf{o}_t, \mathbf{x}_t)$ given $\mathbf{o}_t$, the observation feature vector, and $\mathbf{x}_t$, the information from the prediction DNN, at time t. The prediction DNN predicts future auxiliary information l_{t+n}, in which l can

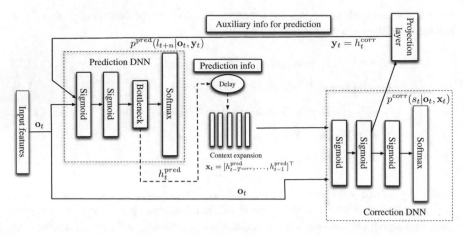

Fig. 11.1 The structure of the PAC-RNN-DNN

be a state s, a phone θ, a noise representation, or other auxiliary information, and n is the number of frames which we look ahead. Note that since $\mathbf{y}_t$, the information from the correction DNN, depends on $\mathbf{x}_t$, the information from the prediction DNN, and vice versa, a recurrent loop is formed.

The information from the prediction DNN, $\mathbf{x}_t$, is extracted from a bottleneck hidden layer h_{t-1}^{pred}. To exploit additional previous predictions, we stack multiple hidden layer values as

$$\mathbf{x}_t = [h_{t-T^{\mathrm{corr}}}^{\mathrm{pred}}, \ldots, h_{t-1}^{\mathrm{pred}}]^T, \tag{11.6}$$

where T^{corr} is the contextual window size used by the correction DNN and was set to 10 in our study. Similarly, we can stack multiple frames to form $\mathbf{y}_t$, the information from the correction DNN, as

$$\mathbf{y}_t = [h_{t-T^{\mathrm{pred}}-1}^{\mathrm{corr}}, \ldots, h_t^{\mathrm{corr}}]^T, \tag{11.7}$$

where T^{pred} is the contextual window size used by the prediction DNN and was set to 1 in our study. In addition, in the specific example shown in Fig. 11.1, the hidden layer output h_t^{corr} is projected to a lower dimension before it is fed into the prediction DNN.

To train the PAC-RNN, we need to provide supervision information to both the prediction and the correction DNNs. As we have mentioned, the correction DNN estimates the state posterior probability, so we provide the state labels, and train it with the frame cross-entropy (CE) criterion. For the prediction DNN, we follow [33], and use the phoneme label as the prediction targets.

The training of the PAC-RNN is a multitask learning problem similar to [2, 29], which also use phoneme targets in addition to the state targets. The two training objectives can be combined into a single one as

$$J = \sum_{t=1}^{T}(\alpha * \ln p^{\text{corr}}(s_t|\mathbf{o}_t, \mathbf{x}_t) + (1 - \alpha) * \ln p^{\text{pred}}(l_{t+n}|\mathbf{o}_t, \mathbf{y}_t)), \qquad (11.8)$$

where α is the interpolation weight, and was set to 0.8 in our study unless otherwise stated, and T is the total number of frames in the training utterance. Note that in the standard PAC-RNN as described so far, both the correction model and the prediction model are DNNs. From this point onwards we will call this particular setup the PAC-RNN-DNN. LSTMs have improved speech recognition accuracy on many tasks over DNNs [10, 11, 22]. To further enhance the PAC-RNN model, we use an LSTM to replace the DNN used in the correction model. The input of this LSTM is the acoustic feature $\mathbf{o}_t$ concatenated with the information from the prediction component, $\mathbf{x}_t$. The prediction component can also be an LSTM, but we did not observe a performance gain in the experiments. To keep it simple, we use the same DNN prediction model as [33].

11.4   Deep Long Short-Term Memory RNN Extensions

In this section we introduce several extensions of DLSTM RNNs that can provide more modeling power yet can be trained effectively. These models include the highway LSTM RNNs, which use a gated function to control the direct information flow from the memory cells in lower layers to those in the higher layers, the grid LSTM RNNs, which use two (or more) separate LSTM RNNs (with separate memory states) to model information flow on different axes (e.g., depth and time), and the residual LSTM RNNs, which feed the output from lower layers to skipping higher layers directly.

11.4.1   Highway RNNs

When the networks become deeper or more complex, accuracy degradation is often observed. Such degradation is not caused by overfitting [13], since the degradation also happens on the training set. For example, in many ASR tasks, three to five LSTM layers are optimal in a deep LSTM. Further increasing the depth leads to higher WER. There are two possible solutions to this degradation problem: (1) to pretrain the network layer by layer; and (2) to modify the network structure so that its optimization can be easier and more effective. In this subsection, we focus on the second approach and propose the HLSTM, which can directly feed information from lower layers to higher layers.

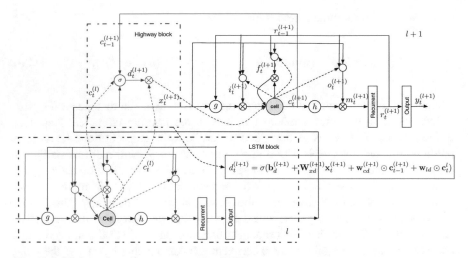

Fig. 11.2 Highway long short-term memory RNNs

The HLSTM [34], as illustrated in Fig. 11.2, improves upon DLSTM RNNs. It has a direct connection (in the block labeled "Highway block") between the memory cells $\mathbf{c}_t^l$ in the lower layer l and the memory cells $\mathbf{c}_t^{l+1}$ in the upper layer $l+1$. The carry gate controls how much information can flow from the lower-layer cells directly to the upper-layer cells. The gate function at layer $l+1$ at time t is

$$\mathbf{d}_t^{(l+1)} = \sigma(\mathbf{b}_d^{(l+1)} + \mathbf{W}_{xd}^{l+1}\mathbf{x}_t^{(l+1)} + \mathbf{w}_{cd}^{l+1} \odot \mathbf{c}_{t-1}^{(l+1)} + \mathbf{w}_{ld}^{(l+1)} \odot \mathbf{c}_t^l), \qquad (11.9)$$

where $\mathbf{b}_d^{(l+1)}$ is a bias term, $\mathbf{W}_{xd}^{(l+1)}$ is the weight matrix connecting the carry gate to the input of this layer, $\mathbf{w}_{cd}^{(L+1)}$ is a weight vector from the carry gate to the past cell state in the current layer, $\mathbf{w}_{ld}^{(L+1)}$ is a weight vector connecting the carry gate to the lower-layer memory cell, and $\mathbf{d}^{(l+1)}$ is the carry gate activation vectors at layer $l+1$.

Using the carry gate, an HLSTM RNN computes the cell state at layer $(l+1)$ according to

$$\mathbf{c}_t^{l+1} = \mathbf{d}_t^{(l+1)} \odot \mathbf{c}_t^l + \mathbf{f}_t^{(l+1)} \odot \mathbf{c}_{t-1}^{(l+1)}$$
$$+ \mathbf{i}_t^{(l+1)} \odot \tanh(\mathbf{W}_{xc}^{(l+1)}\mathbf{x}_t^{(l+1)} + \mathbf{W}_{hc}^{(l+1)}\mathbf{m}_{t-1}^{(l+1)} + \mathbf{b}_c), \qquad (11.10)$$

while all other equations are the same as those for standard LSTM RNNs as described in (11.1), (11.2), (11.4), and (11.5).

Conceptually, the highway connection is a multiplicative modification in analogy to the forget gate. Depending on the output of the carry gates, the highway connection smoothly varies its behavior between that of a plain LSTM layer (no connection) and that of direct linking (i.e., passing the cell memory from the previous layer directly without attenuation). The highway connection between cells in different layers makes the influence of cells in one layer on the other layer more direct and can alleviate the vanishing-gradient problem when training deeper LSTM RNNs.

11.4.2   Bidirectional Highway LSTM RNNs

The unidirectional LSTM RNNs we described above can only exploit past history. In speech recognition, however, future contexts also carry information and should be utilized to further enhance acoustic models. Bidirectional RNNs take advantage of both past and future contexts by processing the data in both directions with two separate hidden layers. It was shown in [5, 10, 11] that bidirectional LSTM RNNs can indeed improve speech recognition accuracy. In this chapter, we also extend HLSTM RNNs from unidirectional to bidirectional. Note that the backward layer follows the same equations as used for the forward layer except that $t-1$ is replaced by $t+1$ to exploit future frames and the model operates from $t = T$ to 1. The outputs of the forward and backward layers are concatenated to form the input to the next layer.

11.4.3   Latency-Controlled Bidirectional Highway LSTM RNNs

Nowadays, graphics processing units (GPUs) are widely used in deep learning due to their massive parallel computation ability. For unidirectional RNN models, multiple sequences (e.g., 40) are often packed into the same minibatch (e.g., in [31]) to better utilize GPUs. When the truncated back-propagation-through-time (BPTT) algorithm is used for parameter updating, this can be easily done since only a small segment (e.g., 20 frames) of each sequence has to be packed into the minibatch. However, when the whole sequence-based BPTT is used, e.g., when doing sequence-discriminative training or when using bidirectional LSTMs (BLSTMs), the GPU's limited memory restricts the number of sequences that can be packed into a minibatch and thus significantly decreases the training and evaluation speed. This problem can be especially severe for LVCSR tasks with long sequences and large model sizes. One way to speed up the training under these conditions is to use asynchronous stochastic gradient descent on a GPU/CPU farm [14], at the cost of low computing resource utilization on each GPU/CPU. In this subsection, we propose a latency-controlled BLSTM that can better utilize computing power on each GPU card during training and incurs much less latency than the basic BLSTM during decoding.

To speed up the training of bidirectional RNNs, the context-sensitive-chunk BPTT (CSC-BPTT) was proposed in [5]. In this method, a sequence is first split into chunks of fixed length N_c. Then N_l past frames and N_r future frames are

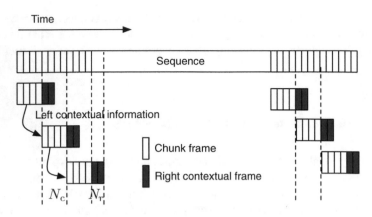

Fig. 11.3 Latency-controlled bidirectional model training

concatenated before and after each chunk as the left and right context, respectively. The appended frames are only used to provide contextual information and do not generate error signals during training. Since each Chunk can be independently drawn and trained, they can be stacked to form large minibatches to speed up training.

Unfortunately, a model trained with CSC-BPTT is no longer a true bidirectional RNN since the history it can exploit is limited by the left and right context concatenated with the chunk. It also introduces additional computation cost during decoding, since both the left and the right contexts need to be recomputed for each chunk.

The latency-controlled bidirectional RNN illustrated in Fig. 11.3 borrows the idea of the CSC-BPTT and improves upon both the CSC-BPTT and the conventional BLSTM. Differently from the CSC-BPTT, in our new model we carry the whole past history while still using a truncated future context. Instead of concatenating and computing N_l left contextual frames for each chunk, we directly carry over the left contextual information from the previous chunk of the same utterance. For every chunk, both the training and the decoding computational cost are reduced by a factor of $N_l/N_l + N_c + N_r$. Moreover, loading the history from the previous minibatch instead of a fixed contextual window makes the context exact when compared to the unidirectional model. Note that the standard BLSTM RNNs come with significant latency since the model can only be evaluated after seeing the whole utterance. In the latency-controlled BLSTM RNNs the latency is limited to N_r, which can be set by the user. In our experiments, we processed 40 utterances in parallel, which is ten times faster than processing the whole utterances without performance loss. Compared to the CSC BPTT, our approach is 1.5 times faster and often leads to better accuracy.

11.4.4   Grid LSTM RNNs

The grid LSTM was proposed in [19] and uses a generic form to add cells along
the depth axis. One grid LSTM block receives N hidden vectors $m_1, \ldots, m_N$ and
N cell (memory) vectors $c_1, \ldots, c_N$ from the N dimensions. The block computes
N transforms denoted by LSTM, for each axis and outputs N hidden vectors and
memory vectors:

$$(\mathbf{m}_1', \mathbf{c}_1') = \text{LSTM}(\mathbf{H}, \mathbf{c}_1, \mathbf{W}_1),$$

$$\cdots$$

$$(\mathbf{m}_N', \mathbf{c}_N') = \text{LSTM}(\mathbf{H}, \mathbf{c}_N, \mathbf{W}_N),$$

$$\mathbf{H} = [\mathbf{m}_1, \ldots, \mathbf{m}_N]^\top, \tag{11.11}$$

where $\mathbf{W}_i$ are the weight matrices for each axis, $\mathbf{H}$ is the concatenation of hidden
outputs, and $\mathbf{c}$ is the cell output (memory) for each axis. All the LSTM formulas used
here are consistent with those in Sect. 11.2.1 and are different from those in [19].
In [19], there is no peephole connection. However, we have found that a peephole
connection is always useful in our experiments.

In the deep neural network model for ASR, there are two axes: time (samples
in the time domain) and depth (multiple layers). For each block, we can define the
LSTM parameters as

$$\{(\mathbf{W}_{(t,d)}^{\text{Time}}, \mathbf{W}_{(t,d)}^{\text{Depth}}) | t = 1, \ldots, T, d = 1, \ldots, D\} \tag{11.12}$$

Note that sharing of weight matrices can be specified along any axis in a grid
LSTM. In our experiments, we always tied all the weight matrices on the temporal
axis:

$$\forall d \in \{1, \ldots, D\}$$

$$\mathbf{W}_{(1,d)}^{\text{Time}} = \mathbf{W}_{(2,d)}^{\text{Time}} \cdots = \mathbf{W}_{(T,d)}^{\text{Time}},$$

$$\mathbf{W}_{(1,d)}^{\text{Depth}} = \mathbf{W}_{(2,d)}^{\text{Depth}} \cdots = \mathbf{W}_{(T,d)}^{\text{Depth}}. \tag{11.13}$$

For the depth axis, we tried both tied ($\mathbf{W}^{\text{Depth}} = \mathbf{W}_{(*,*)}^{\text{Depth}}$) and untied versions.
Differently from the observation in [19], the untied version always gave us better
performance.

11.4.5 Residual LSTM RNNs

The residual network was proposed in [13] and is a special case of the linearly augmented model described in [9]. It defines a building block

$$\mathbf{y} = \mathscr{F}(\mathbf{x}, \mathbf{W}_i) + \mathbf{x}, \tag{11.14}$$

where $\mathbf{x}$ and $\mathbf{y}$ are the input and output vectors of the layers considered. In this study, we replace the convolutional and rectifier linear layer with an LSTM block:

$$\mathbf{m}_{l+1} = \text{LSTM}^r(\mathbf{m}_l) + m_l. \tag{11.15}$$

Here r indicates how many layers we want to skip. In [13], it was reported that it was important to skip more than one layer. However, in our study we didn't find it necessary.

11.5 Experiment Setup

11.5.1 Corpus

11.5.1.1 IARPA-Babel Corpus

The IARPA-Babel program focuses on ASR and spoken-term detection for low-resource languages [18]. The goal of the program is to reduce the amount of time needed to develop ASR and spoken-term detection capabilities for a new language. The data from the Babel program consists of collections of speech from a growing list of languages. For this work, we consider the Full pack (60–80 h of training data) for the 11 languages released in the first 2 years as source languages, while the languages in the third year are the target languages [6]. Some languages also contain a mixture of microphone data recorded at 48 kHz in both training and test utterances. For the purpose of this paper, we downsampled all the wideband data to 8 kHz and treated it the same way as the rest of the recordings. For the target languages, we will focus on the Very Limited Language Pack (VLLP) condition, which includes only 3 h of transcribed training data. This condition excludes any use of a human-generated pronunciation dictionary. Unlike in the previous 2 years of the program, usage of web data is permitted for language modeling and vocabulary expansion.

11.5.1.2 AMI Meeting Corpus

The AMI corpus [4] comprises around 100 h of meeting recordings, recorded in instrumented meeting rooms. Multiple microphones were used, including individual headset microphones (IHMs), lapel microphones, and one or more microphone arrays. In this work, we used the single-distant-microphone (SDM) condition in our experiments. Our systems were trained and tested using the split recommended in the corpus release: a training set of 80 h, and a development set and a test set each of 9 h. For our training, we used all the segments provided by the corpus, including those with overlapping speech. Our models were evaluated on the evaluation set only. NIST's asclite tool [8] was used for scoring.

11.5.2   System Description

Kaldi [21] was used for feature extraction and early-stage triphone training, as well as decoding. A maximum likelihood acoustic-training recipe was used to train a Gaussian-mixture-model—hidden-Markov-model (GMM-HMM) triphone system. Forced alignment was performed on the training data by this triphone system to generate labels for further neural network training.

The Computational Network Toolkit (CNTK) [31] was used for neural network training. We started off by training a six-layer DNN, with 2048 sigmoid units per layer. 40-dimensional filterbank features, together with their corresponding delta and delta–delta features, were used as raw feature vectors. For our DNN training we concatenated 15 frames of raw feature vectors, which leads to a dimension of 1800. This DNN again was used to force-align the training data to generate labels for further LSTM training.

In the PAC-RNN model, the prediction DNN had a 2048-unit hidden layer and an 80-unit bottleneck layer. The correction model had two varieties: a DNN with several 2048-unit hidden layers or an LSTM (with 1024 memory cells) with a projection layer (LSTMP) of 512 nodes. The correction model's projection layer contained 500 units. For the Babel experiments, we used bottleneck features instead of the raw filterbank features as the input to the system.

Our (H)LSTM models, unless explicitly stated otherwise, were added with a projection layer on top of each layer's output (we refer to this as LSTMP here), as proposed in [22], and were trained with 80-dimensional log mel filterbank (FBANK) features. For the LSTMP models, each hidden layer consisted of 1024 memory cells together with a 512-node projection layer. For the BLSTMP models, each hidden layer consisted of 1024 memory cells (512 for forward and 512 for backward) with a 300-node projection layer. Their highway companions shared the same network structure, except for the additional highway connections.

All models were randomly initialized without either generative or discriminative pretraining [23]. A validation set was used to control the learning rate, which was halved when no gain was observed. To train the unidirectional model, truncated BPTT [30] was used to update the model parameters. Each BPTT segment contained 20 frames and we processed 40 utterances simultaneously. To train the latency-controlled bidirectional model, we set $N_c = 22$ and $N_r = 21$ and also processed 40 utterances simultaneously. A start learning rate of 0.2 per minibatch was used and then the learning rate scheduler took action. For frame-level cross-entropy training, $L2$ constraint regularization [16] was used.

11.6 Evaluation

The performance of various models is evaluated using WER in percent below. For the experiments conducted on AMI, the SDM eval set was used if not specified otherwise. Since we did not exclude overlapping speech segments during model training, in addition to results on the full eval set, we also show results on a subset that only contains the nonoverlapping speech segments as in [28].

11.6.1 PAC-RNN

We evaluated the PAC-RNN on two different tasks: LVCSR with low-resource languages using the IARPA-Babel corpus, and distant-talking speech recognition using the AMI corpus.

11.6.1.1 Low-Resource Language

Table 11.1 summarizes the WERs achieved with different models evaluated on the low-resource language setup. The first three rows are the results from stacked bottleneck (SBN) systems; the details can be found in [35]. Both the multilingual and the closest-language system were adapted to the target language for the whole stacked network.[1] For the hybrid systems, the input was the BN features extracted from the first DNN of the adapted multilingual SBN.

The DNN hybrid system outperforms the multilingual SBN but is very similar to the closest-language system. The LSTM improves upon the DNN by around 1%. The PAC-RNN-DNN outperforms the LSTM by another percent across all languages. By simply replacing the correction model with a single-layer LSTM, we observe even further improvements.

We also investigated the effect of multilingual transfer learning for each model. We first used the rich-resource closest language (based on the Language Identification (LID) prediction shown in the table) to train DNN, LSTM, and PAC-RNN models, and then adapted them to the target language. The lower part of Table 11.1 summarizes the ASR results. As shown, the LSTM models perform significantly better than the baseline SBN system. Using the PAC-RNN model yields a noticeable

---

[1] More details, e.g., how to train the multilingual system, can be found in [35].

Table 11.1 WER (%) results for each ASR system

Target language	Cebuano	Kurmanji	Swahili
Closest language	Tagalog	Turkish	Zulu
SBN models			
Monolingual	73.5	86.2	65.8
Adapted multilingual	65.0	75.5	54.9
Closest language	63.7	75.0	54.2
Hybrid models			
DNN	63.9	74.9	54.0
LSTM	63.0	74.0	53.0
PAC-RNN-DNN	62.1	72.9	52.1
PAC-RNN-LSTM	60.6	72.5	51.4
Hybrid models with closest-language initialization			
DNN	62.7	73.1	52.4
LSTM	61.3	72.5	52.2
PAC-RNN-DNN	60.8	71.8	51.6
PAC-RNN-LSTM	59.7	71.4	50.4

SBN is the stacked bottleneck system

Table 11.2 WER (%) results for PAC-RNN on AMI

System	#layers	With overlap	No overlap
DNN	6	57.5	48.4
LSTMP	3	50.7	41.7
PAC-RNN-DNN (no prediction)	3	54.6	45.3
PAC-RNN-DNN	3	53.7	44.6
PAC-RNN-DNN	5	56.8	47.7
PAC-RNN-LSTMP	3	49.5	40.5

The SDM setup was adopted

improvement over the LSTM. Similarly, the PAC-RNN-LSTM can further improve the results.

11.6.1.2 Distant Speech Recognition

Table 11.2 summarizes the WERs achieved with the PAC-RNN model evaluated on the AMI corpus. For the PAC-RNN models, we always fixed the prediction model as a single-layer DNN, and "layers" in Table 11.2 is indicative only of the correction component. The PAC-RNN-DNN is much worse than the LSTM model. We conjecture that the inferior performance is due to two reasons: (1) The PAC-RNN is harder to optimize when more layers are added, since the recursion loop contains both the prediction and the correction components. Row 5 shows that results become much worse when we increase the number of layers to 5. (2) When we have a stronger language model (compared to that in Babel), the gain from the prediction model becomes smaller. Row 3 shows that if we simply remove the prediction softmax operation but keep all the other parts of network the same, we only get 0.7% degradation, which is significantly smaller than that on TIMIT [33].

By simply replacing the correction component with a three-layer LSTM, we observe that the PAC-RNN-LSTMP improves upon the LSTMP by around 1%. However, we have noticed that the PAC-RNN is more sensitive to the learning-rate scheduling than the simple deep LSTMP. We are currently investigating a better PAC-RNN structure that can be optimized easier.

11.6.2 Highway LSTMP

The performance of different RNN structures that can help train deeper networks is evaluated below.

11.6.2.1 Three-Layer Highway (B)LSTMP

Table 11.3 gives the WER performance of the three-layer LSTMP and BLSTMP RNNs, as well as their highway versions, on the AMI corpus. The performance of the DNN network is also listed for comparison. From the table, it's clear that the highway versions of the LSTM RNNs consistently outperform their nonhighway companions, though with a small margin.

11.6.2.2 Highway (B)LSTMP with Dropout

Dropout can be applied to the highway connection to control its flow: a high dropout rate essentially turns off the highway connection, and a small dropout rate, on the other hand, keeps the connection alive. In our experiments, in the early training stages, we used a small dropout rate of 0.1. We increased it to 0.8 after five epochs of training. The performance of highway (B)LSTMP networks with dropout is shown in Table 11.4; as we can see, dropout helps to further bring down the WER for highway networks.

Table 11.3 WER (%) results for highway (B)LSTMP RNNs on the AMI corpus

System	#layers	With overlap	No overlap
DNN	6	57.5	48.4
LSTMP	3	50.7	41.7
HLSTMP	3	50.4	41.2
BLSTMP	3	48.5	38.9
BHLSTMP	3	48.3	38.5

The SDM setup was adopted

Table 11.4 WER (%) results for highway (B)LSTMP RNNs with dropout on the AMI corpus

System	#layers	With overlap	No overlap
LSTMP	3	50.7	41.7
HLSTMP + dropout	3	49.7	40.5
BLSTMP	3	48.5	38.9
BHLSTMP + dropout	3	47.5	37.9

The SDM setup was adopted

Table 11.5 Comparison of shallow and deep networks on the AMI corpus

System	#layers	With overlap	No overlap
LSTMP	3	50.7	41.7
LSTMP	8	52.6	43.8
LSTMP	16	N/A	N/A
HLSTMP	3	50.4	41.2
HLSTMP	8	50.7	41.3
HLSTMP	16	50.7	41.2

The SDM setup was adopted

11.6.2.3 Deeper Highway LSTMP

When a network goes deeper, the training usually becomes more difficult. Table 11.5 compares the performance of shallow and deep networks. From the table we can see that for a normal LSTMP network, when it goes from three layers to eight layers, the recognition performance degrades dramatically. For the highway network, however, the WER only increases a little bit. If we go even deeper, e.g., to 16 layers, a normal LSTMP training would diverge but the highway network can still be trained well. The table suggests that the highway connection between LSTM layers allows the network to go much deeper than the normal LSTM networks. This also indicates that the HLSTMP may gain more when we have much more data, since we could train much deeper models.

11.6.2.4 Grid LSTMP

Section 11.6.2.2 shows that using dropout to regulate on top of the highway connection could further bring down the WER. The grid LSTMP (GLSTMP) can be seen as a special highway block where an LSTM an the depth axis is used to further control the highway connection and thus has the potential to perform better. Table 11.6 compares different variants of grid LSTM RNNs. The vanilla three-layer grid LSTM already outperforms the HLSTM with dropout. By increasing the layers from three to eight we get an additional 1% improvement. It can be observed that using depth as the main axis is more effective than time. This is consistent with the LSTM-DNN structure, which adds a DNN on top of the LSTM before the softmax operation. It also shows that parameter sharing hurts the performance, no matter

Table 11.6 WER (%) results for different grid LSTMP variants on the AMI corpus

System	Priority	Shared	#layers	With overlap	No overlap
GLSTMP	Depth	No	3	49.8	40.5
GLSTMP	Depth	No	8	49.0	39.6
GLSTMP	Time	No	8	51.8	42.8
GLSTMP	Depth	Depth	3	50.0	40.5
GLSTMP	Depth	Depth	8	52.0	42.8
GLSTMP	Depth	Both	8	53.1	44.0

The SDM setup was adopted

Table 11.7 WER (%) results for different residual LSTMP variants on the AMI corpus

System	Skip	#layers	With overlap	No overlap
RLSTMP	No	3	51.3	42.0
RLSTMP	No	8	50.5	40.8
RLSTMP	Yes	16	52.3	43.1
RLSTMP	No	16	49.9	40.4
RLSTMP	No	24	50.3	41.1

The SDM setup was adopted

whether the parameters in depth (row 4) or all the parameters are shared across different layers.

11.6.2.5   Residual LSTMP

Table 11.7 summarizes the results on residual LSTMP (RLSTMP) RNNs. Although the three-layer RLSTMP performs worse than the baseline (41.7%), the accuracy improves as we go deeper. We also compared skipped and nonskipped versions of RLSTMP RNNs. It can be observed that skipping one layer performs much worse than a vanilla version. In addition, we didn't observe further gain when we went even deeper, e.g., to 24 layers, on the AMI corpus, possibly because it contains only 80 h of training data. Evaluating a deeper model on a larger corpus is planned as future work.

11.6.2.6   Summary of Results

Table 11.8 summarizes the WERs across all different models. The GLSTMP performs significantly better than the HLSTMP and RLSTMP when we untie the depth LSTM, for both three- and eight-layer conditions. In that case, the GLSTMP can be viewed as using an LSTM block to control the highway connection. However, when we go up to 16 layers, the GLSTMP cannot be well trained. We attribute this to two possible reasons: the memory from different axes has different attributes and the hyper-parameters may not be set correctly.

Table 11.8 Comparison of the highway, grid, and residual LSTMP

System	#layers	With overlap	No overlap
LSTMP	3	50.7	41.7
LSTMP	8	52.6	43.8
HLSTMP	3	50.4	41.2
HLSTMP (dr)	3	49.7	40.5
HLSTMP	8	50.7	41.3
HLSTMP	16	50.7	41.2
GLSTMP	3	49.8	40.5
GLSTMP	8	49.0	39.6
GLSTMP	16	N/A	N/A
RLSTMP	3	51.3	42.0
RLSTMP	8	50.5	40.8
RLSTMP	16	49.9	40.4

In a deeper setup, the RLSTMP is a better choice than the HLSTMP and GLSTMP: it is faster to train and degrades less in performance on this small corpus. In practice, we also found that the learning rate for the RLSTMP is more stable than for the HLTMP and GLSTMP across different layers. For the HLSTMP and GLSTMP, if we go deeper, we usually need to decrease the learning rate.

11.7 Conclusion

In this chapter, we have reviewed several advanced RNN models for ASR, with a focus on deeper structure. We first applied the PAC-RNN to an LVCSR task and analyzed the reason for the smaller gain compared to relatively smaller tasks. Inspired by recent deeper architectures, we also explored different versions of "highway" networks. The initial experimental results on the AMI corpus showed that:

- The DSR can benefit from a more advanced recurrent neural network architecture.
- The GLSTMP is the best choice if we do not need to go very deep.
- The RLSTMP has more potential in a very deeper setup although it does not perform well in a shallow configuration.

We demonstrated the effectiveness of a deeper model for the AMI SDM task. It is also interesting to evaluate it on a larger tasks, which usually can benefit more from more nonlinearity (more layers) and parameters.

References

1. Abdel-Hamid, O., Mohamed, A., Jiang, H., Deng, L., Penn, G., Yu, D.: Convolutional neural networks for speech recognition. IEEE Trans. Audio Speech Lang. Process. **22**, 1533–1545 (2014). doi:10.1109/TASLP.2014.2339736
2. Bell, P., Renals, S.: Regularization of context-dependent deep neural networks with context-independent multi-task training. In: Proceedings of ICASSP (2015)
3. Bi, M., Qian, Y., Yu, K.: Very deep convolutional neural networks for LVCSR. In: Proceedings of Annual Conference of International Speech Communication Association (INTERSPEECH) (2015)
4. Carletta, J.: Unleashing the killer corpus: experiences in creating the multi-everything AMI meeting corpus. Lang. Resour. Eval. J. **41**(2), 181–190 (2007)
5. Chen, K., Yan, Z.J., Huo, Q.: Training deep bidirectional LSTM acoustic model for LVCSR by a context-sensitive-chunk BPTT approach. In: INTERSPEECH (2015)
6. Chuangsuwanich, E., Zhang, Y., Glass, J.: Multilingual data selection for training stacked bottleneck features. In: Proceedings of ICASSP (2016)
7. Dahl, G.E., Yu, D., Deng, L., Acero, A.: Context-dependent pre-trained deep neural networks for large-vocabulary speech recognition. IEEE Trans. Audio Speech Lang. Process. **20**(1), 30–42 (2012)
8. Fiscus, J., Ajot, J., Radde, N., Laprun, C.: Multiple dimension Levenshtein edit distance calculations for evaluating ASR systems during simultaneous speech. In: LREC (2006)
9. Ghahremani, P., Droppo, J., Seltzer, M.L.: Linearly augmented deep neural network. In: Proceedings of International Conference on Acoustics, Speech and Signal Processing (ICASSP) (2016)
10. Graves, A., Jaitly, N., Mohamed, A.: Hybrid speech recognition with deep bidirectional LSTM. In: Proceedings of IEEE Workshop on Automatic Speech Recognition and Understanding (ASRU), pp. 273–278 (2013)
11. Graves, A., Mohamed, A., Hinton, G.: Speech recognition with deep recurrent neural networks. In: Proceedings of International Conference on Acoustics, Speech and Signal Processing (ICASSP) (2013)
12. Hain, T., Burget, L., Dines, J., Garner, P.N., Grzl, F., Hannani, A.E., Huijbregts, M., Karafit, M., Lincoln, M., Wan, V.: Transcribing meetings with the AMIDA systems. IEEE Trans. Audio Speech Lang. Process. **20**(2), 486–498 (2012)
13. He, K., Zhang, X., Ren, S., Sun, J.: Deep residual learning for image recognition. CoRR abs/1512.03385 (2015). http://arxiv.org/abs/1512.03385
14. Heigold, G., McDermott, E., Vanhoucke, V., Senior, A., Bacchiani, M.: Asynchronous stochastic optimization for sequence training of deep neural networks. In: ICASSP (2014)
15. Hinton, G., Deng, L., Yu, D., Dahl, G., Mohamed, A., Jaitly, N., Senior, A., Vanhoucke, V., Nguyen, P., Sainath, T., Kingsbury, B.: Deep neural networks for acoustic modeling in speech recognition: the shared views of four research groups. IEEE Signal Process. Mag. **29**(6), 82–97 (2012)
16. Hinton, G.E., Srivastava, N., Krizhevsky, A., Sutskever, I., Salakhutdinov, R.: Improving neural networks by preventing co-adaptation of feature detectors. http://arxiv.org/abs/1207.0580 (2012)
17. Hochreiter, S., Schmidhuber, J.: Long short-term memory. Neural Comput. **9**(8), 17351438 (1997)
18. IARPA: Babel program broad agency announcement, IARPA-BAA-11-02 (2011)
19. Kalchbrenner, N., Danihelka, I., Graves, A.: Grid long short-term memory. http://arXiv.org/abs/1507.01526 (2015)
20. Kumatani, K., McDonough, J.W., Raj, B.: Microphone array processing for distant speech recognition: from close-talking microphones to far-field sensors. IEEE Signal Process. Mag. **29**(6), 127–140 (2012)

21. Povey, D., Ghoshal, A., Boulianne, G., Burget, L., Glembek, O., Goel, N., Hannemann, M., Motlíček, P., Qian, Y., Schwarz, P., Silovský, J., Stemmer, G., Veselý, K.: The Kaldi speech recognition toolkit. In: ASRU (2011)
22. Sak, H., Senior, A., Beaufays, F.: Long short-term memory recurrent neural network architectures for large scale acoustic modeling. In: Fifteenth Annual Conference of the International Speech Communication Association (2014)
23. Seide, F., Li, G., Chen, X., Yu, D.: Feature engineering in context-dependent deep neural networks for conversational speech transcription. In: Proceedings of IEEE Workshop on Automatic Speech Recognition and Understanding (ASRU), pp. 24–29 (2011)
24. Seide, F., Li, G., Yu, D.: Conversational speech transcription using context-dependent deep neural networks. In: Proceedings of Annual Conference of International Speech Communication Association (INTERSPEECH), pp. 437–440 (2011)
25. Seltzer, M., Yu, D., Wang, Y.Q.: An investigation of deep neural networks for noise robust speech recognition. In: Proceedings of International Conference on Acoustics, Speech and Signal Processing (ICASSP) (2013)
26. Stolcke, A.: Making the most from multiple microphones in meeting recognition. In: ICASSP (2011)
27. Swietojanski, P., Ghoshal, A., Renals, S.: Hybrid acoustic models for distant and multichannel large vocabulary speech recognition. In: ASRU (2013)
28. Swietojanski, P., Ghoshal, A., Renals, S.: Convolutional neural networks for distant speech recognition. IEEE Signal Process. Lett. 21(9), 1120–1124 (2014). doi:10.1109/LSP.2014.2325781
29. Swietojanski, P., Bell, P., Renals, S.: Structured output layer with auxiliary targets for context-dependent acoustic modelling. In: Proceedings of INTERSPEECH (2015)
30. Williams, R., Peng, J.: An efficient gradient-based algorithm for online training of recurrent network trajectories. Neural Comput. 2, 490501 (1990)
31. Yu, D., Eversole, A., Seltzer, M., Yao, K., Guenter, B., Kuchaiev, O., Zhang, Y., Seide, F., Chen, G., Wang, H., Droppo, J., Agarwal, A., Basoglu, C., Padmilac, M., Kamenev, A., Ivanov, V., Cyphers, S., Parthasarathi, H., Mitra, B., Huang, Z., Zweig, G., Rossbach, C., Currey, J., Gao, J., May, A., Peng, B., Stolcke, A., Slaney, M., Huang, X.: An introduction to computational networks and the computational network toolkit. Microsoft Technical Report (2014)
32. Yu, D., Xiong, W., Droppo, J., Stolcke, A., Ye, G., Li, J., Zweig, G.: Deep convolutional neural networks with layer-wise context expansion and attention. In: Proceedings of Annual Conference of International Speech Communication Association (INTERSPEECH) (2016)
33. Zhang, Y., Yu, D., Seltzer, M., Droppo, J.: Speech recognition with prediction–adaptation–correction recurrent neural networks. In: Proceedings of ICASSP (2015)
34. Zhang, Y., Chen, G., Yu, D., Yao, K., Khudanpur, S., Glass, J.: Highway long short-term memory RNNS for distant speech recognition. In: Proceedings of International Conference on Acoustics, Speech and Signal Processing (ICASSP) (2016)
35. Zhang, Y., Chuangsuwanich, E., Glass, J., Yu, D.: Prediction–adaptation–correction recurrent neural networks for low-resource language speech recognition. In: Proceedings of International Conference on Acoustics, Speech and Signal Processing (ICASSP) (2016)

Chapter 12
Sequence-Discriminative Training of Neural Networks

Guoguo Chen, Yu Zhang, and Dong Yu

Abstract In this chapter we explore sequence-discriminative training techniques for neural-network–hidden-Markov-model (NN-HMM) hybrid speech recognition systems. We first review different sequence-discriminative training criteria for NN-HMM hybrid systems, including maximum mutual information (MMI), boosted, minimum phone error, and state-level minimum Bayes risk (sMBR). We then focus on the sMBR criterion, and demonstrate a few heuristics, such as denominator language model order and frame-smoothing, that may improve the recognition performance. We further propose a two-forward-pass procedure to speed up sequence-discriminative training when memory is the main constraint. Experiments were conducted on the AMI meeting corpus.

12.1 Introduction

We are now in the neural network (NN) era of automatic speech recognition. The basic deep-neural-network–hidden-Markov-model (DNN-HMM) large-vocabulary continuous speech recognition (LVCSR) system involves modeling the HMM state distribution using deep neural networks (DNNs) [6]. In such systems, DNNs are typically optimized to classify each frame into one of the states based on the cross-entropy (CE) criterion, which minimizes the expected frame error rate [29]. Speech recognition, however, is inherently a sequence classification problem. DNNs trained with the CE criterion, therefore, are suboptimal for LVCSR tasks.

Researchers have been investigating sequence-discriminative training techniques since the Gaussian-mixture-model (GMM) era of automatic speech recognition. For example, in [2, 13, 25], the maximum mutual information (MMI) criterion

G. Chen (✉)
Johns Hopkins University, Baltimore, MD, USA
e-mail: guoguo@jhu.edu

Y. Zhang
Massachusetts Institute of Technology, Cambridge, MA, USA

D. Yu
Microsoft Research, Redmond, WA, USA

© Springer International Publishing AG 2017
S. Watanabe et al. (eds.), *New Era for Robust Speech Recognition*,
DOI 10.1007/978-3-319-64680-0_12

was proposed and investigated to improve the recognizer's accuracy. The boosted MMI (BMMI), an improvement over MMI, was later proposed in [19] and further improved the recognition performance. Other sequence-discriminative training criteria, such as minimum phone error (MPE) [16] and minimum Bayes risk (MBR) [8, 12, 14, 17], also earned their reputation for speech recognition in the late 1990s and early 2000s. During that period, a state-of-the-art LVCSR system often consisted of a GMM-HMM architecture, efficiently trained with one of the above-mentioned criteria with statistics collected from lattices.

Sequence-discriminative training for feedforward NN-HMM speech recognition systems also has a long history, well before the resurgence of DNNs in speech recognition systems. It is pointed out in [26] that the "clamped" and "free" posteriors described in [3] are actually the same as the numerator and denominator occupancies described in [25], in which the authors explored the MMI-based sequence-discriminative training technique for GMM-HMM speech recognition systems. In [14], Kingsbury showed that the lattice-based sequence-discriminative training technique originally developed for GMM-HMM systems can improve the recognition accuracy over DNN-HMM systems trained with the CE criterion. This was later confirmed by follow-up works. For example, it was reported in [27] and [15] that discriminatively trained DNNs consistently improved speech recognition accuracy, although differences in criteria and implementation details may lead to slightly different empirical improvements.

The recent resurgence in using recurrent neural networks (RNNs), for example long short-term memory (LSTM) networks [10, 21, 30], in automatic speech recognition systems also motivates the development of sequence-discriminative training for RNN-based speech recognition systems. In [22], Sak et al. compared the MMI and the state-level minimum Bayes risk (sMBR) criteria for training LSTM-based acoustic models. They reported a relative word error rate (WER) reduction of 8.4% on a voice search task. This relative improvement is on a par with that of the DNN-based models. In [30], Zhang et al. applied the sMBR training criterion to a new type of RNN called the highway long short-term memory (HLSTM) network. A similar WER reduction was observed on a distant speech recognition task.

While sequence-discriminative training of neural networks may look trivial at the first glance, as it only requires changing the frame-level CE training criterion to one of the sequence-discriminative training criteria, there are quite some techniques that are needed to make it work in practice. These techniques include criterion selection, frame-smoothing, language model selection, and so on. The best configuration of these techniques depends on the implementation details [26] as well as the dataset used for training and evaluation, and needs to be optimized when building a state-of-the-art discriminatively trained NN-HMM system.

In this chapter, we first review various sequence-discriminative training criteria, including MMI, BMMI, MPE, and sMBR, in Sect. 12.2. We then discuss several techniques that may affect the performance of sequence-discriminative training in practice in Sect. 12.3. In Sect. 12.4, we further propose a two-forward-pass procedure to increase parallelization of sequence-discriminative training on a single graphics processing unit (GPU) when memory is the main constraint. Finally, we demonstrate the performance of the techniques in Sect. 12.6 on a distant speech recognition task.

12.2 Training Criteria

The most commonly used sequence-discriminative training criteria for NN-based speech recognition systems include MMI [2, 13, 25], BMMI [19], MPE [16], and sMBR [8, 12, 14, 17]. Before describing these techniques in detail, we first define several notations used in the following subsections:

- T_m: the total number of frames in the mth utterance;
- N_m: the total number of words in the mth utterance;
- θ: model parameter;
- κ: the acoustic scaling factor;
- $\mathbb{S} = \{(\mathbf{o}^m, \mathbf{w}^m) \,|\, 0 \leq m < M\}$: training set;
- $\mathbf{o}^m = \mathbf{o}_1^m, \ldots, \mathbf{o}_t^m, \ldots, \mathbf{o}_{T_m}^m$: the observation sequence of the mth utterance;
- $\mathbf{w}^m = \mathbf{w}_1^m, \ldots, \mathbf{w}_t^m, \ldots, \mathbf{w}_{N_m}^m$: the correct word transcription of the mth utterance;
- $\mathbf{s}^m = s_1^m, \ldots, s_t^m, \ldots, s_{T_m}^m$: sequence of states corresponding to $\mathbf{w}^m$.

12.2.1 Maximum Mutual Information

MMI criterion [2, 13] used in automatic speech recognition systems aims at maximizing the mutual information between the distributions of the observation sequence and the word sequence, which is highly correlated to minimizing the expected sentence error. The MMI objective function can be written as follows:

$$J_{\text{MMI}} (\theta; \mathbb{S}) = \sum_{m=1}^{M} J_{\text{MMI}} (\theta; \mathbf{o}^m, \mathbf{w}^m)$$

$$= \sum_{m=1}^{M} \log P (\mathbf{w}^m | \mathbf{o}^m; \theta)$$

$$= \sum_{m=1}^{M} \log \frac{p (\mathbf{o}^m | \mathbf{s}^m; \theta)^{\kappa} P (\mathbf{w}^m)}{\sum_{\mathbf{w}} p (\mathbf{o}^m | \mathbf{s}^w; \theta)^{\kappa} P (\mathbf{w})}, \quad (12.1)$$

where the sum in the denominator is supposed to be taken over all possible word sequences. It is, of course, not practical to enumerate all possible word sequences. Therefore, in practice the sum is taken over all possible word sequences in the lattices generated by decoding the mth utterance with the model. In the case of neural networks, computing the gradient of (12.1) with respect to the model parameter θ gives

$$\nabla_\theta J_{\text{MMI}} \left(\theta; \mathbf{o}^m, \mathbf{w}^m\right) = \sum_m \sum_t \nabla_{\mathbf{z}^L_{mt}} J_{\text{MMI}} \left(\theta; \mathbf{o}^m, \mathbf{w}^m\right) \frac{\partial \mathbf{z}^L_{mt}}{\partial \theta}$$

$$= \sum_m \sum_t \ddot{\mathbf{e}}^L_{mt} \frac{\partial \mathbf{z}^L_{mt}}{\partial \theta}, \qquad (12.2)$$

where $\mathbf{z}^L_{mt}$ is the pre-softmax activation for utterance m at frame t, and $\ddot{\mathbf{e}}^L_{mt}$ is the error signal, which can be computed further as

$$\ddot{\mathbf{e}}^L_{mt}(i) = \nabla_{\mathbf{z}^L_{mt}(i)} J_{\text{MMI}} \left(\theta; \mathbf{o}^m, \mathbf{w}^m\right)$$

$$= \sum_r \frac{\partial J_{\text{MMI}} \left(\theta; \mathbf{o}^m, \mathbf{y}^m\right)}{\partial \log p \left(\mathbf{o}^m_t | r\right)} \frac{\partial \log p \left(\mathbf{o}^m_t | r\right)}{\partial \mathbf{z}^L_{mt}(i)}$$

$$= \sum_r \kappa \left(\delta \left(r = s^m_t\right) - \ddot{\gamma}^{\text{DEN}}_{mt}(r)\right) \frac{\partial \log \mathbf{v}^L_{mt}(r)}{\partial \mathbf{z}^L_{mt}(i)}$$

$$= \kappa \left(\delta \left(i = s^m_t\right) - \ddot{\gamma}^{\text{DEN}}_{mt}(i)\right), \qquad (12.3)$$

where $\ddot{\mathbf{e}}^L_{mt}(i)$ is the ith element of the error signal, and $\ddot{\gamma}^{\text{DEN}}_{mt}(r)$ is the posterior probability of being in state r at time t, computed over the denominator lattices of the mth utterance.

12.2.2   Boosted Maximum Mutual Information

A famous variant of the MMI criterion is the BMMI criterion described in [19], where Povey et al. introduced a boosting term to boost the likelihood of paths with more errors. The BMMI criterion can be written as follows:

$$J_{\text{BMMI}} \left(\theta; \mathbb{S}\right) = \sum_{m=1}^M J_{\text{BMMI}} \left(\theta; \mathbf{o}^m, \mathbf{w}^m\right)$$

$$= \sum_{m=1}^M \log \frac{P \left(\mathbf{w}^m | \mathbf{o}^m\right)}{\sum_\mathbf{w} P \left(\mathbf{w} | \mathbf{o}^m\right) e^{-bA(\mathbf{w}, \mathbf{w}^m)}} \qquad (12.4)$$

$$= \sum_{m=1}^M \log \frac{p \left(\mathbf{o}^m | \mathbf{s}^m\right)^\kappa P \left(\mathbf{w}^m\right)}{\sum_\mathbf{w} p \left(\mathbf{o}^m | \mathbf{s}^w\right)^\kappa P \left(\mathbf{w}\right) e^{-bA(\mathbf{w}, \mathbf{w}^m)}},$$

where $e^{-bA(\mathbf{w}, \mathbf{w}^m)}$ is the boosting term. Comparing (12.1) with (12.4), we can see that the only difference between MMI and BMMI is the boosting term in the denominator. Now let's take a closer look at the boosting term. The b in the boosting term is called the boosting factor, which controls the strength of the boosting. The $A(\mathbf{w}, \mathbf{w}^m)$ part in the boosting term defines the accuracy of two sequences $\mathbf{w}$ and $\mathbf{w}^m$. There are several choices of the sequences to compute the accuracy. For example, they can be word sequences, phoneme sequences, or even state sequences. Similarly to MMI, the error signal of the BMMI objective can be derived as

$$\ddot{\mathbf{e}}_{mt}^L(i) = \nabla_{\mathbf{z}_{mt}^L(i)} J_{\text{BMMI}}(\theta; \mathbf{o}^m, \mathbf{w}^m)$$

$$= \kappa \left(\delta \left(i = s_t^m \right) - \ddot{\gamma}_{mt}^{\text{DEN}}(i) \right), \tag{12.5}$$

where $\ddot{\gamma}_{mt}^{\text{DEN}}(i)$ is the posterior probability of being in state r at time t. The difference between the error signal in (12.3) and that in (12.5) lies in the computation of $\ddot{\gamma}_{mt}^{\text{DEN}}(i)$. In the case of BMMI, $\ddot{\gamma}_{mt}^{\text{DEN}}(i)$ also contains the boosting term $e^{-bA(\mathbf{w}, \mathbf{w}^m)}$.

12.2.3 Minimum Phone Error/State-Level Minimum Bayes Risk

The MPE criterion [17], as indicated by its name, attempts to minimize the expected phone error rate. Similarly, the sMBR criterion [14] aims at minimizing the HMM state error rate. Both criteria belong to the more general MBR objective function family [8, 12], which can be written as

$$J_{\text{MBR}}(\theta; \mathbb{S}) = \sum_{m=1}^{M} J_{\text{MBR}}(\theta; \mathbf{o}^m, \mathbf{w}^m)$$

$$= \sum_{m=1}^{M} \sum_{\mathbf{w}} P(\mathbf{w}|\mathbf{o}^m) A(\mathbf{w}, \mathbf{w}^m)$$

$$= \sum_{m=1}^{M} \frac{\sum_{\mathbf{w}} p(\mathbf{o}^m|\mathbf{s}^w)^\kappa P(\mathbf{w}) A(\mathbf{w}, \mathbf{w}^m)}{\sum_{\mathbf{w}'} p(\mathbf{o}^m|\mathbf{s}^{w'})^\kappa P(\mathbf{w}')}, \tag{12.6}$$

where $A(\mathbf{w}, \mathbf{w}^m)$ is the distinguishing factor in the MBR family. It is the accuracy of $\mathbf{w}$ measured against $\mathbf{w}^m$, and essentially defines what kind of "error" the objective function is trying to minimize. In the case of MPE, $\mathbf{w}$ and $\mathbf{w}^m$ should be correct and observed phone sequences, while for sMBR they correspond to state sequences. The error signal of the general MBR objective is

$$\ddot{\mathbf{e}}_{mt}^L(i) = \nabla_{\mathbf{z}_{mt}^L(i)} J_{\text{MBR}}(\theta; \mathbf{o}^m, \mathbf{w}^m)$$

$$= \sum_r \frac{\partial J_{\text{MBR}}(\theta; \mathbf{o}^m, \mathbf{w}^m)}{\partial \log p(\mathbf{o}_t^m|r)} \frac{\partial \log p(\mathbf{o}_t^m|r)}{\partial \mathbf{z}_{mt}^L(i)}$$

$$= \sum_r \kappa \dddot{\gamma}_{mt}^{\text{DEN}}(r) \left(\bar{A}^m \left(r = s_t^m \right) - \bar{A}^m \right) \frac{\partial \log \mathbf{v}_{mt}^L(r)}{\partial \mathbf{z}_{mt}^L(i)}$$

$$= \kappa \dddot{\gamma}_{mt}^{\text{DEN}}(i) \left(\bar{A}^m \left(i = s_t^m \right) - \bar{A}^m \right), \tag{12.7}$$

where $\bar{A}^m$ is the average accuracy of all paths in the lattice, $\bar{A}^m \left(r = s_t^m \right)$ is the average accuracy of all paths in the lattice for utterance m that pass through state r at time t, $\mathbf{z}_{mt}^L$ is the pre-softmax activation for utterance m at frame t, and $\dddot{\gamma}_{mt}^{\text{DEN}}(r)$ is the MBR posterior.

12.3 Practical Training Strategy

Building a state-of-the-art discriminatively trained NN-HMM speech recognition system requires, during the training process, optimizing various configurations such as criterion selection, frame-smoothing, and lattice generation. While some of these techniques only help on certain tasks or datasets, others, for example frame-smoothing, generally help to stabilize the training and thus improve the recognition accuracy. We review several training strategies that are effective in practice in this section.

12.3.1 Criterion Selection

Different observations have been made on different speech recognition tasks with regard to the relative effectiveness of various sequence-discriminative training criteria. For example, the sMBR criterion was shown to be superior in [14] and [15], while the authors of [27] suggested that the MMI criterion outperforms the MPE criterion on their particular task. In general, most of the observations seem to suggest that the sMBR training criterion usually provides the best performance.

Table 12.1 is summarized from [26], where Veselý et al. compared all the sequence-discriminative training criteria mentioned in Sect. 12.2 for DNNs, evaluated on the 300-h Switchboard conversational telephone speech transcription task. From this table, it is clear that DNNs trained with the sMBR criterion give the best

Table 12.1 Performance (% WER) comparison of DNNs trained with different sequence-discriminative training criteria, on the 300-h Switchboard conversational telephone speech task (summarized from [26])

System	Hub5'00 SWB	Hub5'01 SWB
DNN CE	14.2	14.5
DNN MMI	12.9	13.3
DNN BMMI	12.9	13.2
DNN sMBR	12.6	13.0
DNN MPE	12.9	13.2

Table 12.2 Performance (% WER) comparison of LSTMs trained with different sequence-discriminative training criteria, starting from different CE models, on a voice search task (summarized from [22])

System	Model 1	Model 2	Model 3	Model 4	Model 5
DNN CE	15.9	14.9	12.0	11.2	10.7
DNN MMI	13.8	12.0	10.8	10.8	10.5
DNN sMBR	–	–	10.7	10.3	9.8

performance on both evaluation sets, although the performance difference between different criteria is small.

Table 12.2 is summarized from [22], where Sak et al. developed discriminatively trained LSTMs for acoustic modeling. They compared MMI and sMBR sequence-discriminative training on top of CE LSTM models trained at different stages (with a switch early from CE to sequence level), and showed that the sMBR criterion consistently outperformed the MMI criterion on their internal voice search task, especially when the CE model had a lower error rate.

Since most of the observations seem to suggest that the sMBR criterion is superior to other criteria, we recommend to use the sMBR criterion as the default and compare it with the MMI criterion on your specific task if resources permit. For the same reason, we mainly report results in later sections using the sMBR training criterion in our experiments.

12.3.2 Frame-Smoothing

One issue that often occurs in the training of neural networks with sequence-discriminative training criteria is the overfitting problem. It is often demonstrated as having an improving sequence-discriminative training objective, and a dramatically decreasing frame accuracy.

In [24] Su et al. observed that even the fattest lattices could generate denominator supervision with only about 300 senones per frame, out of 9304. This suggests that the overfitting problem might be caused by the sparse lattice. In [29], however, Yu and Deng argued that the problem might be rooted in the fact that the sequences are in a higher-dimensional space than the frames, which makes the posterior distribution estimated from the training set deviate from that of the testing set.

In any case, in order to alleviate this problem, one can make the sequence-discriminative training criterion closer to the cross entropy training criterion, for example, by using a weaker language model such as a unigram language model when generating the lattices. In [24] Su et al. proposed a technique called frame-smoothing (F-smoothing), which essentially introduced a new training objective that interpolates the sequence-discriminative training objective with the cross-entropy

Table 12.3 Performance (% WER) comparison of DNNs trained with MMI criterion, with and without frame-smoothing (F-smoothing), on the 300-h Switchboard conversational telephone speech training set, and the Hub5'00 evaluation set (summarized from [24])

System	Number of MMI epochs				
	1	2	3	4	5
DNN CE	15.6				
DNN MMI	14.4	14.3	14.3	14.2	14.3
DNN MMI + F-smoothing	13.9	13.7	13.8	13.6	13.8

objective as follows:

$$J_{\text{FS-SEQ}}(\theta; \mathbb{S}) = (1 - \alpha) J_{\text{CE}}(\theta; \mathbb{S}) + \alpha J_{\text{SEQ}}(\theta; \mathbb{S}), \quad (12.8)$$

where α is the smoothing factor. Similar ideas have also been developed for discriminative training of GMM models, for example, the H-criterion technique in [9] and the I-smoothing technique in [18].

Table 12.3 is summarized from [24], where the authors compared MMI training with and without frame-smoothing, on the 300-h Switchboard conversational telephone speech training set and the Hub5'00 evaluation set. It is clear from the table that frame-smoothing helps improve the speech recognition performance. In [24] Su et al. found it helpful to use a frame/sequence ratio of 1:4 or 1:10. In the later sections, we set the frame/sequence ratio to 1:10 as that gave the best performance on the development set in our experiments.

12.3.3 Lattice Generation

The lattice generation process plays an important role in sequence-discriminative training of neural networks. The conventional wisdom is to use the best available CE neural network model to generate both the numerator lattice and the denominator lattice, and then train the CE model with sequence-discriminative training criteria [24].

12.3.3.1 Numerator Lattice

In practice, the numerator lattice often reduces to the forced alignment of the training transcription. It was demonstrated an several occasions that both CE training and sequence-discriminative training can benefit from using better alignments.

Table 12.4 is extracted from [24], where the authors compared sequence-discriminative training performance with alignments of different qualities. In this table, "DNN1 CE" indicates a DNN CE model trained with alignment with a GMM model, and "DNN2 CE" indicates a DNN CE model trained with alignment with

Table 12.4 Performance (% WER) comparison of DNN CE and MMI models trained with different alignments, on the 300-h Switchboard conversational telephone speech task (summarized from [24])

System	WER
GMM	–
DNN1 CE (alignment generated from GMM)	16.2
DNN2 CE (alignment generated from DNN1 CE)	15.6
DNN1 MMI (alignment DNN1 CE)	14.1
DNN2 MMI (alignment DNN2 CE)	13.5

Table 12.5 Performance (% WER) comparison of LSTM CE and sMBR models trained with different alignments, on a voice search task (summarized from [22])

System	WER
DNN CE	–
LSTM1 CE (alignment generated from DNN CE)	10.7
LSTM2 CE (alignment generated from LSTM1 CE)	10.7
LSTM1 sMBR (alignment generated online)	10.1
LSTM2 sMBR (alignment generated online)	10.0

a DNN1 CE model. The "DNN1 MMI" model is a model trained with the MMI criterion using alignment with the DNN1 CE model as numerator lattice, and the "DNN2 MMI" model is a model trained with the MMI criterion using alignment with the DNN2 CE model. It is clear from the table that both CE and MMI training benefit from forced alignment generated with better acoustic models.

Table 12.5 is summarized from [22], where Sak et al. conducted a similar set of experiments, but this time for LSTMs instead of DNNs. Since the authors implemented an asynchronous stochastic gradient descent framework for parameter updating, they generated the numerator lattice (alignment) online right before lattice computation and parameter updating. The observation is similar to what the authors reported in [24], that sMBR training benefits from forced alignment generated with better acoustic models, but very marginally.

12.3.3.2   Denominator Lattice

It was suggested in [16] that it is important to generate denominator lattices with a unigram language model trained on a training transcription for sequence-discriminative training of GMM-HMM systems. It was reiterated in [29] that the same weak language model should be used to generate denominator lattices in the case of sequence-discriminative training of neural networks. Recent work in the literature, however, indicates that the selection of language models for generating the denominator lattice is also task dependent.

Table 12.6 is summarized from [22], where the authors explored language models with different modeling power when generating denominator lattices for sequence-discriminative training. The table suggests that a bigram language model leads to the best performance in their particular task. In our experiments in later sections, we will also explore language models with different modeling powers for denominator lattice generation.

Table 12.6 Performance (% WER) comparison of LSTMs trained with the sMBR criterion, with denominator lattices generated using different language models, on a voice search task (summarized from [22])

System	Denominator language model		
	Unigram	Bigram	Trigram
LSTM CE	10.7		
LSTM sMBR	10.9	10.0	10.1

In most sequence-discriminative training setups, denominator lattices are only generated once and are reused across training epochs, since it is relatively expensive to generate lattices. In [26], the authors indicated that further improvement can be achieved by regenerating lattices after one or two sequence-discriminative training epochs. In our experiments in later sections, we also explore the benefits of lattice regeneration.

12.4 Two-Forward-Pass Method for Sequence Training

Nowadays, GPUs are widely used in deep learning due to their massive parallel computation power. For the cross-entropy training of basic DNN models, since back-propagation does not depend on future or past data samples, the parallelization ability of GPUs can be fully utilized. For unidirectional RNNs, the truncated back-propagation-through-time (BPTT) algorithm is commonly used for parameter updating, and multiple (e.g., 40) sequences are often packed into the same mini-batch (e.g., in [1]) to better utilize the GPU's parallelization ability. This is possible because only a small segment (e.g., 20 frames) of each sequence has to be packed into the same minibatch. For training bidirectional RNNs, or sequence-discriminative training of memory-demanding neural networks, however, since whole-sequence-based BPTT is often used, the number of sequences that can be packed into the same minibatch is usually quite restricted due to the GPU's memory limit. This significantly decreases the training and evaluation speed.

One way to speed up the training in this case is to use asynchronous SGD on a GPU/CPU farm [11], at the cost of low computing resource utilization on each a GPU/CPU. This solution is of course not ideal, and GPU/CPU farm can be quite expensive to build and maintain.

For bidirectional RNNs, various techniques have been proposed to improve resource utilization on GPUs, for example, the context-sensitive-chunk BPTT (CSC-BPTT) proposed in [5] and the latency-controlled method proposed in [30]. Those techniques, however, cannot be directly applied to the sequence-discriminative training of RNNs and other memory-hungry models such as deep convolutional neural networks (CNNs), since the signal computation itself requires having posteriors for the whole sequence.

Algorithm 1 Two-forward-pass sequence training

1: **procedure** TwoForwardPassSequenceTraining()
2:    $\mathscr{S} \leftarrow$ Sequences
3:    $\mathscr{A} \leftarrow$ Alignments corresponding to $\mathscr{S}$
4:    $\mathscr{D} \leftarrow$ Denominator lattices corresponding to $\mathscr{U}$
5:    $M \leftarrow MinibatchReader(\mathscr{S})$ ▷ (E.g., 40 sequences, each with 20 frames)
6:    $P \leftarrow SequencePool(\mathscr{S}, \mathscr{A}, \mathscr{D})$
7:    **for** *all* $m \in M$ **do**
8:        **if** *P.HasGradient*(*m*) **then**
9:            $g \leftarrow P.Gradient(m)$
10:           *forward_pass*(*m*)
11:           *set_output_node_gradient*(*g*)
12:           *backward_pass*(*m*)
13:           *parameter_update*()
14:       **else**
15:           $m\_p \leftarrow M.CurrentMinibatchPointer()$
16:           **while** *P.NeedMoreMinibatch*() **do**
17:               $m_1 \leftarrow m\_p.ReadMinibatch()$
18:               $p \leftarrow forward\_pass(m_1)$ ▷ Posterior from forward pass
19:               $P.ComputeGradient(m_1, p)$
20:           $M.ResetMinibatchPointer(m\_p)$

In this section, we propose a two-forward-pass method for efficient sequence-discriminative training of memory-hungry models. Algorithm 1 demonstrates the pseudocode of the proposed method. The general idea is to enable partial sequences in each minibatch for sequence-discriminative training. For this to happen, we have to maintain a sequence pool, and compute the gradient for those sequences from their corresponding lattices in advance. This requires an additional forward pass so that gradients from lattices can be computed at sequence level and be stored in the sequence pool, thus the name "two-forward-pass." After preparing the gradients in the sequence pool, sequences can again be split into small segments (e.g., 20 frames), and segments from multiple sequences (e.g., 40) can be packed into the same minibatch for efficient parameter updating.

12.5   Experiment Setup

We used the AMI [4] corpus for all our experiments, and Kaldi [20] and the Computational Network Toolkit (CNTK) [1] for system building. Details of the corpus and system descriptions are as follows.

12.5.1 Corpus

The AMI corpus comprises around 100 h of meeting recordings, recorded in instrumented meeting rooms. Multiple microphones were used, including individual headset microphones (IHMs), lapel microphones, and one or more microphone arrays. In this work, we used the single-distant-microphone (SDM) condition in our experiments. Our systems were trained and tested using the split recommended in the corpus release: a training set of 80 h, and a development set and a test set each of 9 h. For our training, we used all the segments provided by the corpus, including those with overlapping speech. Our models were evaluated on the evaluation set only. NIST's asclite tool [7] was used for scoring.

12.5.2 System Description

Kaldi [20] was used for feature extraction, and early-stage triphone training, as well as decoding. A maximum likelihood acoustic training recipe was used to train a GMM-HMM triphone system. Forced alignment was performed on the training data by this triphone system to generate labels for further neural network training.

CNTK [1] was used for neural network training. We started off by training a 6-layer DNN, with 2048 sigmoid units per layer. 40-dimensional filter bank features, together with their corresponding delta and delta–delta features, were used as raw feature vectors. For our DNN training we concatenated 15 frames of raw feature vectors, which leads to a dimension of 1800. This DNN was again used to force-align the training data to generate labels for further LSTM training.

Our (H)LSTM models, unless explicitly stated otherwise, were added with a projection layer on top of each layer's output (we refer to this as LSTMP here), as proposed in [21], and were trained with 80-dimensional log mel filterbank (FBANK) features. For LSTMP models, each hidden layer consisted of 1024 memory cells together with a 512-node projection layer. For the bidirectional LSTMP (BLSTMP) models, each hidden layer consisted of 1024 memory cells (512 for forward and 512 for backward) with a 300-node projection layer. Their highway companions shared the same network structure, except for the additional highway connections as proposed in [30, 31].

All models were randomly initialized without either generative or discriminative pretraining [23]. A validation set was used to control the learning rate, which was halved when no gain was observed. To train the unidirectional recurrent model, the truncated BPTT [28] was used to update the model parameters. Each BPTT segment contained 20 frames, and we processed 40 utterances simultaneously. To train the bidirectional model, the latency-controlled method proposed in [30] was applied. We set $N_c = 22$ and $N_r = 21$ and also processed 40 utterances simultaneously. To train the recurrent model with the sMBR criterion, we adopted the two-forward-pass method described in Sect. 12.4, and processed 40 utterances simultaneously.

12.6   Evaluation

The performance of various models was evaluated using WER in percent as below. Experiments were conducted on the full evaluation set, including utterances with overlapping speech segments.

12.6.1   Practical Strategy

We evaluated different strategies described in Sect. 12.3 for a DNN acoustic model, trained with the sMBR criterion. Table 12.7 illustrates the WER performance of various techniques on the AMI SDM task. From this table, we can see that we get an additional 3.9% relative WER reduction by adding F-smoothing to the sMBR training objective. This is consistent with what Su et al. observed in [24]. We also achieved another 2% WER reduction by switching from the unigram language model to the bigram language model when generating denominator lattices. This echoes the results in [22], although a unigram language model was suggested in [16] and [29].

In the table, "realign" refers to the case where we realigned the training transcription with the DNN CE model, and performed further CE training before sMBR training, as described in Sect. 12.3.3.1. "Regenerate" refers to the case where denominator lattices were regenerated by the sMBR model after sweeping the data once, as mentioned in Sect. 12.3.3.2. Unfortunately, these two strategies only lead to minor improvement in terms of WER reduction.

12.6.2   Two-Forward-Pass Method

We evaluated the proposed two-forward-pass method for the sMBR training of (B)LSTMP and (B)HLSTMP recurrent neural networks. For a detailed description

Table 12.7 Performance (% WER) comparison of DNN sMBR models trained with different techniques described in Sect. 12.3, on AMI SDM task

System	WER
DNN CE	55.9
DNN + sMBR + unigram LM	54.4
DNN + sMBR + unigram LM + F-smoothing	52.4
DNN + sMBR + bigram LM + F-smoothing	51.3
DNN + sMBR + bigram LM + F-smoothing + realign	51.2
DNN + sMBR + bigram LM + F-smoothing + realign + regenerate	51.2

LM language model

of those recurrent neural networks, readers are referred to [30]. Following the observations in Sect. 12.6.1, we used a bigram language model trained on a training transcription for denominator lattice generation, and we applied F-smoothing to the objective function. Since "realign" and "regenerate" are quite expensive processes, and only lead to tiny improvements in our particular setup, we did not apply them in the following evaluation.

12.6.2.1 Speed

The motivation of the two-forward-pass method is to allow more utterance parallelization in each minibatch when performing sequence-discriminative training for recurrent neural networks. The conventional whole-utterance approach to sequence-discriminative training of recurrent neural networks often limits the number of utterances that can be processed in the same minibatch due to the GPU's memory limit. For example, on the NVIDIA Grid K520 GPUs that we experimented with, we were only able to parallelize at most four utterances in the same minibatch for our given LSTMP network structure.

Table 12.8 compares the training time of the conventional whole-utterance approach without multiutterance parallelization and our proposed two-forward-pass method with 40 utterances processed in the same minibatch. Since the training in the conventional whole-utterance approach is quite time-consuming, we conducted the comparison on a $10K$ utterance subset of the SDM task. From the table, we can see that we can get an $18\times$ speedup by using our proposed two-forward-pass method. Further speed improvement is possible by increasing the number of utterances processed in the same minibatch. For example, in our experiments, we have processed 80 utterances in the same minibatch without hurting the performance.

12.6.2.2 Performance

Table 12.9 reports the WER of sMBR models trained with the proposed two-forward-pass method. The "LSTMP" in this table refers to the LSTM model with projection layers, while "BLSTMP" is its bidirectional version. The "HLSTMP" in the table indicates the highway LSTMP model proposed in [30], while "BHLSTMP" is its bidirectional companion.

Table 12.8 Speed performance (hours per epoch) comparison of LSTMP sMBR training with and without parallelization in each minibatch

	#utterances in each minibatch	
System	1	40
LSTMP sMBR	13.7	0.75

The experiment was conducted on a $10K$ utterance subset of the AMI SDM task, with NVIDIA Grid K520 GPUs

Table 12.9 Performance
(% WER) comparison of
DNN sMBR models trained
with different techniques
described in Sect. 12.3, on
AMI SDM task

System	WER
LSTMP CE	50.7
LSTMP sMBR	49.3
BLSTMP CE	47.3[a]
BLSTMP sMBR	45.6[a]
HLSTMP CE	49.7
HLSTMP sMBR	47.7
BHLSTMP CE	47.9[a]
BHLSTMP sMBR	45.4[a]

[a]Experiments were conducted after the JSALT15 workshop with the latest CNTK, which may give slightly better results than what we obtained at the workshop

It is clear from the table that the sMBR sequence-discriminative training criterion consistently improves upon the cross-entropy model.

12.7   Conclusion

In this chapter, we have reviewed several criteria for sequence-discriminative training. We also conducted a survey regarding practical strategies that might help improve the sequence training in practice. The combination of the survey and what we observed from our own experiments on the AMI SDM task seems to suggest that:

- Sequence-discriminative training often benefits from frame-smoothing, since it helps to prevent overfitting and also stabilizes the training.
- The language model used to generate the denominator lattices may have an effect on the performance, but this is usually task dependent.

We further proposed a two-forward-pass method for sequence-discriminative training of memory-hungry neural networks, which enables more utterance parallelization in each minibatch and dramatically decreases the training time. We demonstrated the effectiveness of this method on the AMI SDM task with sMBR training of various recurrent neural networks.

References

1. Agarwal, A., Akchurin, E., Basoglu, C., Chen, G., Cyphers, S., Droppo, J., Eversole, A., Guenter, B., Hillebrand, M., Hoens, R., Huang, X., Huang, Z., Ivanov, V., Kamenev, A., Kranen, P., Kuchaiev, O., Manousek, W., May, A., Mitra, B., Nano, O., Navarro, G., Orlov, A., Parthasarathi, H., Peng, B., Padmilac, M., Reznichenko, A., Seide, F., Seltzer, M.L., Slaney, M., Stolcke, A., Wang, Y., Wang, H., Yao, K., Yu, D., Zhang, Y., Zweig, G.: An introduction to computational networks and the computational network toolkit. Technical Report MSR-TR-2014-112, Microsoft Research (2014)
2. Bahl, L., Brown, P.F., De Souza, P.V., Mercer, R.L.: Maximum mutual information estimation of hidden Markov model parameters for speech recognition. In: Proceedings of the International Conference on Acoustics, Speech and Signal Processing (ICASSP), vol. 86, pp. 49–52 (1986)
3. Bridle, J., Dodd, L.: An Alphanet approach to optimising input transformations for continuous speech recognition. In: Proceedings of the International Conference on Acoustics, Speech and Signal Processing (ICASSP), pp. 277–280. IEEE (1991)
4. Carletta, J.: Unleashing the killer corpus: experiences in creating the multi-everything AMI Meeting Corpus. Lang. Resour. Eval. **41**(2), 181–190 (2007)
5. Chen, K., Huo, Q.: Training deep bidirectional LSTM acoustic model for LVCSR by a context-sensitive-chunk BPTT approach. IEEE/ACM Trans. Audio Speech Lang. Process. **24**(7), 1185–1193 (2016)
6. Dahl, G.E., Yu, D., Deng, L., Acero, A.: Context-dependent pre-trained deep neural networks for large-vocabulary speech recognition. IEEE Trans. Audio Speech Lang. Process. **20**(1), 30–42 (2012)
7. Fiscus, J.G., Ajot, J., Radde, N., Laprun, C.: Multiple dimension Levenshtein edit distance calculations for evaluating automatic speech recognition systems during simultaneous speech. In: Proceedings of the International Conference on Language Resources and Evaluation (LERC) (2006)
8. Gibson, M., Hain, T.: Hypothesis spaces for minimum Bayes risk training in large vocabulary speech recognition. In: Proceedings of INTERSPEECH (2006)
9. Gopalakrishnan, P., Kanevsky, D., Nadas, A., Nahamoo, D., Picheny, M.: Decoder selection based on cross-entropies. In: Proceedings of the International Conference on Acoustics, Speech and Signal Processing (ICASSP), pp. 20–23. IEEE, New York (1988)
10. Graves, A., Jaitly, N., Mohamed, A.R.: Hybrid speech recognition with deep bidirectional LSTM. In: Proceedings of Automatic Speech Recognition and Understanding (ASRU), pp. 273–278. IEEE, New York (2013)
11. Heigold, G., McDermott, E., Vanhoucke, V., Senior, A., Bacchiani, M.: Asynchronous stochastic optimization for sequence training of deep neural networks. In: Proceedings of the International Conference on Acoustics, Speech and Signal Processing (ICASSP), pp. 5587–5591. IEEE, New York (2014)
12. Kaiser, J., Horvat, B., Kacic, Z.: A novel loss function for the overall risk criterion based discriminative training of HMM models. In: Proceedings of the Sixth International Conference on Spoken Language Processing (2000)
13. Kapadia, S., Valtchev, V., Young, S.: MMI training for continuous phoneme recognition on the TIMIT database. In: Proceedings of the International Conference on Acoustics, Speech and Signal Processing (ICASSP), vol. 2, pp. 491–494. IEEE, New York (1993)
14. Kingsbury, B.: Lattice-based optimization of sequence classification criteria for neural-network acoustic modeling. In: Proceedings of the International Conference on Acoustics, Speech and Signal Processing (ICASSP), pp. 3761–3764. IEEE, New York (2009)
15. Kingsbury, B., Sainath, T.N., Soltau, H.: Scalable minimum Bayes risk training of deep neural network acoustic models using distributed Hessian-free optimization. In: Proceedings of INTERSPEECH (2012)

16. Povey, D.: Discriminative training for large vocabulary speech recognition. Ph.D. thesis, University of Cambridge (2005)
17. Povey, D., Kingsbury, B.: Evaluation of proposed modifications to MPE for large scale discriminative training. In: Proceedings of the International Conference on Acoustics, Speech and Signal Processing (ICASSP), vol. 4, pp. IV-321. IEEE, New York (2007)
18. Povey, D., Woodland, P.C.: Minimum phone error and I-smoothing for improved discriminative training. In: Proceedings of the International Conference on Acoustics, Speech and Signal Processing (ICASSP), vol. 1, pp. I-105. IEEE, New York (2002)
19. Povey, D., Kanevsky, D., Kingsbury, B., Ramabhadran, B., Saon, G., Visweswariah, K.: Boosted MMI for model and feature-space discriminative training. In: Proceedings of the International Conference on Acoustics, Speech and Signal Processing (ICASSP), pp. 4057–4060. IEEE, New York (2008)
20. Povey, D., Ghoshal, A., Boulianne, G., Burget, L., Glembek, O., Goel, N., Hannemann, M., Motlicek, P., Qian, Y., Schwarz, P., et al.: The Kaldi speech recognition toolkit. In: Proceedings of Automatic Speech Recognition and Understanding (ASRU), EPFL-CONF-192584. IEEE Signal Processing Society, Piscataway (2011)
21. Sak, H., Senior, A., Beaufays, F.: Long short-term memory based recurrent neural network architectures for large vocabulary speech recognition (2014). arXiv preprint arXiv:1402.1128
22. Sak, H., Vinyals, O., Heigold, G., Senior, A., McDermott, E., Monga, R., Mao, M.: Sequence discriminative distributed training of long short-term memory recurrent neural networks. In: Proceedings of INTERSPEECH (2014)
23. Seide, F., Li, G., Chen, X., Yu, D.: Feature engineering in context-dependent deep neural networks for conversational speech transcription. In: Proceedings of Automatic Speech Recognition and Understanding (ASRU), pp. 24–29. IEEE, New York (2011)
24. Su, H., Li, G., Yu, D., Seide, F.: Error back propagation for sequence training of context-dependent deep networks for conversational speech transcription. In: Proceedings of the International Conference on Acoustics, Speech and Signal Processing (ICASSP), pp. 6664–6668. IEEE, New York (2013)
25. Valtchev, V., Odell, J., Woodland, P.C., Young, S.J.: MMIE training of large vocabulary recognition systems. Speech Commun. 22(4), 303–314 (1997)
26. Veselý, K., Ghoshal, A., Burget, L., Povey, D.: Sequence-discriminative training of deep neural networks. In: Proceedings of INTERSPEECH, pp. 2345–2349 (2013)
27. Wang, G., Sim, K.C.: Sequential classification criteria for NNs in automatic speech recognition. In: Proceedings of INTERSPEECH (2011)
28. Williams, R.J., Peng, J.: An efficient gradient-based algorithm for on-line training of recurrent network trajectories. Neural Comput. 2(4), 490–501 (1990)
29. Yu, D., Deng, L.: Automatic Speech Recognition, pp. 137–153. Springer, London (2015)
30. Zhang, Y., Chen, G., Yu, D., Yao, K., Khudanpur, S., Glass, J.: Highway long short-term memory RNNs for distant speech recognition. In: Proceedings of the International Conference on Acoustics, Speech and Signal Processing (ICASSP). IEEE, New York (2016)
31. Zilly, J.G., Srivastava, R.K., Koutník, J., Schmidhuber, J.: Recurrent highway networks (2016). arXiv preprint arXiv:1607.03474

Chapter 13
End-to-End Architectures for Speech Recognition

Yajie Miao and Florian Metze

Abstract Automatic speech recognition (ASR) has traditionally integrated ideas from many different domains, such as signal processing (mel-frequency cepstral coefficient features), natural language processing (*n*-gram language models), or statistics (hidden markov models). Because of this "compartmentalization," it is widely accepted that components of an ASR system will largely be optimized individually and in isolation, which will negatively influence overall performance. End-to-end approaches attempt to solve this problem by optimizing components jointly, and using a single criterion. This can also reduce the need for human experts to design and build speech recognition systems by painstakingly finding the best combination of several resources—which is still somewhat of a "black art." This chapter will first discuss several recent deep-learning-based approaches to end-to-end speech recognition. Next, we will present the EESEN framework, which combines connectionist-temporal-classification-based acoustic models with a weighted finite state transducer decoding setup. EESEN achieves state-of-the art word error rates, while at the same time drastically simplifying the ASR pipeline.

13.1 Introduction

At the heart of the problem are the different optimization criteria that are being used during system development. In the standard formulation of the speech recognition problem [6], the search for the "best" word sequence W' on an observation $\mathbf{X} = \{\mathbf{x}_1, \mathbf{x}_2, \mathbf{x}_3, \ldots, \mathbf{x}_T\}$ is broken down as follows (the "fundamental equation" of speech recognition):

$$W' = \arg\max_W P(W|O) \equiv \arg\max_W p(O|W)P(W). \tag{13.1}$$

Y. Miao • F. Metze (✉)
Carnegie Mellon University, 5000 Forbes Ave, Pittsburgh, PA, USA
e-mail: fmetze@cs.cmu.edu

© Springer International Publishing AG 2017
S. Watanabe et al. (eds.), *New Era for Robust Speech Recognition*,
DOI 10.1007/978-3-319-64680-0_13

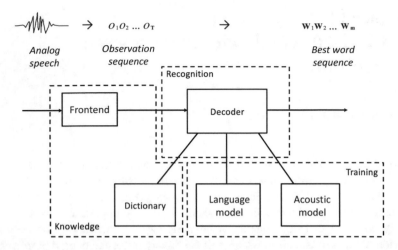

Fig. 13.1 Typical components of conventional speech-to-text systems. In most cases, the preprocessing and lexicon are knowledge-based, without receiving any training. The acoustic model is
often trained to optimize likelihood, frame cross-entropy, or some discriminative criterion. The
language model optimizes perplexity. Standard Viterbi decoding optimizes *sentence* error rate,
while the evaluation is based on *word* error rate

The prior $P(W)$ and the likelihood $p(O|W)$ are called the *language model*,
and the *acoustic model* respectively. Note that this formulation seeks to minimize
the "sentence" error rate (SER), i.e., it describes the word sequence W' with the
highest expectation of being correct in its entirety. Most speech-to-text systems,
however, are being evaluated using the word error rate (WER), which corresponds
to the sequence $\{w_1, w_2, \ldots, w_m\}$ of words that have the highest expectation of
being correct (individually)—a different, and maybe more "forgiving" criterion. In
practice, SER and WER are usually correlated, and improving one also tends to
improve the other. Still, this seemingly small discrepancy in how we chose to set up
the speech recognition problem is part of the reason why consensus decoding [47],
sequence training [37], and many other (discriminative) techniques are required to
achieve state-of-the-art automatic speech recognition (ASR) results, bringing with
them a slew of parameters, fudge factors, and heuristics.

Figure 13.1 shows the main components of such a standard ASR pipeline. Note
that not all the parts of the system are being trained, and that different criteria are
used to optimize each of them.

13.1.1 Complexity and Suboptimality of the Conventional ASR Pipeline

Figure 13.2 illustrates the development process of such a conventional, state-of-the
art "hybrid" ASR system. This consists of three different types of acoustic models,

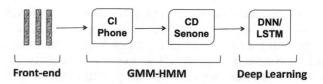

Fig. 13.2 Pipeline for building a state-of-the-art ASR system using deep learning acoustic modeling. Context-dependent (CD) subphonetic states (senones) are learned using a succession of Gaussian mixture models (GMMs), which are required for alignment and clustering only

only the last of which is ultimately used. Such a pipeline is intrinsically complex, and even if it is built without failures, it is prone to fall into local minima during optimization for multiple reasons:

- *Multiple training stages.* The pipeline is initialized by constructing a hidden Markov model/Gaussian mixture model (HMM/GMM). It generates frame-level alignments of the training data, which are used as targets for the deep-neural-network (DNN) training. The deep learning model can be trained only after the HMM/GMM model has been completely built. Unfortunately, training a good HMM/GMM itself requires execution of a series of steps, often starting with a context-independent (CI) model and then moving on to a context-dependent (CD) model, with senones as HMM states. In practice, each step is typically repeated multiple times, slowly growing the number of parameters, updating the alignments, and often integrating more advanced training techniques, such as speaker-adaptive training (SAT) and discriminative training (DT).
- *Various types of resources.* A conventional pipeline requires careful preparation of dictionaries and often phonetic information, in order to perform acoustic modeling. These resources are not always available, especially for low-resource languages/conditions. The lack of these resources can hamper the deployment of ASR systems in these scenarios. Using preexisting alignments or other available resources can speed up development, but introduces additional dependencies which make it hard to reproduce a training from scratch.
- *Tuning of hyper-parameters.* Another layer of complexity lies in the efforts involved in hyper-parameter tuning. There exist quite a few hyper-parameters in the pipeline, e.g., the number of HMM states and the number of Gaussian components inside each state. Deciding the values of these hyper-parameters requires intensive empirical tuning and relies on the knowledge and experience of ASR experts.
- *Multiple optimization functions.* The main drawback of the conventional ASR pipeline is the separate optimization of various components towards individual objective functions. In the speech recognition community, obtaining an appropriate feature representation and learning an effective model have mostly been treated as separate problems. Speech feature design follows the auditory mechanism of how humans perceive speech signals (e.g., the mel frequency scale), but this audiologically inspired feature extraction does not necessarily fit

the training of acoustic models best. The inconsistency of optimization objectives becomes more manifest in acoustic modeling. For example, the HMM/GMM model is normally trained by maximum likelihood estimation (MLE), or various discriminative criteria, whereas deep learning models are trained by optimizing discriminative objectives such as cross-entropy (CE). In almost all cases, these criteria are based on frame classification, while word recognition is a sequence classification problem. Moreover, during decoding, language models are tuned towards lower perplexity, which is not guaranteed to correspond to lower WERs. The independent optimization of these inconsistent components certainly hurts the performance of the resulting ASR systems.

- *Model mismatch.* It has been recognized early on [57] that hidden Markov Models do not fit the observed phone durations well. While the probability of staying in a given state of a first-order HMM decays exponentially for $t > 0$, observed distributions of phone durations have a distinct peak for $t > 0$, and could thus be approximated with, e.g., a gamma distribution. Using subphonetic units as the states of the HMM alleviates this fundamental problem somewhat, but most state-of-the-art ASR systems simply ignore phone durations, and set all state transition probabilities to fixed values, often 1. While this seems to work OK in practice, this inconsequential violation of the conservation of probability is an indication that the model is fundamentally mismatched, and that alternative solutions should be investigated.

13.1.2 Simplification of the Conventional ASR Pipeline

The aforementioned complexity of the conventional ASR pipeline motivates researchers to come up with various simplifications. A major complication involves the reliance on well-trained HMM/GMM models. It is then favorable to remove the GMMs-building stage and flat-start deep learning models directly. In [69], the authors proposed an approach to training DNNs acoustic models with no dependence on the HMM/GMM. Starting with initial uniform alignments (every state is assigned an equal duration in the utterance), their approach refines the alignments progressively by multiple iterations of network training and alignment regeneration. Phonetic state tying is then performed in the space of hidden activations from the trained network. Another similar proposal was developed simultaneously in [62], where other types of network outputs were investigated in context-dependent state clustering. The proposed GMM-free DNN training was further scaled up to a distributed asynchronous training infrastructure and larger datasets [2].

Still, these approaches inherit the fundamental disadvantages of an HMM-based approach, which "end-to-end" systems attempt to avoid.

13.1.3   End-to-End Learning

In recent years, with the advancement of deep learning, end-to-end solutions have emerged in many areas. A salient example is the wide application of deep convolutional neural networks (CNNs) to the image classification task. Conventionally, image classification starts with extracting hand-engineered features (e.g., SIFT [42]), and then applies classifiers such as support vector machines (SVMs) to the extracted features. With deep CNNs, image classification can be performed in a purely end-to-end manner. Inputs into the CNN are simply raw pixel values with proper preprocessing. After being trained on labeled data, the CNN generates classification results directly from its softmax layer. This end-to-end paradigm has been applied further to a variety of computer vision tasks, e.g., object detection [22], face recognition [39], scene labeling [70], and video classification [34]. Another area where end-to-end approaches have accomplished tremendous success is machine translation (MT). Building a conventional statistical MT system [38] contains a series of intermediate steps, potentially suffering from separate optimization. An encoder–decoder architecture [12, 63] has been proposed to achieve end-to-end machine translation. An elegant encoder–decoder variant, called attention models, was developed in [3]. Beyond machine translation, this encoder–decoder paradigm has been used in image captioning [67], video captioning [68], and many other tasks.

Using attention-based encoder–decoder models, the end-to-end idea can be naturally ported to speech recognition [9, 14, 43]. In principle, building an ASR system involves learning a mapping from speech feature vectors to a transcript (e.g., words, phones, characters, etc.), both of which are in sequence, so that no reordering is expected to take place. If we can learn such a mapping directly, all the components are optimized under a unified objective, which can enhance the final recognition performance—obliterating the need for separate acoustic and language models.

In practice, however, it may be pragmatic to maintain a separation of the language model (which describes "what" is being said) and the acoustic model ("how" it is being said). Even though this separation defeats the original motivation behind end-to-end learning, such approaches, most notably those based on connectionist temporal classification (CTC, [26]), have been accepted as "end-to-end" by the community, because their objective function is inherently sequence based.

13.2   End-to-End ASR Architectures

At present, two main end-to-end approaches dominate the speech-processing field. They differ in the way in which they make the alignment between observations and output symbols explicit, and in how they order dependencies between output symbols. Examples in the first category, which create explicit alignments and treat symbols as independent units, use algorithms like CTC, and will be discussed in Sect. 13.2.1. Encoder–decoder models dominate the second category, and compute no alignments at all, unless attention mechanisms are used; see Sect. 13.2.2. In Sect. 13.2.4, we will discuss other end-to-end approaches, for tasks other than speech-to-text, while Sect. 13.2.3 presents efforts to learn the front end of a recognizer.

13.2.1 Connectionist Temporal Classification

Frame-based neural networks require training targets for every segment or time step in the input sequence, and produce equally "dense" outputs. This has two important consequences: training data must be presegmented to provide the targets, and any dependency between successive labels must be modelled externally. In ASR, however, our goal is not necessarily a segmentation of the data, but rather a labeling of the sequence of states in the data, which allows us to define and train a network structure that can be directly optimized to predict a label sequence, and marginalizes over all possible alignments of that label sequence to the observations.

The CTC [26] loss function is defined over the target symbols, and introduces an additional "blank" label, which the network can predict at any point in time without influencing the output sequence. The introduction of the blank label enables the projection of the label sequences to frame-level independent labels. An important consequence of CTC training is that the sequence of labels is monotonically mapped to the observations (i.e., speech frames), which removes the need for additional constraints (e.g., the monotonic alignment constraint imposed in [13]). Importantly, the CTC loss function assumes that neighboring symbols are independent of each other, so that it can be used as a "tokenization" of the input features, and language models can be applied easily.

Most CTC-trained ASR systems have been built using stacked layers of long short-term memory networks (LSTMs) [30], e.g., in [25, 26, 61]. Other work used other types of recurrent neural networks [28], including simplifications such as rectified linear units (ReLUs [53]) in [27], or other improvements such as online learning [32], or more complex structures [58].

Recently, [71] introduced several improvements to "all-neural" end-to-end speech recognition, including an iterated CTC method, which boosts the performance of lexicon-free CTC-based systems to around 10% WER on the Switchboard 300 h setup. CTC training will be discussed in more detail in Sect. 13.3.2.

13.2.2 Encoder–Decoder Paradigm

Alternatively, it is possible to compress the entire sequence (i.e., values and order) of input features into a single vector, which then "encodes" all the input information in a single entity. This is achieved by using an encoder recurrent neural network

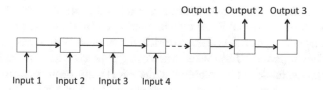

Fig. 13.3 Illustration of the encoder–decoder paradigm in which the model absorbs four input symbols (e.g., English words) and emits three output symbols (e.g., German words). The compressed input representation is depicted by the *dashed arrow*

(RNN) to read the input sequence, one time step at a time, to obtain a fixed-dimensional vector representation. Then, a decoder RNN is used to generate the output sequence from that vector representation. The decoder RNN is essentially an RNN-based language model [8], except that it is conditioned on the input sequence. In many practical implementations, LSTM networks (LSTMs, [30]) have acted as the building block in the encoder/decoder network. The LSTM's ability to successfully learn long-range dependencies [35] makes it a natural choice for this application, although other units such as gated recurrent units (GRUs) [11] can be used as well. Figure 13.3 illustrates the encoder-decoder idea, where the–encoder encodes three input symbols into a fixed-length representation. This input representation is then propagated to the decoder, which emits four output symbols sequentially.

The idea of mapping an entire input sequence into a vector was first presented in [33]. Machine translation presents an obvious extension of this idea [12]. The performance of this simple and elegant model can be further improved by introducing a mechanism [3] by which the model can rely on parts of the input sequence that are relevant to predicting a target output symbol, without having to identify these parts as a hard segment explicitly. This attention approach seems to better allow the network to recover from errors during the generation step, but does not imply monotonicity as is the case for CTC.

The first application of this technique to speech recognition was reported in [13, 14], where the authors found that only a slight modification to the attention mechanism was required in order to achieve state-of-the-art results on the TIMIT task [18]. While the unaltered MT model reached an overall 18.7% phoneme error rate (PER), its performance degraded quickly for longer, or concatenated utterances. This model seems to track the absolute location of a frame in the input sequence, which might help on a small task such as TIMIT, but would not generalize well.

The proposed modification explicitly takes into account both the location of the focus from the previous time step and the input feature sequence. This is achieved by adding as inputs to the attention mechanism auxiliary convolutional features, which are extracted by convolving the attention weights from the previous step with trainable filters. Together with other minor modifications to make the attention mechanism cope better with the noisy and frame-based characteristics of speech data, this model performed significantly better at 17.6% PER. Moreover, almost no degradation is observed on concatenated utterances. In follow-up work, the authors applied a deep bidirectional RNN encoder and attention-based recurrent sequence generator (ARSG) [5] as a decoder network to the English Wall Street Journal (WSJ) task [55]. They reported 10.8% WER when using characters as units and including a trigram language model. It is notable that the ARSG approach performs better than CTC approaches when no language models are used, mainly due to the fact that the decoder RNN has learned a language model implicitly.

A similar approach, named "listen, attend, and spell" (LAS), was implemented on a large-scale voice search corpus [10], achieving 14.1% WER without a language model and 10.3% WER with language model rescoring. These results get close to the 8.0% WER baseline, which consists of a combination of convolutional and (unidirectional) LSTM networks [58]. The LAS system does not use a pronunciation lexicon, but models characters directly. The listener is a 3-layer recurrent network encoder that accepts filterbank spectra as inputs. Each layer of the stacked bidirectional LSTM encoder reduces the time resolution by a factor of 2. This pyramid architecture is found to result in training speedup and better recognition results. The speller is an attention-based recurrent network decoder that emits each character conditioned on all the previous characters and the entire acoustic sequence. Beam search is used similarly to [63], and the top N (typically 32) hypotheses can be rescored with a language model, at which time a small bias towards shorter utterances is also corrected. One special property of the LAS implementation is that it can produce multiple spelling variants for the same acoustics: the model can, for example, produce both "triple A" and "AAA" (within the top four beams). As there is no conditional-independence assumption between output symbols as with CTC, the decoder will score both variants highly for matching acoustics. In comparison, conventional HMM/DNN systems would require both spellings to be in the pronunciation dictionary to generate both transcriptions.

Most recently, Lu et al. [45] investigated training strategies for end-to-end systems on the Switchboard task, and achieved reasonable overall accuracies with GRUs as the network building block. A multilayer decoder model was employed, which shows improved long-term memory behavior. This work also confirmed the beneficial effect of hierarchically subsampling the input frames in learning the encoder–decoder model.

13.2.3 Learning the Front End

Another line of work attempts to remove feature extraction, and use time samples for neural network training. Standard deep learning models take as inputs hand-engineered features, for example, log mel filterbank magnitudes. Various attempts have now been made to train a DNN or CNN model directly on the raw speech waveform [31, 54, 64]. Apart from the network, filters are placed over the raw waveform and learned jointly with the rest of the network. This has the advantage of optimizing feature extraction towards the objective at hand (network training) and unifying the entire training process. In [59], used together with a powerful deep learning model [58], the raw-waveform features (with filter learning applied) were observed to match the performance of log mel filterbanks.

It should be easily possible to combine, e.g., CTC or encoder–decoder models with a learned front end; however, we are not aware of any work that has implemented this as yet.

13.2.4 Other Ideas

An extension of CTC towards so-called *recurrent neural network transducers* was presented in [24], and evaluated on TIMIT. This approach defines a distribution over output sequences of all lengths, and jointly models both input–output and output–output dependencies. A recent paper [44] introduced a segmental version of CTC, by adding another marginalization step over all possible segmentations, which resulted in the currently best PER on the TIMIT dataset.

Keyword spotting is another task for which an end-to-end approach may be beneficial. Fernández et al. [17] implemented an end-to-end keyword-spotting system with LSTMs and CTC, with the disadvantage that new keywords cannot easily be added. This limitation was partly resolved in [66] which applied dynamic Bayesian networks (DBNs) and relies on phones detected by CTC. A related approach was presented in [36], which attempted to use neural networks to optimize the fusion of multiple sources of information at the input or model level. It shows how neural networks can be used to combine the acoustic model and the language model, or how neural networks can optimize the combination of multiple input features.

"Wav2Letter" [15] is another recent approach, which goes from waveforms directly to letters, using only convolutional neural networks. The advantage of this approach is that it uses nonrecursive models only, which can be faster to train and evaluate. At present, the results of completely end-to-end (i.e., wavefile to letter) approaches are a bit behind systems trained on mel-frequency cepstral coefficient (MFCC) or power spectrum features.

13.3 The EESEN Framework

In this section, we exemplify end-to-end ASR using the "EESEN" toolkit [50]. We will describe the model structure and training objective together with a decoding approach based on weighed finite state transducers (WFSTs). EESEN shares data preparation scripts and other infrastructure with the popular "Kaldi" toolkit [56], and has been released[1] as open source under the Apache license. More related toolkits can be found in Chap. 17.

13.3.1 Model Structure

The acoustic models in EESEN are deep bidirectional RNN networks. The basic RNN model and its LSTM variant have been introduced in Chaps. 7 and 11. Here we restate their principles for complete formulation. Compared to the standard feedforward networks, RNNs have the advantage of learning complex temporal dynamics on sequences. Given an input sequence $X = (x_1, \ldots, x_T)$ which totally contains T frames, a recurrent layer computes the *forward* sequence of hidden states $\overrightarrow{H} = (\overrightarrow{h}_1, \ldots, \overrightarrow{h}_T)$ by iterating from $t = 1$ to T:

$$\overrightarrow{h}_t = \sigma(\overrightarrow{W}_{hx} x_t + \overrightarrow{W}_{hh} \overrightarrow{h}_{t-1} + \overrightarrow{b}_h) \tag{13.2}$$

where t represents the current speech frame, $\overrightarrow{W}_{hx}$ is the input-to-hidden weight matrix, and $\overrightarrow{W}_{hh}$ is the hidden-to-hidden weight matrix. In addition to the inputs x_t, the hidden activation h_{t-1} from the previous time step are fed to influence the hidden outputs at the current time step. In a bidirectional RNN, an additional recurrent layer computes the *backward* sequence of hidden outputs $\overleftarrow{H}$ from $t = T$ to 1:

$$\overleftarrow{h}_t = \sigma(\overleftarrow{W}_{hx} x_t + \overleftarrow{W}_{hh} \overleftarrow{h}_{t-1} + \overleftarrow{b}_h). \tag{13.3}$$

The acoustic model is a deep architecture, in which we stack multiple bidirectional recurrent layers. At each frame t, the concatenation of the forward and backward hidden outputs $[\overrightarrow{h}_t, \overleftarrow{h}_t]$ from the current layer are treated as inputs into the next recurrent layer.

Learning of RNNs can be done using back-propagation through time (BPTT). In practice, training RNNs to learn long-term temporal dependency can be difficult due to the vanishing-gradients problem [7]. To overcome this issue, we apply LSTM units [30] as the building blocks of RNNs. An LSTM contains memory cells with self-connections to store the temporal states of the network. Also, multiplicative gates are added to control the flow of information. Figure 13.4 depicts the structure of the LSTM units we use. The light grey represents peephole connections [20] that link the memory cells to the gates to learn the precise timing of the outputs. The computation at the time step t can be formally written as follows. We omit the

---

[1]Code and latest recipes and results can be found at https://github.com/srvk/eesen.

Fig. 13.4 A memory block
of an LSTM

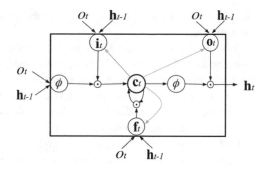

arrow → for an uncluttered formulation:

$$i_t = \sigma(W_{ix}x_t + W_{ih}h_{t-1} + W_{ic}c_{t-1} + b_i), \tag{13.4a}$$

$$f_t = \sigma(W_{fx}x_t + W_{fh}h_{t-1} + W_{fc}c_{t-1} + b_f), \tag{13.4b}$$

$$c_t = f_t \odot c_{t-1} + i_t \odot \phi(W_{cx}x_t + W_{ch}h_{t-1} + b_c), \tag{13.4c}$$

$$o_t = \sigma(W_{ox}x_t + W_{oh}h_{t-1} + W_{oc}c_t + b_o), \tag{13.4d}$$

$$h_t = o_t \odot \phi(c_t), \tag{13.4e}$$

where i_t, o_t, f_t, c_t are the activations of the input gates, output gates, forget gates and memory cells, respectively. The $W_{.x}$ weight matrices connect the inputs with the units, whereas the $W_{.h}$ matrices connect the *previous* hidden states with the units. The $W_{.c}$ terms are diagonal weight matrices for peephole connections. Also, σ is the logistic sigmoid nonlinearity, and ϕ is the hyperbolic tangent nonlinearity. The computation of the *backward* LSTM layer can be represented similarly.

13.3.2   Model Training

Unlike in the hybrid approach, the RNN model in the EESEN framework is not trained using frame-level labels with respect to the CE criterion. Instead, following [25, 28, 46], the CTC objective is adopted [26] to automatically learn the alignments between speech frames and their label sequences (e.g., phones or characters). Assume that the label sequences in the training data contain K unique labels. Normally K is a relatively small number, e.g., around 45 for English when the labels are phones. An additional *blank* label $\varnothing$, which means no labels being emitted, is added to the labels. For simplicity of formulation, we denote every label using its index in the label set. Given an utterance $O = (x_1, \ldots, x_T)$, its label sequence is denoted as $z = (z_1, \ldots, z_U)$. The blank is always indexed as 0. Therefore z_u is an integer ranging from 1 to K. The length of z is constrained to be no greater than the length of the utterance, i.e., $U \leq T$. CTC aims to maximize $\ln p(z|O)$, the log-likelihood of the label sequence given the inputs, by optimizing the RNN model parameters.

The final layer of the RNN is a softmax layer which has $K + 1$ nodes that correspond to the $K + 1$ labels (including $\varnothing$). At each frame t, we get the output vector $\boldsymbol{y}_t$ whose kth element y_t^k is the posterior probability of the label k. However, since the labels z are not aligned with the frames, it is difficult to evaluate the likelihood of z given the RNN outputs. To bridge the RNN outputs with label sequences, an intermediate representation, the *CTC path*, was introduced in [26]. A CTC path $\boldsymbol{p} = (p_1, \ldots, p_T)$ is a sequence of labels at the frame level. It differs from z in that the CTC path allows occurrences of the blank label and repetitions of nonblank labels. The total probability of the CTC path is decomposed into the probability of the label p_t at each frame:

$$p(\boldsymbol{p}|O) = \prod_{t=1}^{T} y_t^{p_t}. \tag{13.5}$$

The label sequence z can then be mapped to its corresponding CTC paths. This is a one-to-many mapping because multiple CTC paths can correspond to the same label sequence. For example, both "A A $\varnothing$ $\varnothing$ B C $\varnothing$" and "$\varnothing$ A A B $\varnothing$ C C" are mapped to the label sequence "A B C". We denote the set of CTC paths for z as $\Phi(z)$. Then, the likelihood of z can be evaluated as a sum of the probabilities of its CTC paths:

$$p(z|X) = \sum_{p \in \Phi(z)} p(\boldsymbol{p}|O). \tag{13.6}$$

However, summing over all the CTC paths is computationally intractable. A solution is to represent the possible CTC paths compactly as a trellis. To allow blanks in CTC paths, we add "0" (the index of $\varnothing$) to the beginning and the end of z, and also insert "0" between every pair of the original labels in z. The resulting *augmented label sequence* $l = (l_1, \ldots, l_{2U+1})$ is leveraged in a forward–backward algorithm for efficient likelihood evaluation. Specifically, in a forward pass, the variable α_t^u represents the total probability of all CTC paths that end with label l_u at frame t. As in the case of HMMs [57], α_t^u can be recursively computed from the variable values that have been obtained in the previous frame $t - 1$. Similarly, a backward variable β_t^u carries the total probability of all CTC paths that start with label l_u at t and reach the final frame T. The likelihood of the label sequence z can then be computed as

$$p(z|O) = \sum_{u=1}^{2U+1} \alpha_t^u \beta_t^u, \tag{13.7}$$

where t can be any frame $1 \leq t \leq T$. The objective $\ln p(z|O)$ now becomes differentiable with respect to the RNN outputs $\boldsymbol{y}_t$. We define an operation on the augmented label sequence $\Upsilon(l, k) = \{u | l_u = k\}$ that returns the elements of l which have the value k. The derivative of the objective with respect to y_t^k can be derived as

$$\frac{\partial \ln p(z|X)}{\partial y_t^k} = \frac{1}{p(z|X)} \frac{1}{y_t^k} \sum_{u \in \Upsilon(l,k)} \alpha_t^u \beta_t^u. \tag{13.8}$$

These errors are back-propagated through the softmax layer and further into the RNN to update the model parameters.

EESEN implements this model-training stage on graphics processing unit (GPU) devices. To fully exploit the capacity of GPUs, multiple utterances are processed at a time in parallel. This parallel processing speeds up model training by replacing matrix–vector multiplication over single frames with matrix–matrix multiplication over multiple frames. Within a group of parallel utterances, every utterance is padded to the length of the longest utterance in the group. These padding frames are excluded from gradient computation and parameter updating. For further acceleration, the training utterances are sorted by their lengths, from the shortest to the longest. The utterances in the same group then have approximately the same length, which minimizes the number of padding frames. CTC evaluation is also expensive because the forward and backward vectors (α_t and β_t) have to be computed sequentially, either from $t = 1$ to T or from $t = T$ to 1. Like RNNs, CTC implementation in EESEN also processes multiple utterances at the same time. Moreover, at a specific frame t, the elements of α_t (and β_t) are independent and thus can be computed in parallel.

13.3.3 Decoding

Previous work has introduced a variety of methods [25, 28, 46] to decode CTC-trained models. These methods, however, either fail to integrate word-level language models [46] or achieve the integration under constrained conditions (e.g., n-best list rescoring in [25]). A distinctive feature of EESEN is a generalized decoding approach based on WFSTs [52, 56]. A WFST is a finite state acceptor (FSA) in which each transition has an input symbol, an output symbol, and a weight. A path through the WFST takes a sequence of input symbols and emits a sequence of output symbols. EESEN's decoding method represents the CTC labels, lexicons, and language models as separate WFSTs. Using highly optimized FST libraries such as OpenFst [1], the WFSTs are fused efficiently into a single search graph. The Building of the individual WFSTs is described as follows. Although exemplified in the scenario of English, the same procedures hold for other languages.

13.3.3.1 Grammar

A grammar WFST encodes the permissible word sequences in a language/domain. The WFST shown in Fig. 13.5 represents a toy language model which permits two sentences, "how are you" and "how is it." The WFST symbols are the words, and the arc weights are the language model probabilities. With this WFST representation, CTC decoding in principle can leverage any language models that can be converted into WFSTs. Following conventions in the literature [56], the language model WFST is denoted as G.

13.3.3.2 Lexicon

A lexicon WFST encodes the mapping from sequences of lexicon units to words. Depending on what labels our RNN has modeled, there are two cases to consider. If the labels are phones, the lexicon is a standard dictionary as we normally have in the hybrid approach. When the labels are characters, the lexicon simply contains the spellings of the words. A key difference between these two cases is that the *spelling lexicon* can be easily expanded to include any out-of-vocabulary (OOV) words. In contrast, expansion of the *phone lexicon* is not so straightforward. It relies on some grapheme-to-phoneme rules/models, and is potentially subject to errors. The lexicon WFST is denoted as L. Figures 13.6 and 13.7 illustrate these two cases of building L.

For the spelling lexicon, there is another complication to deal with. With characters as CTC labels, an additional *space* character is usually inserted between every pair of words, in order to model word delimiting in the original transcripts. Decoding allows the space character to optionally appear at the beginning and end of a word. This complication can be handled easily by the WFST shown in Fig. 13.7.

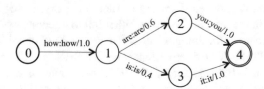

Fig. 13.5 A toy example of the grammar (language model) WFST. The arc weights are the probability of emitting the next word when given the previous word. The node 0 is the start node, and the double-circled node is the end node

Fig. 13.6 The WFST for the phone-lexicon entry "is IH Z". The "<eps>" symbol means no inputs are consumed or no outputs are emitted

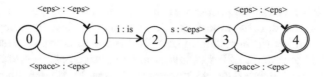

Fig. 13.7 The WFST for the spelling of the word "is". We allow the word to optionally start and end with the space character "<space>"

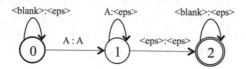

Fig. 13.8 An example of the token WFST which depicts the phone "IH". We allow occurrences of the blank label "<blank>" and repetitions of the nonblank label "IH"

13.3.3.3   Token

The third WFST component maps a sequence of frame-level CTC labels to a single lexicon unit (phone or character). For a lexicon unit, its *token WFST* is designed to subsume all of its possible label sequences at the frame level. Therefore, this WFST allows occurrences of the blank label ∅, as well as repetitions of any nonblank labels. For example, after processing five frames, the RNN model may generate three possible label sequences, "AAAAA", "∅ ∅ A A ∅", "∅ A A A ∅". The token WFST maps all these three sequences into a singleton lexicon unit "A". Figure 13.8 shows the WFST structure for the phone "IH". The token WFST is denoted as T.

13.3.3.4   Search Graph

After compiling the three individual WFSTs, we compose them into a comprehensive search graph. The lexicon and grammar WFSTs are firstly composed. Two special WFST operations, *determinization* and *minimization*, are performed over the composition of them, in order to compress the search space and thus speed up decoding. The resulting WFST LG is then composed with the token WFST, which finally generates the search graph. Overall, the order of the FST operations is

$$S = T \circ \min(\det(L \circ G)),\tag{13.9}$$

where ○, det and min denote composition, determinization, and minimization, respectively. The search graph S encodes the mapping from a sequence of CTC labels emitted on speech frames to a sequence of words.

When decoding with the conventional ASR pipeline, the state posteriors from deep learning models are normally scaled using state priors. The priors are estimated from forced alignments of the training data [16]. In contrast, decoding of CTC-trained models requires no posterior scaling, as the posteriors of the entire label sequences can be directly evaluated. However, in practice, posterior scaling is observed to still benefit decoding in EESEN. Instead of collecting statistics from frame-level alignments, EESEN estimates more robust label priors from the label sequences in the training data. As mentioned in Sect. 13.3.2, the label sequences actually used in CTC training are the augmented label sequences, which insert a blank at the beginning, at the end, and between every label pair in the original label sequences. The priors are computed from the augmented label sequences (e.g., "∅ IH ∅ Z ∅") instead of the original ones (e.g., "IH Z"), through simple counting. This simple method is empirically found to give better recognition accuracy than both using priors derived from frame-level alignments and the proposal described in [61].

13.3.4   Experiments and Analysis

In this section, we present experiments conducted on a variety of benchmark ASR tasks, and an analysis regarding the advantages and disadvantages of EESEN.

13.3.4.1   Wall Street Journal

The EESEN framework was first of all verified on the WSJ corpus [55], which can be obtained from LDC under the catalog numbers LDC93S6B and LDC94S13B. Data preparation gave us 81 h of transcribed speech, from which we selected 95% as the training set and the remaining 5% for cross-validation. As discussed in Sect. 13.3, we applied deep RNNs as the acoustic models. The inputs of the RNNs were 40-dimensional filterbank features together with their first- and second-order derivatives. The features were normalized via mean subtraction and variance normalization on the speaker basis.

The RNN model had four bidirectional LSTM layers. In each layer, both the forward and the backward sublayers contained 320 memory cells. Initial values of the model parameters were randomly drawn from a uniform distribution in the range [−0.1, 0.1]. The model was trained with BPTT, in which the errors were back-propagated from CTC. Utterances in the training set were sorted by their lengths, and ten utterances were processed in parallel at a time. Model training adopted an initial learning rate of 0.00004, which was decayed based on the change of the accuracy of the hypothesis labels with respect to the reference label sequences.

Our decoding followed the WFST-based approach in Sect. 13.3.3. We applied the WSJ standard pruned trigram language model in the ARPA format (which we will consistently refer to as *standard*). To be consistent with previous work [25, 28], we report our results on the *eval92* set. Our experimental setup has been released together with EESEN, which should enable readers to reproduce the numbers reported here.

When phones were taken as CTC labels, we employed the CMU dictionary[2] as the lexicon. From the lexicon, we extracted 72 labels including phones, noise marks, and the blank. On the eval92 testing set, the EESEN end-to-end system finally achieved a WER of 7.87%. As a comparison, we also constructed a conventional ASR system using the Kaldi toolkit [56]. The system employed a hybrid HMM/DNN as its acoustic model which contained slightly more parameters (9.2 vs. 8.5 million) than EESEN's RNN model. The DNN was fine-tuned to optimize the CE objective with respect to 3421 clustered context-dependent states. From Table 13.1, we observe that the performance of the EESEN system is still behind the conventional system. As will be discussed in Sect. 13.3.4.2, EESEN is able to outperform the conventional pipeline on larger-scale tasks.

When switching to characters as CTC labels, we took the word list from the CMU dictionary as our vocabulary, ignoring the word pronunciations. CTC training dealt with 59 labels including letters, digits, punctuation marks, etc. Table 13.1 shows that with the standard language model, the character-based system gets a WER of 9.07%. CTC experiments in past work [25] have adopted an *expanded* vocabulary, and retrained the language model using text data released together with the WSJ corpus. For fair comparison, we follow an identical configuration. OOV words that occurred at least twice in the language model training texts were added to the vocabulary. A new trigram language model was built (and then pruned) with the language model training texts. Under this setup, the WER of the EESEN character-based system was reduced to 7.34%.

In Table 13.1, we also list results for end-to-end ASR systems that have been reported in previous work [25, 28] on the same dataset. EESEN outperforms both [25] and [28] in terms of WERs on the testing set. It is worth pointing out that the 8.7% WER reported in [25] was not obtained in a purely end-to-end manner. Instead, the authors of [25] generated an *n*-best list of hypothesis from a conventional system, and applied the CTC model to rescore the hypotheses candidates. The EESEN numbers shown here, in contrast, carry no dependency on any existing systems.

Table 13.1 Performance of EESEN systems, and comparison with conventional systems built with Kaldi and results reported in previous work

Targets type	LM setup	Model	# parameters	WER (%)
Phone-based	Standard	EESEN	8.5 million	7.87
Phone-based	Standard	Kaldi HMM/DNN	9.2 million	7.14
Character-based	Standard	EESEN	8.5 million	9.07
Character-based	Expanded	EESEN	8.5 million	7.34
Character-based	Expanded	Graves et al. [25]	–	8.7
Character-based	Standard	Hannun et al. [28]	–	14.1
Character-based	Standard	ARSG [5]	–	10.8

[2]http://www.speech.cs.cmu.edu/cgi-bin/cmudict.

Moreover, in the last row of Table 13.1, we show the result from ARSG, an attention-based encoder–decoder model, reported in [5]. We can see that with the same configuration, the CTC model constructed in our EESEN framework outperforms the encoder–decoder model.

13.3.4.2 Switchboard

EESEN was applied to the Switchboard conversational telephone transcription task [23]. We used Switchboard-1 Release 2 (LDC97S62) as the training set, which contained over 300 h of speech. For fast turnarounds, we also selected 110 h from the training set and created a lighter setup. On the 110-h and 300-h setups, the LSTM network consisted of four and five bidirectional LSTM layers, respectively. Other training configurations (layer size, learning rate, etc.) remained the same as those for WSJ. The CTC training modeled 46 labels including phones, noise marks, and the blank. For decoding, a trigram language model was trained on the training transcripts, and was then interpolated with another language model trained on the Fisher English Part 1 transcripts (LDC2004T19). We report WERs on the Switchboard part of the Hub5'00 (LDC2002S09) test set.

Our baseline systems were conventional Kaldi systems where both DNNs and LSTMs were used as acoustic models. For the 110-h setup, the DNN had five hidden layers, each of which contained 1200 neurons. The LSTM model had two unidirectional LSTM layers, where linear projection layers were applied over the hidden outputs [60]. Each LSTM layer had 800 memory cells and 512 output units. The parameters of both the DNN and LSTM models were randomly initialized. For the 300-h setup, the DNN model had 6 hidden layers, each of which contained 2048 neurons. The LSTM model had 2 projected LSTM layers, where each LSTM layer had 1024 memory cells and 512 output units. The DNN was initialized with restricted Boltzmann machines (RBMs) [29], while the LSTM model was randomly initialized. More details about these hybrid models can be found in [49].

Table 13.2 shows that on the 110-h setup, the EESEN system performs slightly better than the conventional HMM/DNN system, but worse than the HMM/LSTM system.[3] In contrast, when we switch to the 300-h setup, EESEN outperforms both conventional systems. This comparison indicates that CTC training becomes more advantageous when the amount of training data increases. This observation is understandable because in the conventional systems, deep learning models (DNNs or LSTMs) are trained as frame-level classifiers, classifying speech frames into their corresponding state labels. CTC-based training in EESEN aims for sequence-to-sequence learning, which is apparently more complicated than frame-level

---

[3]Note that the HMM/LSTM system employed a unidirectional LSTM while the EESEN system applied a bidirectional LSTM. Readers should take this discrepancy into account when evaluating the results.

Table 13.2 Comparisons of EESEN and conventional baseline systems on the Switchboard 110-h and 300-h setups

Dataset	Model	# parameters	WER (%)
110-h	EESEN	8 million	19.9
110-h	Kaldi HMM/DNN	12 million	20.2
110-h	Kaldi HMM/LSTM	8 million	19.2
300-h	EESEN	11 million	15.0
300-h	Kaldi HMM/DNN	40 million	16.9
300-h	Kaldi HMM/LSTM	12 million	15.8

Both DNNs and LSTMs were exploited as acoustic models in the conventional systems

Table 13.3 Comparison of decoding speed between EESEN and conventional systems on the Switchboard 300-h setup

Model	Decoding graph	Decoding-graph size	Real-time factor
EESEN	TLG	123M	0.71
Kaldi DNN-HMM	HCLG	216M	1.43
Kaldi LSTM-HMM	HCLG	216M	1.12

The decoding graph size is measured in terms of megabytes. The *real-time factor* is the ratio of the time consumed by decoding to the duration of the testing speech. For example, a real-time factor of 1.5 indicates that decoding 1 h of speech takes 1.5 h of decoding time

classification. Therefore, in order to learn high-quality CTC models, we need to pool more training sequences into the training data.

Compared with conventional ASR systems, a major advantage of EESEN lies in the decoding speed. The acceleration comes from the drastic reduction of the number of states, i.e., from thousands of clustered context-dependent states to tens of purely context-independent phones/characters. To verify this, Table 13.3 compares the decoding speed of EESEN and the conventional systems with their best decoding settings. Due to the reduction of states, the decoding graph in EESEN is significantly smaller than the graph used by the conventional systems, which saves disk space for storing the graphs. From the real-time factors in Table 13.3, we observe that decoding in EESEN is nearly two times faster than decoding in conventional systems.

13.3.4.3 HKUST Mandarin Chinese

So far, we have evaluated EESEN on English. Here, we continue by applying it to the HKUST Mandarin Chinese conversational telephone ASR task [41]. The training and testing sets contained 174 and 5 h of speech, respectively. The acoustic model contained five bidirectional LSTM layers, each of which had 320 memory cells in both the forward and the backward sublayers. Instead of phones, CTC in this setup modeled characters directly. Data preparation gave us 3667 labels, including English characters, Mandarin characters, noise marks and the blank. A trigram language model was employed in the WFST-based decoding. From Table 13.4, we

Table 13.4 Comparison of
EESEN and the conventional
Kaldi DNN-HMM system on
the HKUST Mandarin corpus

Model	Features	CER (%)
Kaldi DNN-HMM	Filterbank	39.42
EESEN	Filterbank + pitch	39.70
EESEN	Filterbank + pitch	38.67

The evaluation metric is the character error rate (CER)

can see that EESEN achieves a character error rate (CER) of 39.70%. This number is comparable to a competitive Kaldi HMM/DNN system (39.42%) which was trained over speaker-adaptive (SA) features, as reported in the Kaldi repository [56]. This observation is contrary to [40] where CTC was found to perform much worse than hybrid models, due to the lack of word language models in decoding. Finally, we enriched the speech front end with pitch features, which have been observed to benefit ASR on tonal languages. The pitch features were extracted by following, the implementation adopted by Kaldi [21]. We observe that, as expected, adding Pitch features brings additional gains for the EESEN system.

13.4 Summary and Future Directions

This chapter has presented several approaches to automatic speech recognition from an end-to-end perspective. End-to-end ASR aims to learn a direct mapping from speech to transcript (or indeed any "meaning representation"), without composing individual components and using intermediate optimization functions, as is the case in a conventional pipeline. Using approaches such as connectionist temporal classification and attention-based encoder–decoder models, speech recognition becomes a joint learning process and conceptually a very simple task. We have demonstrated the simplicity of end-to-end ASR using the "EESEN" open-source framework [50], which combines a CTC objective function with WFST-based decoding. This retains the useful separation of acoustic model and language model, and enables the efficient incorporation of lexicons and language models into decoding. The EESEN framework achieves competitive recognition accuracy, while simplifying the existing ASR pipeline drastically.

One caveat of the existing end-to-end approaches is that there are still hyper-parameters (e.g., network architecture, learning rate schedule, etc.) which need to be prespecified. We expect future work to make this an easier task (especially for non-ASR experts [48]) or a task that can be fully automated [65]. Also, mostly to limit computational complexity, most current practical work still uses conventional acoustic features (e.g., filterbanks).

It seems safe to predict that we will soon see work that exploits multitask learning for end-to-end approaches, e.g., multilingual speech recognition.

A natural extension is to combine model learning and feature learning together, which enables us to learn the mapping from raw waveforms to transcripts directly. Moreover, it will be exciting to see how end-to-end approaches can be extended to tasks of grander scales, in which speech recognition is merely a subtask. Examples of these tasks include dialog systems, dialog state tracking, parsing and slot filling, speech summarization, lecture captioning, speech translation, etc. These tasks are being treated as a cascade of independent modules these days. Therefore, they should benefit enormously from joint optimization with speech recognizers. Different forms of loss functions for different tasks have already been proposed [4].

Also, while most current work on end-to-end ASR is using some form of recurrent neural networks, it is not clear if recurrence and long-term memory are required, at least for the strictly linear speech-to-text task. Some work indicates that this may not be the case [19, 51].

In fact, given the current rate of progress on many of these tasks, it is highly likely that by the time this book chapter appears, many of these ideas will already have been realized. We apologize in advance for providing an out-of-date discussion in these cases.

References

1. Allauzen, C., Riley, M., Schalkwyk, J., Skut, W., Mohri, M.: OpenFST: a general and efficient weighted finite-state transducer library. In: Holub, J., Žd'ávn, J. (eds.) Implementation and Application of Automata, pp. 11–23. Springer, Heidelberg (2007)
2. Bacchiani, M., Senior, A., Heigold, G.: Asynchronous, online, GMM-free training of a context dependent acoustic model for speech recognition. In: Fifteenth Annual Conference of the International Speech Communication Association (INTERSPEECH). ISCA, Singapore (2014)
3. Bahdanau, D., Cho, K., Bengio, Y.: Neural machine translation by jointly learning to align and translate (2014). arXiv preprint arXiv:1409.0473
4. Bahdanau, D., Serdyuk, D., Brakel, P., Ke, N.R., Chorowski, J., Courville, A.C., Bengio, Y.: Task loss estimation for sequence prediction. CoRR abs/1511.06456 (2015). http://arxiv.org/abs/1511.06456
5. Bahdanau, D., Chorowski, J., Serdyuk, D., Brakel, P., Bengio, Y.: End-to-end attention-based large vocabulary speech recognition. In: Seventeenth Annual Conference of the International Speech Communication Association (INTERSPEECH) (2016)
6. Bahl, L.R., Jelinek, F., Mercer, R.L.: A maximum likelihood approach to continuous speech recognition. IEEE Trans. Pattern Anal. Mach. Intell. 5(2), 179–190 (1983)
7. Bengio, Y., Simard, P., Frasconi, P.: Learning long-term dependencies with gradient descent is difficult. IEEE Trans. Neural Netw. 5(2), 157–166 (1994)
8. Bengio, Y., Ducharme, R., Vincent, P., Jauvin, C.: A neural probabilistic language model. J. Mach. Learn. Res. 3, 1137–1155 (2003)
9. Chan, W., Jaitly, N., Le, Q.V., Vinyals, O.: Listen, attend and spell (2015). arXiv preprint arXiv:1508.01211
10. Chan, W., Jaitly, N., Le, Q.V., Vinyals, O.: Listen, attend and spell: a neural network for large vocabulary conversational speech recognition. In: 2016 IEEE International Conference on Acoustics, Speech and Signal Processing (ICASSP). IEEE, New York (2016)
11. Cho, K., van Merrienboer, B., Bahdanau, D., Bengio, Y.: On the properties of neural machine translation: encoder–decoder approaches. CoRR abs/1409.1259 (2014). http://arxiv.org/abs/1409.1259
12. Cho, K., Van Merriënboer, B., Gulcehre, C., Bahdanau, D., Bougares, F., Schwenk, H., Bengio, Y.: Learning phrase representations using RNN encoder–decoder for statistical machine translation (2014). arXiv preprint arXiv:1406.1078

13. Chorowski, J., Bahdanau, D., Cho, K., Bengio, Y.: End-to-end continuous speech recognition using attention-based recurrent NN: first results (2014). arXiv preprint arXiv:1412.1602
14. Chorowski, J.K., Bahdanau, D., Serdyuk, D., Cho, K., Bengio, Y.: Attention-based models for speech recognition. In: Advances in Neural Information Processing Systems, pp. 577–585 (2015)
15. Collobert, R., Puhrsch, C., Synnaeve, G.: Wav2Letter: an end-to-end convNet-based speech recognition system. CoRR abs/1609.03193 (2016). http://arxiv.org/abs/1609.03193
16. Dahl, G.E., Yu, D., Deng, L., Acero, A.: Context-dependent pre-trained deep neural networks for large-vocabulary speech recognition. IEEE Trans. Audio Speech Lang. Process. **20**(1), 30–42 (2012)
17. Fernández, S., Graves, A., Schmidhuber, J.: An application of recurrent neural networks to discriminative keyword spotting. In: Artificial Neural Networks–ICANN 2007, pp. 220–229. Springer, Heidelberg (2007)
18. Garofolo, J.S., Lamel, L.F., Fisher, W.M., Fiscus, J.G., Pallett, D.S.: DARPA TIMIT acoustic-phonetic continuous speech corpus CD-ROM. NIST speech disc 1-1.1. NASA STI/Recon Technical Report N 93 (1993)
19. Geras, K.J., Mohamed, A.R., Caruana, R., Urban, G., Wang, S., Aslan, O., Philipose, M., Richardson, M., Sutton, C.: Blending LSTMS into CNNS (2015). arXiv preprint arXiv:1511.06433
20. Gers, F.A., Schraudolph, N.N., Schmidhuber, J.: Learning precise timing with LSTM recurrent networks. J. Mach. Learn. Res. **3**, 115–143 (2003)
21. Ghahremani, P., BabaAli, B., Povey, D., et al.: A pitch extraction algorithm tuned for automatic speech recognition. In: 2014 IEEE International Conference on Acoustics, Speech and Signal Processing (ICASSP), pp. 2494–2498. IEEE, New York (2014)
22. Girshick, R., Donahue, J., Darrell, T., Malik, J.: Rich feature hierarchies for accurate object detection and semantic segmentation. In: Proceedings of the IEEE Conference on Computer Vision and Pattern Recognition, pp. 580–587 (2014)
23. Godfrey, J.J., Holliman, E.C., McDaniel, J.: Switchboard: telephone speech corpus for research and development. In: 1992 IEEE International Conference on Acoustics, Speech, and Signal Processing, ICASSP-92, vol. 1, pp. 517–520. IEEE, New York (1992)
24. Graves, A.: Sequence transduction with recurrent neural networks (2012). arXiv preprint arXiv:1211.3711
25. Graves, A., Jaitly, N.: Towards end-to-end speech recognition with recurrent neural networks. In: Proceedings of the 31st International Conference on Machine Learning (ICML-14), pp. 1764–1772 (2014)
26. Graves, A., Fernández, S., Gomez, F., Schmidhuber, J.: Connectionist temporal classification: labelling unsegmented sequence data with recurrent neural networks. In: Proceedings of the 23rd International Conference on Machine Learning (ICML-06), pp. 369–376 (2006)
27. Hannun, A., Case, C., Casper, J., Catanzaro, B., Diamos, G., Elsen, E., Prenger, R., Satheesh, S., Sengupta, S., Coates, A., et al.: Deepspeech: scaling up end-to-end speech recognition (2014). arXiv preprint arXiv:1412.5567
28. Hannun, A.Y., Maas, A.L., Jurafsky, D., Ng, A.Y.: First-pass large vocabulary continuous speech recognition using bi-directional recurrent DNNs. arXiv preprint arXiv:1408.2873 (2014)
29. Hinton, G.E.: A practical guide to training restricted Boltzmann machines. In: Montavon, G., Orr, G., Müller, K.R. (eds.) Neural Networks: Tricks of the Trade, pp. 599–619. Springer, Heidelberg (2012)
30. Hochreiter, S., Schmidhuber, J.: Long short-term memory. Neural Comput. **9**(8), 1735–1780 (1997)
31. Hoshen, Y., Weiss, R.J., Wilson, K.W.: Speech acoustic modeling from raw multichannel waveforms. In: 2015 IEEE International Conference on Acoustics, Speech and Signal Processing (ICASSP), pp. 4624–4628. IEEE, New York (2015)

32. Hwang, K., Sung, W.: Online sequence training of recurrent neural networks with connectionist temporal classification (2015). arXiv preprint arXiv:1511.06841
33. Kalchbrenner, N., Blunsom, P.: Recurrent convolutional neural networks for discourse compositionality. CoRR abs/1306.3584 (2013). http://arxiv.org/abs/1306.3584
34. Karpathy, A., Toderici, G., Shetty, S., Leung, T., Sukthankar, R., Fei-Fei, L.: Large-scale video classification with convolutional neural networks. In: Proceedings of the IEEE Conference on Computer Vision and Pattern Recognition, pp. 1725–1732 (2014)
35. Karpathy, A., Johnson, J., Li, F.F.: Visualizing and understanding recurrent networks (2015). arXiv preprint arXiv:1506.02078
36. Kilgour, K.: Modularity and neural integration in large-vocabulary continuous speech recognition. Ph.D. thesis, Karlsruhe Institute of Technology (2015)
37. Kingsbury, B.: Lattice-based optimization of sequence classification criteria for neural-network acoustic modeling. In: 2009 IEEE International Conference on Acoustics, Speech and Signal Processing, pp. 3761–3764. IEEE, New York (2009)
38. Koehn, P., Och, F.J., Marcu, D.: Statistical phrase-based translation. In: Proceedings of the 2003 Conference of the North American Chapter of the Association for Computational Linguistics on Human Language Technology, vol. 1, pp. 48–54. Association for Computational Linguistics, Stroudsburg (2003)
39. Li, H., Lin, Z., Shen, X., Brandt, J., Hua, G.: A convolutional neural network cascade for face detection. In: Proceedings of the IEEE Conference on Computer Vision and Pattern Recognition, pp. 5325–5334 (2015)
40. Li, J., Zhang, H., Cai, X., Xu, B.: Towards end-to-end speech recognition for Chinese Mandarin using long short-term memory recurrent neural networks. In: Sixteenth Annual Conference of the International Speech Communication Association (INTERSPEECH). ISCA, Dresden (2015)
41. Liu, Y., Fung, P., Yang, Y., Cieri, C., Huang, S., Graff, D.: HKUST/MTS: a very large scale Mandarin telephone speech corpus. In: Chinese Spoken Language Processing, pp. 724–735 (2006)
42. Lowe, D.G.: Object recognition from local scale-invariant features. In: Proceedings of the Seventh IEEE International Conference on Computer Vision, 1999, vol. 2, pp. 1150–1157. IEEE, New York (1999)
43. Lu, L., Zhang, X., Cho, K., Renals, S.: A study of the recurrent neural network encoder-decoder for large vocabulary speech recognition. In: Sixteenth Annual Conference of the International Speech Communication Association (2015)
44. Lu, L., Kong, L., Dyer, C., Smith, N.A., Renals, S.: Segmental recurrent neural networks for end-to-end speech recognition. CoRR abs/1603.00223 (2016). http://arxiv.org/abs/1603.00223
45. Lu, L., Zhang, X., Renals, S.: On training the recurrent neural network encoder–decoder for large vocabulary end-to-end speech recognition. In: 2016 IEEE International Conference on Acoustics, Speech and Signal Processing (ICASSP). IEEE, New York (2016)
46. Maas, A.L., Xie, Z., Jurafsky, D., Ng, A.Y.: Lexicon-free conversational speech recognition with neural networks. In: Proceedings of the 2015 Conference of the North American Chapter of the Association for Computational Linguistics: Human Language Technologies (2015)
47. Mangu, L., Brill, E., Stolcke, A.: Finding consensus in speech recognition: word error minimization and other applications of confusion networks. Comput. Speech Lang. 14(4), 373–400 (2000)
48. Metze, F., Fosler-Lussier, E., Bates, R.: The speech recognition virtual kitchen. In: Proceedings of INTERSPEECH. ISCA, Lyon, France (2013). https://github.com/srvk/eesen-transcriber
49. Miao, Y., Metze, F.: On speaker adaptation of long short-term memory recurrent neural networks. In: Sixteenth Annual Conference of the International Speech Communication Association (INTERSPEECH). ISCA, Dresden (2015)
50. Miao, Y., Gowayyed, M., Metze, F.: EESEN: end-to-end speech recognition using deep RNN models and WFST-based decoding. In: 2015 IEEE Workshop on Automatic Speech Recognition and Understanding (ASRU). IEEE, New York (2015)

51. Mohamed, A.R., Seide, F., Yu, D., Droppo, J., Stoicke, A., Zweig, G., Penn, G.: Deep bidirectional recurrent networks over spectral windows. In: 2015 IEEE Workshop on Automatic Speech Recognition and Understanding (ASRU), pp. 78–83. IEEE, New York (2015)
52. Mohri, M., Pereira, F., Riley, M.: Weighted finite-state transducers in speech recognition. Comput. Speech Lang. **16**(1), 69–88 (2002)
53. Nair, V., Hinton, G.E.: Rectified linear units improve restricted Boltzmann machines. In: Proceedings of the 27th International Conference on Machine Learning (ICML-10), pp. 807–814 (2010)
54. Palaz, D., Collobert, R., Doss, M.M.: Estimating phoneme class conditional probabilities from raw speech signal using convolutional neural networks (2013). arXiv preprint arXiv:1304.1018
55. Paul, D.B., Baker, J.M.: The design for the wall street journal-based CSR corpus. In: Proceedings of the Workshop on Speech and Natural Language, pp. 357–362. Association for Computational Linguistics, Morristown (1992)
56. Povey, D., Ghoshal, A., Boulianne, G., Burget, L., Glembek, O., Goel, N., Hannemann, M., Motlíček, P., Qian, Y., Schwarz, P., Silovský, J., Stemmer, G., Veselý, K.: The Kaldi speech recognition toolkit. In: 2011 IEEE Workshop on Automatic Speech Recognition and Understanding (ASRU), pp. 1–4. IEEE, New York (2011)
57. Rabiner, L.R.: A tutorial on hidden Markov models and selected applications in speech recognition. Proc. IEEE **77**(2), 257–286 (1989)
58. Sainath, T.N., Vinyals, O., Senior, A., Sak, H.: Convolutional, long short-term memory, fully connected deep neural networks. In: 2015 IEEE International Conference on Acoustics, Speech and Signal Processing (ICASSP), pp. 4580–4584. IEEE, New York (2015)
59. Sainath, T.N., Weiss, R.J., Senior, A., Wilson, K.W., Vinyals, O.: Learning the speech front-end with raw waveform CLDNNs. In: Sixteenth Annual Conference of the International Speech Communication Association (INTERSPEECH). ISCA, Dresden (2015)
60. Sak, H., Senior, A., Beaufays, F.: Long short-term memory recurrent neural network architectures for large scale acoustic modeling. In: Fifteenth Annual Conference of the International Speech Communication Association (INTERSPEECH). ISCA, Singapore (2014)
61. Sak, H., Senior, A., Rao, K., Irsoy, O., Graves, A., Beaufays, F., Schalkwyk, J.: Learning acoustic frame labeling for speech recognition with recurrent neural networks. In: 2015 IEEE International Conference on Acoustics, Speech and Signal Processing (ICASSP), pp. 4280–4284. IEEE, New York (2015)
62. Senior, A., Heigold, G., Bacchiani, M., Liao, H.: GMM-free DNN training. In: 2014 IEEE International Conference on Acoustics, Speech and Signal Processing (ICASSP), pp. 5639–5643. IEEE, New York (2014)
63. Sutskever, I., Vinyals, O., Le, Q.V.: Sequence to sequence learning with neural networks. In: Advances in Neural Information Processing Systems, pp. 3104–3112 (2014)
64. Tüske, Z., Golik, P., Schlüter, R., Ney, H.: Acoustic modeling with deep neural networks using raw time signal for LVCSR. In: Fifteenth Annual Conference of the International Speech Communication Association (INTERSPEECH), pp. 890–894. ISCA, Singapore (2014)
65. Watanabe, S., Le Roux, J.: Black box optimization for automatic speech recognition. In: 2014 IEEE International Conference on Acoustics, Speech and Signal Processing (ICASSP), pp. 3256–3260. IEEE, New York (2014)
66. Wöllmer, M., Eyben, F., Schuller, B., Rigoll, G.: Spoken term detection with connectionist temporal classification: a novel hybrid CTC-DBN decoder. In: 2010 IEEE International Conference on Acoustics, Speech and Signal Processing (ICASSP), pp. 5274–5277. IEEE, New York (2010)
67. Xu, K., Ba, J., Kiros, R., Courville, A., Salakhutdinov, R., Zemel, R., Bengio, Y.: Show, attend and tell: neural image caption generation with visual attention (2015). arXiv preprint arXiv:1502.03044
68. Yao, L., Torabi, A., Cho, K., Ballas, N., Pal, C., Larochelle, H., Courville, A.: Describing videos by exploiting temporal structure. In: Proceedings of the IEEE International Conference on Computer Vision, pp. 4507–4515 (2015)

69. Zhang, C., Woodland, P.C.: Standalone training of context-dependent deep neural network acoustic models. In: 2014 IEEE International Conference on Acoustics, Speech and Signal Processing (ICASSP), pp. 5597–5601. IEEE, New York (2014)
70. Zhou, B., Lapedriza, A., Xiao, J., Torralba, A., Oliva, A.: Learning deep features for scene recognition using places database. In: Advances in Neural Information Processing Systems, pp. 487–495 (2014)
71. Zweig, G., Yu, C., Droppo, J., Stolcke, A.: Advances in all-neural speech recognition (2016). arXiv:1609.05935

Part III
Resources

Chapter 14
The CHiME Challenges: Robust Speech Recognition in Everyday Environments

Jon P. Barker, Ricard Marxer, Emmanuel Vincent, and Shinji Watanabe

Abstract The CHiME challenge series has been aiming to advance the development of robust automatic speech recognition for use in everyday environments by encouraging research at the interface of signal processing and statistical modelling. The series has been running since 2011 and is now entering its 4th iteration. This chapter provides an overview of the CHiME series, including a description of the datasets that have been collected and the tasks that have been defined for each edition. In particular, the chapter describes novel approaches that have been developed for producing simulated data for system training and evaluation, and conclusions about the validity of using simulated data for robust-speech-recognition development. We also provide a brief overview of the systems and specific techniques that have proved successful for each task. These systems have demonstrated the remarkable robustness that can be achieved through a combination of training data simulation and multicondition training, well-engineered multi-channel enhancement, and state-of-the-art discriminative acoustic and language modelling techniques.

14.1 Introduction

Speech recognition technology is becoming increasingly pervasive. In particular, it is now being deployed in home and mobile consumer devices, where it is expected to work reliably in noisy, everyday listening environments. In many of these applications the microphones are at a significant distance from the user, so the captured speech signal is corrupted by interfering noise sources and reverberation.

J.P. Barker (✉) • R. Marxer
University of Sheffield, Regent Court, 211 Portobello, Sheffield S1 4DP, UK
e-mail: j.p.barker@sheffield.ac.uk; r.marxer@sheffield.ac.uk

E. Vincent
Inria, 615 rue du Jardin Botanique 54600 Villers-lès-Nancy, France
e-mail: emmanuel.vincent@inria.fr

S. Watanabe
Mitsubishi Electric Research Laboratories (MERL), Cambridge, MA, USA

© Springer International Publishing AG 2017
S. Watanabe et al. (eds.), *New Era for Robust Speech Recognition*,
DOI 10.1007/978-3-319-64680-0_14

Delivering reliable recognition performance in these conditions remains a challenging engineering problem.

One of the commonest approaches to distant speech recognition is to use a multichannel microphone array. Beamforming algorithms can then be used to capture the signal from the direction of the target talker while suppressing spatially distinct noise interferers. Although beamforming is a mature technique, the design and evaluation of algorithms is often performed by signal-processing researchers optimising speech enhancement objectives. Conversely, builders of speech recognition systems are often disappointed when they try to use beamforming algorithms 'off the shelf' with little idea how to properly optimise them for recognition.

The CHiME challenges were designed with the goal of building a community of researchers that would span signal processing and statistical speech recognition and make progress to robust distant-microphone speech recognition through closer collaboration. They were also prompted by a perceived gap in the speech recognition challenge landscape. Most challenges were being designed around lecture hall or meeting room scenarios where, although there might be considerable reverberation, the environment is essentially quiet, e.g. [18, 23, 24]. Other challenges model more extreme noise levels, but these typically use artificially mixed-in noise and pre-segmented test utterances, thus providing no opportunity to learn the structure of the noise background or to observe the noise context prior to the utterance, e.g. [14, 20, 25]. In contrast, the CHiME challenges were designed to draw attention towards the noise background by providing speech embedded in continuous recordings and accompanied by considerable quantities of matched noise-only training material.

The 1st CHiME challenge was launched in 2011 and the series is now entering its 4th iteration. Over that time the challenges have developed from small highly controlled tasks towards more complex scenarios with multiple dimensions of difficulty and greater commercial realism. This chapter provides an account of this development, providing a full description of the task design for each iteration and an overview of findings arising from analysis of challenge systems.

14.2   The 1st and 2nd CHiME Challenges (CHiME-1 and CHiME-2)

The 1st and 2nd CHiME challenges [4, 31] were conducted between 2011 and 2013 and were both based on a 'home automation' scenario, involving the recognition of command-like utterances using distant microphones in a noisy domestic environment. They both used simulated mixing, allowing the choice of speech and background materials to be separately controlled.

14.2.1   Domestic Noise Background

The noise backgrounds for the 1st and 2nd CHiME challenges were taken from the CHiME Domestic Audio dataset [6]. This data consists of recordings made in a family home using a B & K head and torso simulator type 4128 C. The head has built-in ear simulators that record a left- and a right-ear signal that approximate the signals that would be received by an average adult listener.

The CHiME challenge used recordings taken from a single room—a family living room—over the course of several weeks. The living room recordings were made during 22 separate morning and evening sessions, typically lasting around 1 h each and totalling over 20 h. The manikin remained in the same location throughout. The major noise sources were those that are typical of a family home: television, computer games console, children playing, conversations, some street noise from outside and noises from adjoining rooms, including washing machine noise and general kitchen noise.

The Domestic Audio dataset is also distributed with binaural room impulse response (BRIR) measurements that were made in the same recording room. The BRIRs were estimated using the sine sweep method [11] from a number of locations relative to the manikin. For each location, several BRIR estimates were made. For the particular location 2 m directly in front of the manikin (i.e. at an azimuthal angle of 0°), estimates were made with variable 'room settings': with a set of floor-length bay window curtains opened or closed, and with the door to the adjoining hallway open or closed.

14.2.2   The Speech Recognition Task Design

The 1st and 2nd CHiME challenges (CHiME-1 and CHiME-2) both employed artificial mixing of the speech and the noise background in order to carefully control the target signal-to-noise ratio. CHiME-1 used a small-vocabulary task and a fixed speaker location. CHiME-2 had two tracks, extending CHiME-1 in two separate directions: speaker motion and vocabulary size. In all tasks, utterances were embedded within complete unsegmented CHiME Domestic Audio recording sessions. Participants were supplied with the start and end times of each test utterance (i.e. speech activity detection was not part of the task) and they were also allowed to make use of knowledge of the surrounding audio before and after the utterance (e.g. to help estimate the noise component of the speech and noise mixture).

14.2.2.1   CHiME-1: Small Vocabulary

CHiME-1 was based on the small-vocabulary Grid corpus task [7]. This is a simple command sentence task that was initially designed for measuring the robustness of *human* speech recognition in noisy conditions. The corpus consists of 34 speakers (18 male and 16 female) each uttering 1000 unique 6-word commands with a simple fixed grammar. Each utterance contains a letter–digit grid reference. These two words were considered as the target keywords and performance was reported in terms of keyword correctness.

The Grid data was split such that 500 utterances per speaker were designated as training data and the remaining utterances were set aside as test data. From the test data, test sets of 600 utterances (about 20 utterances per speaker) were defined. To form noisy test utterances, Grid test set speech was convolved with the CHiME BRIRs and then added to a 14 h subset of the CHiME background audio. Temporal locations were selected such that the 600 utterances did not overlap and such that the mixtures had a fixed target signal-to-noise ratio (SNR). By varying the temporal locations, it was possible to achieve test sets with SNRs of −6, −3, 0, 3 and 6 dB. Separate test sets were produced for development and final evaluation. The final evaluation test set was released close to the deadline for submitting final results.

For training purposes, participants were supplied with a reverberated version of the 17,000-utterance CHiME training set, plus a further 6 h of background recording. The background audio was from the same room but made up from different recording sessions to those that had been used for the test data. Likewise, a different instance of the 2 m and 0° BRIR was used. No restrictions were placed on how this data could be used for system training.

14.2.2.2   CHiME-2 Track 1: Simulated Motion

CHiME-2 Track 1 was designed in response to criticism that the fixed impulse responses used in CHiME-1 made the task too artificial. To test this claim, variability was introduced into the training and test set BRIRs. Specifically, the effect of small speaker movements was simulated. To do this, a new set of BRIRs were recorded on a grid of locations around the 2 m and 0° location used for CHiME-1. The grid had a size of 20 cm by 20 cm and a 2 cm resolution, requiring a total of 121 (i.e. 11×11) BRIR measurements.

To simulate motion, first interpolation was used to increase the resolution of the BRIR grid in the left–right direction down to a 2.5 mm step size. Then, for each utterance, a random straight-line trajectory was produced such that the speaker moved at a constant speed of at most 15 cm/s over a distance of at most 5 cm within the grid. Then each sample of the clean utterances was convolved with the impulse response from the grid location that was closest to the speaker at that instant.

As with CHiME-1, a 17,000-utterance training set was provided and separate 600-utterance development and final test sets. All utterances had simulated motion. The test sets were produced with the same range of SNRs as CHiME-1.

14.2.2.3   CHiME-2 Track 2: Medium Vocabulary

CHiME-2 Track 2 extended CHiME-1 by replacing the small-vocabulary Grid task with the medium-vocabulary 5000-word Wall Street Journal (WSJ) task. The data were mixed in the same way as per CHiME-1, with a fixed BRIR at 2 m directly in front of the manikin. As with CHiME-1, different instances of the BRIR were used for the training, development and final test sets. The SNRs were defined as the median value of segmental SNRs computed over 200 ms windows to be compatible with the SNRs used in other WSJ tasks. It was found that because the WSJ utterances are longer than Grid utterances, there were fewer periods of CHiME background where low SNRs could be sustained. Hence, some signal rescaling had to be employed to obtain the lowest SNRs. Also, it was not possible to follow the rule that temporal locations should be chosen such that test utterances would not share some portions of the background.

The training data consisted of 7138 reverberated utterances from 83 speakers forming the WSJ0 SI-84 training set. The development data was 409 utterances from the 10 speakers forming the "no verbal punctuation" part of the WSJ0 speaker-independent 5k development set. The final test set was 330 utterances from 12 different speakers forming the Nov92 ARPA WSJ evaluation set.

14.2.3   Overview of System Performance

The CHiME-1 challenge attracted participation from 13 teams. A broad range of strategies were employed that could be grouped under target enhancement, robust feature extraction and robust decoding. A full review of the systems can be found in [4]. Generally, systems that delivered the best performance successfully combined an enhancement stage (exploiting both spatial and spectral diversity) and a robust decoder, using either some form of uncertainty propagation, an adapted training objective (e.g. MLLR, MAP, bMMI) or simply a multicondition training strategy using speech plus background mixtures.

For comparison, the challenge was published with a 'vanilla' hidden-Markov-model/Gaussian-mixture-model (HMM/GMM) baseline system trained on the reverberated speech with mel-frequency cepstral coefficient (MFCC) features. This non-robust system scored 82% keyword correctness at 9 dB, with performance falling to 30% at −6 dB. Listening tests established human performance to be 98% at 9 dB and falling to 90% at −6 dB. Scores for the submitted systems were broadly spread between the non-robust baseline system and the human performance. The overall best-performing system [8] made only 57% more errors than the human, with correctness varying between 86% and 96% across the SNR range. Analysis of the top-performing systems indicated that the most important strategies were multicondition training, spatial-diversity-based enhancement and robust training.

The CHiME-2 outcomes are reviewed in [30]. CHiME-2 Track 1 attracted participation from 11 teams, with some overlap with the earlier CHiME-1 challenge. The top-performing team achieved a score very similar to the best performance achieved on CHiME-1. Further, teams that made a direct comparison between CHiME-1 and CHiME-2 achieved equal scores on both tasks. It was concluded that the simulated small speaker movements caused little extra difficulty. Track 2 received only four entrants with one clear top performer [27] achieving word error rates (WERs) ranging from 14.8% to 44.1% from 9 to −6 dB SNR. Achieving this performance required a highly optimised system using spatial enhancement, a host of feature-space transformations, a decoder employing discriminative acoustic and language models, and ROVER combination of system variants. Spectral-diversity-based enhancement, which had performed extremely well in the small-vocabulary task, was found to be less useful in Track 2.

14.2.4 Interim Conclusions

The CHiME-1 and CHiME-2 challenges clearly demonstrated that distant microphone automatic speech recognition (ASR) systems need careful optimisation of both the signal processing and the statistical back end. However, the challenge design left several questions unanswered.

Are results obtained on artificially mixed speech representative of performance on real tasks? The artificial mixing is useful in allowing SNRs to be carefully controlled, but it raises questions about the realism of the data. First, the challenges used studio-recorded speech from the Grid and WSJ corpora. Although this speech was convolved with room impulse responses to model the effects of reverberation, speech read in a studio environment will differ in other significant ways from speech spoken and recorded live in noise. Second, it is likely that the range of the SNRs used is not representative of SNRs observed in real distant-microphone speech applications. Third, the simulation does not capture the channel variability of real acoustic mixing, where many factors will have an impact on the BRIRs.

How can evaluation be designed so as to allow fairer cross-team comparisons? One problem with both CHiME-1 and CHiME-2 was that the lack of a state-of-the-art baseline left every team to develop systems from the ground up. This led to an interesting diversity of approaches but reduced the opportunity for scientifically controlled comparison. Further, although the noise background training data was specified, there were no restrictions on how it could be used. Systems that employed multicondition training and generated larger noisy training datasets had higher performance. Tighter control of the training conditions could have allowed for more meaningful comparison.

14.3   The 3rd CHiME Challenge (CHiME-3)

The 3rd CHiME challenge was designed in response to feedback from the earlier challenges. Several priorities were identified. First, the domestic setting of the earlier challenge had been considered rather narrow and there was a desire to broaden the range of noise environments. Second, it was decided to move away from the binaural microphone configuration towards a more traditional microphone array setting that would have greater commercial relevance. The CHiME-3 scenario was therefore chosen to be that of an automatic speech recognition application running on a mobile device that would be used in noisy everyday settings. In order to make the task challenging, a lap-held tablet computer was selected as the target device, for which it was estimated microphone distances would be in the range of 30–40 cm (i.e. considerably greater than the typical distance for mobile phone usage). Finally, to answer questions that had been raised about the validity of using simulated mixing for training and testing systems, a direct 'simulated versus real' data comparison was built into the task design.

14.3.1   The Mobile Tablet Recordings

The CHiME-3 speech recordings were made using a 6-channel microphone array constructed by embedding Audio-Technica ATR3350 omnidirectional lavalier microphones around the edge of a frame designed to hold a Samsung Galaxy tablet computer. The array was designed to be held in landscape orientation with three microphones positioned along the top and bottom edges, as indicated in Fig. 14.1. All microphones faced forward except the top-central microphone, which faced backwards and was mounted flush with the rear of the frame.

The microphone signals were recorded sample-synchronously using a 6-channel TASCAM DR-680 portable digital recorder. A second TASCAM DR-680 was used to record a signal from a Beyerdynamic condenser close-talking microphone (CTM). The recorders were daisy-chained together to allow their transports to be controlled via a common interface. There was a variable delay between the units of up to 20 ms. All recordings were made with 16 bits at 48 kHz and later downsampled to 16 kHz.

Speech was recorded for training, development and test sets. Four native US talkers were recruited for each set (two male and two female). Speakers were instructed to read sentences that were presented on the tablet PC while holding the device in any way that felt natural. Each speaker recorded utterances first in an IAC single-walled acoustically isolated booth and then in each of the following environments: on a bus (BUS), on a street junction (STR), in a café (CAF) and in a pedestrian area (PED). Speakers were prompted to change their seating/standing position after every ten utterances. Utterances that were misread or read with disfluency were re-read until a satisfactory rendition had been recorded.

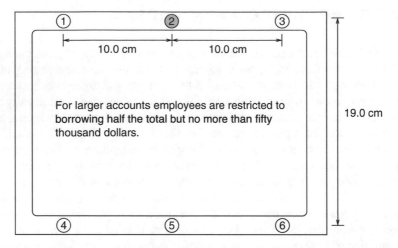

Fig. 14.1 The geometry of the 6-channel CHiME-3 microphone array. All microphones are forward facing except for channel 2 (*shaded grey*), which faces backwards and is flush with the rear of the 1 cm thick frame

14.3.2   The CHiME-3 Task Design: Real and Simulated Data

The task was based on the WSJ0 5K ASR task, i.e. it remained comparable with CHiME-2 Track 2. For the training data, 100 utterances were recorded by each speaker in each environment, totalling 1600 utterances selected at random from the full 7138 WSJ0 SI-84 training set. Speakers assigned to the 409-utterance development set or the 330-utterance final test set each spoke 1/4 of each set in each environment, resulting in 1636 (4 × 409) and 1320 (4 × 330) utterances for development and final testing, respectively.

The live-recorded training data was supplemented with 7138 simulated noisy utterances constructed by artificially adding the WSJ training set to a separately recorded 8 h of noise background (2 h from each of the environments). Techniques for simulation were included as part of the baseline described in the next section. Participants were encouraged to try and improve on the baseline simulation technique under the assumption that reducing the mismatch between simulated training data and real test data would lead to better ASR performance. In order to extend the scientific outcomes of the challenge, a simulated development and test set was also produced. Given that previous CHiME challenges had used only simulated data, it was important to know whether the performance of a system evaluated using simulated data was a good predictor of performance on real data.

Additional rules were imposed in order to keep systems as comparable as possible. Chiefly, participants were asked to tune system parameters using only the development data and to report results on the final test data. Any language model was allowed as long as it was trained with official WSJ language model training data. New simulation techniques for training data were not allowed to expand the amount of training data and had to keep the same pairing between utterances and segments of noise background. A constraint of 5 s was placed on the amount of audio context that could be used preceding an utterance.

14.3.3   The CHiME-3 Baseline Systems

The CHiME-3 challenge was distributed alongside baseline systems for training data simulation, multichannel speech enhancement and automatic speech recognition. These systems are outlined below and are described in greater detail in [5].

14.3.3.1   Simulation

The simulation baseline software was designed for adding clean WSJ speech to microphone array noise recordings in such a way as to model the effects of speaker and tablet motion. The procedure for mixing was performed in two stages.

First, a set of short-time Fourier transform (STFT)-domain time-varying impulse responses (IRs) between the close-talking microphone (considered to be clean speech) and each of the other microphones were estimated in the least-squares sense. Estimates were made in each frequency bin and in blocks of frames partitioned such that each partition contained a similar amount of speech. The SNR at each tablet microphone could then be estimated.

In a second stage, the spatial position of the speaker was tracked in each of the CHiME training data recordings. To do this, signals were first represented in the complex-valued STFT domain using 1024-sample, half-overlapped sine windows. The position of the speaker was encoded by a nonlinear SRP-PHAT pseudo-spectrum. The peaks of the pseudo-spectrum were tracked using the Viterbi algorithm. A time-varying filter modelling direct sound between the speaker and the microphones was then constructed.

Original WSJ training utterances were then convolved with filters estimated from CHiME training utterances. An additional equalisation filter was applied that was estimated as the ratio of the average power spectrum of the CHiME booth recordings and the average power spectrum of the WSJ training data. Finally, the equalised recordings were rescaled to match the estimated real training data SNRs and were then mixed with noise backgrounds taken from the separate 8 h set of noise-only recordings.

14.3.3.2   Enhancement

The baseline enhancement system was designed to take the 6-channel array recordings and produce a single-channel output with reduced background noise, suitable for input into the ASR system.

The baseline system was based on a minimum variance distortionless response (MVDR) beamforming approach. The target talker was tracked using the peaks in the nonlinear SRP-PHAT pseudo-spectrum (as used in the simulation component). The multichannel noise covariance matrix was estimated from 400 to 800 ms of context prior to the utterance. MVDR with diagonal loading was then employed to estimate the target speech spectrum.

Some of the CHiME test recordings were subject to microphone failures. These could be caused by microphone occlusion during handling, or vibrations leading to intermittent connection failures (particularly in the BUS environment). The baseline system applied a simple energy-based criterion to detect microphone failure and ignore failed channels.

14.3.3.3   ASR

Two Kaldi-based ASR baseline systems were provided: a lightweight GMM/HMM system for rapid experimentation and a state-of-the-art deep-neural-network (DNN) baseline for final benchmarking.

The GMM/HMM system employed 13th-order MFCCs to represent individual frames. Feature vectors were then formed by concatenating three frames of left and right context and compressing to 40 dimensions using linear discriminative analysis, with classes being one of 2500 tied tri-phone HMM states. A total of 15,000 Gaussians were used to model the tied states. The system also employed maximum likelihood linear transformation and feature-space maximum likelihood linear regression with speaker-adaptive training.

The DNN baseline employed a network with 7 layers and 2048 units per hidden layer. Input was based on the 40-dimensional filterbank frames with 5 frames of left and right context (i.e. a total of $11 \times 40 = 440$ input units). The DNN was trained using standard procedures described in [29]: pre-training using restricted Boltzmann machines, cross-entropy training and sequence-discriminative training using the state-level minimum Bayes risk criterion.

14.4   The CHiME-3 Evaluations

A total of 26 systems were submitted to the CHiME-3 challenge, all of which achieved a lower test set WER than the 33.4% scored by the baseline DNN system. This section presents the performance of the top systems and provides an overview of the strategies that were most effective for reducing WERs.

Table 14.1 Overview of the top ten systems submitted to the CHiME-3 challenge

System	Tr	ME	SE	FE	FT	AM	LM	SC	BUS	CAF	PED	STR	Ave.
Yoshioka et al. [34]	X	X		X		X	X		7.4	4.5	6.2	5.2	5.8
Hori et al. [15]		X	X	X	X		X	X	13.5	7.7	7.1	8.1	9.1
Du et al. [9]		X		X	X	X		X	13.8	11.4	9.3	7.8	10.6
Sivasankaran et al. [26]	X	X		X	X		X		16.2	9.6	12.3	7.2	11.3
Moritz et al. [17]	X			X	X		X		13.5	13.5	10.6	9.2	11.7
Fujita et al. [12]		X	X	X	X			X	16.6	11.8	10.0	8.8	11.8
Zhao et al. [35]	X	X		X	X				14.5	11.7	11.5	10.0	11.9
Vu et al. [32]		X	X		X		X		17.6	12.1	8.5	9.6	11.9
Tran et al.	X	X		X	X		X		18.6	10.7	9.7	9.6	12.1
Heymann et al. [13]	X	X					X		17.5	10.5	11.0	10.0	12.3
DNN baseline v2		X			X		X		19.1	11.4	10.3	10.3	12.8
DNN baseline									51.8	34.7	27.2	20.1	33.4

The left side of the table summarizes the key features of each system, indicating where the systems have differed from the baseline with respect to training (Tr), multichannel enhancement (ME), single channel enhancement (SE), feature extraction (FE), feature transformation (FT), acoustic modelling (AM), language modelling (LM) and system combination (SC). The right-hand side reports WERs for the final test set overall (Ave.) and for each environment individually. Results are shown for the real-data test set only. For performance on the simulated data see [5]

14.4.1 An Overview of CHiME-3 System Performance

Table 14.1 presents the results of the top ten overall best systems. Most of the best systems achieved WERs in the range from 13% down to 10%. The overall best system achieved a WER of just 5.8%, significantly better than the 2nd-placed system. The table also shows WERs broken down by noise environment. For most systems the highest WERs were observed in the BUS environment and the lowest in STR, with CAF and PED lying somewhere in between. However, there are notable exceptions; for example, the best system has a WER of just 4.5% in the CAF environment.

14.4.2 An Overview of Successful Strategies

Analysis of the results shows that no single technique is sufficient for success. Systems near the top of the table modified multiple components, whereas systems that improved one or two components performed consistently poorly. Best performance required a combination of improved multichannel processing, good feature normalisation and improvement of the baseline language model. The most commonly employed strategies are reviewed briefly below.

14.4.2.1 Strategies for Improved Signal Enhancement

Good target enhancement is crucial for success, and nearly all teams attempted to improve this component of the baseline. Many teams replaced the baseline's super-directive MVDR beamformer with a conventional delay-and-sum beamformer, e.g. [15, 22, 26]. Others kept the MVDR framework but tried to improve the estimates of the steering vector [34], or the speech and noise covariances [13]. Another popular strategy was to add a postfilter stage, for example spatial coherence filtering [3, 19] or dereverberation [10, 34]. A smaller number of teams used additional single-channel enhancement stages after the array processing, e.g. NMF-based source separation [2, 32], but these approaches were found to have a more marginal benefit.

14.4.2.2 Strategies for Improved Statistical Modelling

Most teams adopted the same feature design as the baseline design, i.e. MFCC features for the initial alignment stages followed by filterbank features for the DNN pass. However, good speaker/environment normalisation was found to be important. Whereas the baseline only applied explicit speaker-normalising transforms in the HMM/GMM training, it was found that it was also advantageous to improve normalisation for DNN training. Strategies included performing utterance-based feature mean and variance normalisation [9, 12, 33, 35] and augmenting DNN inputs with pitch-based features [9, 16, 33]. The most successful strategies were found to be feature space maximum likelihood linear regression (fMLLR) [15, 17, 26, 32] and augmentation of DNN inputs with either i-vectors [17, 36] or bottleneck features extracted from a speaker classification DNN [28]. Using both fMLLR and feature vector augmentation provided additive benefits [19, 28, 36].

For acoustic modelling, most teams adopted the DNN architecture provided by the baseline system. Notable alternatives included convolutional neural networks, e.g. [2, 16, 33], and long short-time memory (LSTM) networks, e.g. [2, 19]. A comparison of submission performance did not demonstrate any clear advantage for any particular architecture, and, indeed, some of the best systems employed the baseline architecture. Where alternative architectures were employed they were often used in combination, e.g. [9, 34, 36].

Most teams implemented a language-model rescoring stage using a more sophisticated model than the 3-gram model used by the baseline decoder. All teams doing so were able to achieve significant performance enhancements. Language models used for rescoring included DNN-LMs [32], LSTM-LMs [10] or, most commonly, recurrent neural network language models (RNN-LMs) [21, 26, 28, 34].

14.4.2.3   Strategies for Improved System Training

The CHiME-3 challenge was designed so as to let teams experiment with training data simulation. It was stressed that the simulation technique used to make the simulated training data was to be considered as a baseline, and the MATLAB source code was made available to all participants. The rules allowed the WSJ and noise background to be remixed as long as each training utterance remained paired with the same segment of noise background.

Despite encouragement, few teams attempted to experiment with alterations to the training data. The only exceptions were [13] and [33], who achieved significant performance improvements by remixing the training data at a range of SNRs. Although within the rules, this increases the training-data quantity rather than just the quality. One other team [26] generated simulated training data in the feature domain using a condition-restricted Boltzmann machine but failed to achieve better results. A number of teams generated an expanded training set by simply applying feature extraction directly to the individual channels (i.e. rather than first combining them into a single enhanced signal) [17, 34–36]. Surprisingly, this produced consistent improvements in performance despite the mismatch between the individual channels and the enhanced signals used for testing.

Techniques for improved training data simulation have remained largely unexplored. Given the relative simplicity of the baseline simulation, there is potential for significant advancements in this area.

14.4.3   Key Findings

The analysis of CHiME-3 systems indicates that to achieve the highest scores requires complex systems applying multiple recognition passes and the possible combination of multiple feature extractors and classifiers. However, the largest consistently observed gains over the baseline came from three commonly applied techniques. First, replacing the MVDR beamformer with a delay-and-sum beamformer. (The teams taking this step used the BeamformIt toolkit beamformer implementation [1] and therefore the improvement in WER may be partly due to the manner in which BeamformIt implicitly weights microphones according to their correlation, hence making it robust to microphone failures, in addition to the difference between MVDR and delay-and-sum.) Second, providing better speaker and environment normalisation by employing fMLLR transformed features for training the DNN. Third, adding a language-model-rescoring stage using a more complex language model, e.g. either a 5-gram model or an RNN language model.

After the challenge, a new baseline system was built that incorporated these three changes. This reduced the baseline WER from 33.4% to 12.8% making it competitive with the top ten systems (see the row labelled 'DNN Baseline v2' in Table 14.1). This system has now replaced the original baseline as the official CHiME-3 baseline distributed with Kaldi.

A secondary goal of the challenge was to investigate the utility of simulated multichannel data either for training systems or for evaluation. Regarding acoustic modelling of noisy data, where comparisons were made it was found that using the simulated data always improved results compared to using real data alone, despite possible mismatch. However, some care is needed with the microphone array processing of the simulated data. The simple nature of the mixing means that array processing that has been optimised for the simulated data can produce overly optimistic enhancements, i.e. enhancements in which the SNRs are not representative of the SNRs that will be achieved when enhancing the real data. This mismatch can lead to poorer system performance and may explain why remixing the simulated training with a broader range of SNRs was beneficial. The problem could be fixed in a more principled fashion by improving the simulation itself; however, few teams attempted this, so there is more work to be done before conclusions can be drawn.

Finally, considering simulated test data, Barker et al. [5] presents the correlation between system performance on the real and simulated test sets across all 26 systems submitted to the challenge. Although the correlation is strong, there were observed to be many outlier systems, in particular, systems which achieved very low WERs on the simulated data but proportionally poorer WERs on the real data. This result suggests that extreme caution is needed when interpreting the results of fully simulated challenges.

14.5   Future Directions: CHiME-4 and Beyond

Although significant progress has been made, distant-microphone speech recognition still remains a significant challenge. For modern everyday applications that are expected to work in a wide variety of noise environments, the root of the problem is the potential mismatch between training and test data: it is not possible to anticipate the acoustic environment in which the device will be used when training the system. The CHiME challenges reviewed in this chapter have highlighted two key distant-microphone ASR strategies that can address this mismatch problem. First, microphone array processing, which reduces the potential for mismatch by the degree to which it successfully removes noise from the signal. Second, multicondition training, which reduces mismatch to the extent that the noise environment can be successfully anticipated.

Table 14.2 A summary of the CHiME challenge tasks

Edition		Channels	Noise	Task	Mixing	SNR (dB)	Distance
CHiME-1		Binaural	Domestic	Grid	Simu static	−6 to 9	2 m
CHiME-2	Track 1	Binaural	Domestic	Grid	Simu moving	−6 to 9	2 m
	Track 2	Binaural	Domestic	WSJ 5K	Simu static	−6 to 9	2 m
CHiME-3		6	Urban	WSJ 5K	Real/simu	−5 to 0	30–40 cm
CHiME-4	1-CH	1	Urban	WSJ 5K	Real/simu	−5 to 0	30–40 cm
	2-CH	2	Urban	WSJ 5K	Real/simu	−5 to 0	30–40 cm
	6-CH	6	Urban	WSJ 5K	Real/simu	−5 to 0	30–40 cm

The solutions to mismatch seen in the CHiME systems have proved remarkably effective, particularly in CHiME-3. However, although efforts were made to increase the realism of the evaluation, the challenge design significantly under-represents the degree of mismatch that real systems will have to handle. First, the training and test speech both come from the same narrow and well-represented domain for which it is possible to build well-matched language and acoustic models. Second, the training data has been recorded on the exact same device that is used for testing. This means that not only is the microphone array geometry matched, but so too are the individual microphone channels. Third, the noise environments, although more varied than those employed in many challenges, still only represent four rather narrow situations. Again, the same noise environments were employed in both the training and the test data.

One of the aims of the Fred Jelinek Workshop presented in this book was to develop novel solutions to the mismatch problem. In order to emphasise mismatch, novel evaluation protocols were developed, in particular cross-corpora evaluation in which the training data from one corpus (e.g. AMI) would be used for building systems to be tested with data from another (e.g. CHiME-3). Inspired by this work, a new iteration of the CHiME challenge (CHiME-4) is now in progress. This iteration will use the same datasets that were constructed for CHiME-3 but has taken two steps towards increasing the mismatch challenge. First, 1- and 2-channel tracks are being introduced that will reduce the opportunity for noise removal in the enhancement stage. Second, the 1- and 2-channel tasks will employ different channel subsets for training and testing.

To conclude, a summary of all the datasets and tasks comprising the complete CHiME challenge series is presented in Table 14.2. The datasets for all CHiME editions are publicly available, and state-of-the-art baseline systems are distributed with the Kaldi speech recognition toolkit.[1]

---

[1]Instructions for obtaining CHiME datasets can be found at http://spandh.dcs.shef.ac.uk/chime.

References

1. Anguera, X., Wooters, C., Hernando, J.: Acoustic beamforming for speaker diarization of meetings. IEEE Trans. Audio Speech Lang. Process. **15**(7), 2011–2023 (2007)
2. Baby, D., Virtanen, T., Van Hamme, H.: Coupled dictionary-based speech enhancement for CHiME-3 challenge. Technical Report KUL/ESAT/PSI/1503, KU Leuven, ESAT, Leuven (2015)
3. Barfuss, H., Huemmer, C., Schwarz, A., Kellermann, W.: Robust coherence-based spectral enhancement for distant speech recognition (2015). arXiv:1509.06882
4. Barker, J., Vincent, E., Ma, N., Christensen, H., Green, P.: The PASCAL CHiME speech separation and recognition challenge. Comput. Speech Lang. **27**(3), 621–633 (2013)
5. Barker, J., Marxer, R., Vincent, E., Watanabe, S.: The third 'CHiME' speech separation and recognition challenge: dataset, task and baselines. In: 2015 IEEE Workshop on Automatic Speech Recognition and Understanding, ASRU 2015, Scottsdale, AZ, December 13–17, 2015, pp. 504–511 (2015). doi:10.1109/ASRU.2015.7404837
6. Christensen, H., Barker, J., Ma, N., Green, P.: The CHiME corpus: a resource and a challenge for computational hearing in multisoure environments. In: Proceedings of the 11th Annual Conference of the International Speech Communication Association (Interspeech 2010), Makuhari (2010)
7. Cooke, M., Barker, J., Cunningham, S., Shao, X.: An audio-visual corpus for speech perception and automatic speech recognition. J. Acoust. Soc. Am. **120**(5), 2421–2424 (2006). doi:10.1121/1.2229005
8. Delcroix, M., Kinoshita, K., Nakatani, T., Araki, S., Ogawa, A., Hori, T., Watanabe, S., Fujimoto, M., Yoshioka, T., Oba, T., Kubo, Y., Souden, M., Hahm, S.J., Nakamura, A.: Speech recognition in the presence of highly non-stationary noise based on spatial, spectral and temporal speech/noise modeling combined with dynamic variance adaptation. In: Proceedings of the 1st CHiME Workshop on Machine Listening in Multisource Environments, Florence, pp. 12–17 (2011)
9. Du, J., Wang, Q., Tu, Y.H., Bao, X., Dai, L.R., Lee, C.H.: An information fusion approach to recognizing microphone array speech in the CHiME-3 challenge based on a deep learning framework. In: 2015 IEEE Workshop on Automatic Speech Recognition and Understanding, ASRU 2015, Scottsdale, AZ, December 13–17, 2015, pp. 430–435 (2015)
10. El-Desoky Mousa, A., Marchi, E., Schuller, B.: The ICSTM+TUM+UP approach to the 3rd CHiME challenge: single-channel LSTM speech enhancement with multi-channel correlation shaping dereverberation and LSTM language models (2015). arXiv:1510.00268
11. Farina, A.: Simultaneous measurement of impulse response and distortion with a swept sine technique. In: Proceedings of the 108th AES Convention, Paris (2000)
12. Fujita, Y., Takashima, R., Homma, T., Ikeshita, R., Kawaguchi, Y., Sumiyoshi, T., Endo, T., Togami, M.: Unified ASR system using LGM-based source separation, noise-robust feature extraction, and word hypothesis selection. In: 2015 IEEE Workshop on Automatic Speech Recognition and Understanding ASRU 2015, Scottsdale, AZ, December 13–17, 2015, pp. 416–422 (2015)
13. Heymann, J., Drude, L., Chinaev, A., Haeb-Umbach, R.: BLSTM supported GEV beamformer front-end for the 3rd CHiME challenge. In: 2015 IEEE Workshop on Automatic Speech Recognition and Understanding ASRU 2015, Scottsdale, AZ, December 13–17, 2015, pp. 444–451 (2015)
14. Hirsch, H.G., Pearce, D.: The Aurora experimental framework for the performance evaluation of speech recognition systems under noisy conditions. In: Proceedings of the 6th International Conference on Spoken Language Processing (ICSLP), vol. 4, pp. 29–32 (2000)
15. Hori, T., Chen, Z., Erdogan, H., Hershey, J.R., Le Roux, J., Mitra, V., Watanabe, S.: The MERL/SRI system for the 3rd CHiME challenge using beamforming, robust feature extraction, and advanced speech recognition. In: 2015 IEEE Workshop on Automatic Speech Recognition and Understanding ASRU 2015, Scottsdale, AZ, December 13–17, 2015, pp. 475–481 (2015)

16. Ma, N., Marxer, R., Barker, J., Brown, G.J.: Exploiting synchrony spectra and deep neural networks for noise-robust automatic speech recognition. In: 2015 IEEE Workshop on Automatic Speech Recognition and Understanding ASRU 2015, Scottsdale, AZ, December 13–17, 2015, pp. 490–495 (2015)

17. Moritz, N., Gerlach, S., Adiloglu, K., Anemüller, J., Kollmeier, B., Goetze, S.: A CHiME-3 challenge system: long-term acoustic features for noise robust automatic speech recognition. In: 2015 IEEE Workshop on Automatic Speech Recognition and Understanding ASRU 2015, Scottsdale, AZ, December 13–17, 2015, pp. 468–474 (2015)

18. Mostefa, D., Moreau, N., Choukri, K., Potamianos, G., Chu, S.M., Tyagi, A., Casas, J.R., Turmo, J., Cristoforetti, L., Tobia, F., , Pnevmatikakis, A., Mylonakis, V., Talantzis, F., Burger, S., Stiefelhagen, R., Bernardin, K., Rochet, C.: The CHIL audiovisual corpus for lecture and meeting analysis inside smart rooms. Lang. Resour. Eval. **41**(3–4), 389–407 (2007)

19. Pang, Z., Zhu, F.: Noise-robust ASR for the third 'CHiME' challenge exploiting time–frequency masking based multi-channel speech enhancement and recurrent neural network (2015). arXiv:1509.07211

20. Parihar, N., Picone, J., Pearce, D., Hirsch, H.G.: Performance analysis of the Aurora large vocabulary baseline system. In: Proceedings of the 2004 European Signal Processing Conference (EUSIPCO), Vienna, pp. 553–556 (2004)

21. Pfeifenberger, L., Schrank, T., Zöhrer, M., Hagmüller, M., Pernkopf, F.: Multi-channel speech processing architectures for noise robust speech recognition: 3rd CHiME challenge results. In: 2015 IEEE Workshop on Automatic Speech Recognition and Understanding, ASRU 2015, Scottsdale, AZ, December 13–17, 2015, pp. 452–459 (2015)

22. Prudnikov, A., Korenevsky, M., Aleinik, S.: Adaptive beamforming and adaptive training of DNN acoustic models for enhanced multichannel noisy speech recognition. In: 2015 IEEE Workshop on Automatic Speech Recognition and Understanding, ASRU 2015, Scottsdale, AZ, December 13–17, 2015, pp. 401–408 (2015)

23. Renals, S., Hain, T., Bourlard, H.: Interpretation of multiparty meetings: the AMI and AMIDA projects. In: Proceedings of the 2nd Joint Workshop on Hands-free Speech Communication and Microphone Arrays (HSCMA), pp. 115–118 (2008)

24. RWCP meeting speech corpus (RWCP-SP01) (2001). http://research.nii.ac.jp/src/en/RWCP-SP01.html

25. Segura, J.C., Ehrette, T., Potamianos, A., Fohr, D., Illina, I., Breton, P.A., Clot, V., Gemello, R., Matassoni, M., Maragos, P.: The HIWIRE database, a noisy and non-native English speech corpus for cockpit communication (2007). http://islrn.org/resources/934-733-835-065-0/

26. Sivasankaran, S., Nugraha, A.A., Vincent, E., Morales-Cordovilla, J.A., Dalmia, S., Illina, I.: Robust ASR using neural network based speech enhancement and feature simulation. In: 2015 IEEE Workshop on Automatic Speech Recognition and Understanding, ASRU 2015, Scottsdale, AZ, December 13–17, 2015, pp. 482–489 (2015)

27. Tachioka, Y., Watanabe, S., Le Roux, J., Hershey, J.R.: Discriminative methods for noise robust speech recognition: a CHiME challenge benchmark. In: Proceedings of the 2nd CHiME Workshop on Machine Listening in Multisource Environments, Vancouver (2013)

28. Tachioka, Y., Kanagawa, H., Ishii, J.: The overview of the MELCO ASR system for the third CHiME challenge. Technical Report SVAN154551, Mitsubishi Electric (2015)

29. Veselý, K., Ghoshal, A., Burget, L., Povey, D.: Sequence-discriminative training of deep neural networks. In: Proceedings of INTERSPEECH, pp. 2345–2349 (2013)

30. Vincent, E., Barker, J., Watanabe, S., Le Roux, J., Nesta, F., Matassoni, M.: The second 'CHiME' speech separation and recognition challenge: an overview of challenge systems and outcomes. In: Proceedings of the 2013 IEEE Workshop on Automatic Speech Recognition and Understanding (ASRU), pp. 162–167 (2013)

31. Vincent, E., Barker, J., Watanabe, S., Le Roux, J., Nesta, F., Matassoni, M.: The second 'CHiME' speech separation and recognition challenge: datasets, tasks and baselines. In: Proceedings of ICASSP (2013)

32. Vu, T.T., Bigot, B., Chng, E.S.: Speech enhancement using beamforming and non negative matrix factorization for robust speech recognition in the CHiME-3 challenge. In: 2015 IEEE

Workshop on Automatic Speech Recognition and Understanding, ASRU 2015, Scottsdale, AZ, December 13–17, 2015, pp. 423–429 (2015)

33. Wang, X., Wu, C., Zhang, P., Wang, Z., Liu, Y., Li, X., Fu, Q., Yan, Y.: Noise robust IOA/CAS speech separation and recognition system for the third 'CHiME' challenge (2015). arXiv:1509.06103

34. Yoshioka, T., Ito, N., Delcroix, M., Ogawa, A., Kinoshita, K., Fujimoto, M., Yu, C., Fabian, W.J., Espi, M., Higuchi, T., Araki, S., Nakatani, T.: The NTT CHiME-3 system: advances in speech enhancement and recognition for mobile multi-microphone devices. In: 2015 IEEE Workshop on Automatic Speech Recognition and Understanding, ASRU 2015, Scottsdale, AZ, December 13–17, 2015, pp. 436–443 (2015)

35. Zhao, S., Xiao, X., Zhang, Z., Nguyen, T.N.T., Zhong, X., Ren, B., Wang, L., Jones, D.L., Chng, E.S., Li, H.: Robust speech recognition using beamforming with adaptive microphone gains and multichannel noise reduction. In: 2015 IEEE Workshop on Automatic Speech Recognition and Understanding, ASRU 2015, Scottsdale, AZ, December 13–17, 2015, pp. 460–467 (2015)

36. Zhuang, Y., You, Y., Tan, T., Bi, M., Bu, S., Deng, W., Qian, Y., Yin, M., Yu, K.: System combination for multi-channel noise robust ASR. Technical Report SP2015-07, Department of Computer Science and Engineering, Shanghai Jiao Tong University, Shanghai (2015)

Chapter 15
The REVERB Challenge: A Benchmark Task for Reverberation-Robust ASR Techniques

Keisuke Kinoshita, Marc Delcroix, Sharon Gannot, Emanuël A.P. Habets, Reinhold Haeb-Umbach, Walter Kellermann, Volker Leutnant, Roland Maas, Tomohiro Nakatani, Bhiksha Raj, Armin Sehr, and Takuya Yoshioka

Abstract The REVERB challenge is a benchmark task designed to evaluate reverberation-robust automatic speech recognition techniques under various conditions. A particular novelty of the REVERB challenge database is that it comprises both real reverberant speech recordings and simulated reverberant speech, both of which include tasks to evaluate techniques for 1-, 2-, and 8-microphone situations. In this chapter, we describe the problem of reverberation and characteristics of the REVERB challenge data, and finally briefly introduce some results and findings useful for reverberant speech processing in the current deep-neural-network era.

K. Kinoshita (✉) • M. Delcroix • T. Nakatani • T. Yoshioka
NTT Communication Science Laboratories, NTT Corporation, 2-4, Hikaridai, Seika-cho, Kyoto, Japan
e-mail: kinoshita.k@lab.ntt.co.jp

S. Gannot
Bar-Ilan University, Ramat Gan, Israel

E.A.P. Habets
International Audio Laboratories Erlangen, Erlangen, Germany

R. Haeb-Umbach
University of Paderborn, Paderborn, Germany

W. Kellermann • R. Maas
Friedrich-Alexander University of Erlangen-Nuremberg, Erlangen, Germany

V. Leutnant
Amazon Development Center Germany GmbH, Aachen, Germany

B. Raj
Carnegie Mellon University, Pittsburgh, PA, USA

A. Sehr
Ostbayerische Technische Hochschule Regensburg, Regensburg, Germany

© Springer International Publishing AG 2017
S. Watanabe et al. (eds.), *New Era for Robust Speech Recognition*,
DOI 10.1007/978-3-319-64680-0_15

345

15.1 Introduction

Speech signal-processing technologies have advanced significantly in the last few decades, and now play various important roles in our daily lives. Especially, speech recognition technology has advanced rapidly, and is increasingly coming into practical use, enabling a wide spectrum of innovative and exciting voice-driven applications. However, most applications consider a microphone located near the talker as a prerequisite for reliable performance, which prohibits further progress in automatic speech recognition (ASR) applications.

Speech signals captured with distant microphones inevitably contain interfering noise and reverberation, which severely degrade the speech intelligibility of the captured signals [16] and the performance of ASR systems [4, 18]. A noisy reverberant observed speech signal $y(t)$ at time t can be expressed as

$$y(t) = h(t) * s(t) + n(t), \tag{15.1}$$

where $h(t)$ corresponds to the room impulse response between the speaker and the microphone, $s(t)$ to the clean speech signal, $n(t)$ to the background noise, and $*$ to the convolution operator. Note that the primary focus of interest in the REVERB challenge is on reverberation, i.e., the effect of $h(t)$ on $s(t)$, and techniques which address it.

Research on reverberant speech processing has made significant progress in recent years [11, 19], mainly driven by multidisciplinary approaches that combine ideas from room acoustics, optimal filtering, machine learning, speech modeling, enhancement, and recognition. The motivation behind the REVERB challenge was to provide a common evaluation framework, i.e., tasks and databases, to assess and collectively compare algorithms and gain new insights regarding the potential future research directions for reverberant speech-processing technology.

This chapter summarizes the REVERB challenge, which took place in 2014 as a community-wide evaluation campaign for speech enhancement (SE) and ASR techniques [6, 7, 13]. While other benchmark tasks and challenges [1, 12, 17] mainly focus on the noise-robustness issue and sometimes only on a single-channel scenario, the REVERB challenge was designed to test robustness against *reverberation* under moderately noisy environments. The evaluation data of the challenge contains both *single-channel* and *multichannel* recordings, both of which comprise *real recordings* and *simulated data*, which has similar characteristics to real recordings. Although the REVERB challenge contains two tasks, namely SE and ASR tasks, we focus only on the latter task in this chapter.

The remainder of this chapter is organized as follows. In Sect. 15.2, we describe the scenario assumed in the challenge and details of the challenge data. Section 15.3 introduces results for baseline systems and top-performing systems. Section 15.4 provides a summary of the chapter and potential research directions to further develop reverberation-robust ASR techniques.

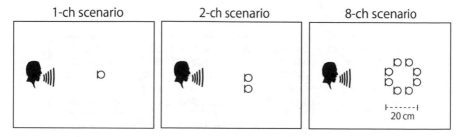

Fig. 15.1 Scenarios assumed in the REVERB challenge

15.2 Challenge Scenarios, Data, and Regulations

15.2.1 Scenarios Assumed in the Challenge

Figure 15.1 shows the three scenarios considered in this challenge [6, 7], in which an utterance spoken by a single spatially stationary speaker is captured with single-channel (1-ch), two-channel (2-ch), or eight-channel (8-ch) circular microphone arrays in a moderately noisy reverberant room. In practice, we commonly encounter this kind of acoustic situation when, e.g., we attend a presentation given in a small lecture room or a meeting room. In fact, the real recordings used in the challenge were recorded in an actual university meeting room, closely simulating the acoustic conditions of a lecture hall [10]. The 1-ch and 2-ch data are simply a subset of the 8-ch circular-microphone-array data. The 1-ch data were generated by randomly picking up one of eight microphones, while the 2-ch data were generated by randomly picking up adjacent two microphones from the eight microphones. For more details of the recording setting, refer to a document in the "download" section of the challenge webpage [13].

15.2.2 Data

For the challenge, the organizers provided a dataset which consisted of training data and test data. The test data comprised a development (Dev) test set and an evaluation (Eval) test set. All the data was provided as 1-ch, 2-ch, and 8-ch reverberant speech recordings at a sampling frequency of 16 kHz, and is available through the challenge webpage [13] via its "download" section. An overview of all the datasets is given in Fig. 15.2. Details of the test and training data are given in the following subsections.

15.2.2.1 Test Data: Dev and Eval Test Sets

By having the test data (i.e., the Dev and Eval test sets) consisting of both real recordings (RealData) and simulated data (SimData), the REVERB challenge

Test data

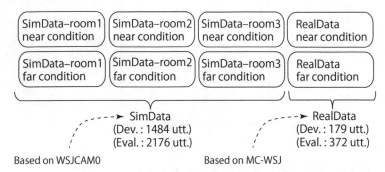

Training data

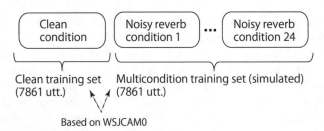

Fig. 15.2 Overview of datasets used in the REVERB challenge

provided researchers with an opportunity to thoroughly evaluate their algorithms for (1) practicality in realistic conditions and (2) robustness against a wide range of reverberant conditions.

- *SimData* is composed of reverberant utterances generated based on the WSJ-CAM0 British English corpus [9, 14]. These utterances were artificially distorted by convolving clean WSJCAM0 signals with measured room impulse responses (RIRs) and subsequently adding measured stationary ambient noise signals with a signal-to-noise ratio (SNR) of 20 dB. SimData simulated six different reverberation conditions: three rooms with different volumes (small, medium, and large) and two distances between a speaker and a microphone array (near = 50 cm and far = 200 cm). Hereafter, the rooms are referred to as SimData-room1, -room2, and -room3. The reverberation times (i.e., T60) of SimData-room1, -room2, and -room3 were about 0.3, 0.6, and 0.7 s, respectively. The RIRs and added noise were recorded in the corresponding reverberant room with an 8-ch circular array with a diameter of 20 cm. The recorded noise was stationary diffuse background noise, which was mainly caused by the air conditioning systems in the rooms, and thus has relatively large energy at lower frequencies.

- *RealData*, which comprises utterances from the MC-WSJ-AV British English corpus [8, 10], consists of utterances spoken by human speakers in a noisy and reverberant meeting room. RealData contains two reverberation conditions: one room and two distances between the speaker and the microphone array (near, i.e., about 100 cm, and far, i.e., about 250 cm). The reverberation time of the room was about 0.7 s [10]. Judging by the reverberation time and the distance between the microphone array and the speaker, the characteristics of RealData resemble those of the SimData-room-3-far condition. The text prompts for the utterances used in RealData and in part of SimData were the same. Therefore, we can use the same language and acoustic models for both SimData and RealData. For RealData recordings, a microphone array which had the same array geometry as the one used for SimData was employed.

For both SimData and RealData, we assumed that the speakers stayed in the same room for each test condition. However, within each condition, the relative speaker–microphone position changed from utterance to utterance. The term "test condition" in this chapter refers to one of the eight reverberation conditions that comprise two conditions in RealData and six conditions in SimData (see Fig. 15.2).

15.2.2.2   Training Data

As shown in Fig. 15.2, the training data consisted of (1) a clean training set taken from the original WSJCAM0 training set and (2) a multicondition (MC) training set. The MC training set was generated from the clean WSJCAM0 training data by convolving the clean utterances with 24 measured room impulse responses and adding recorded background noise at an SNR of 20 dB. The reverberation times of the 24 measured impulse responses for this dataset range roughly from 0.2 to 0.8 s. Different recording rooms were used for the test data and the training data, while the same set of microphone arrays was used for the training data and SimData.

15.2.3   Regulations

The ASR task in the REVERB challenge was to recognize each noisy reverberant test utterance without a priori information about the speaker identity/label, room parameters such as the reverberation time, the speaker–microphone distance and the speaker location, and the correct transcription. Therefore, systems had to perform recognition without knowing which speaker was talking in which acoustic condition.

Although the relative speaker–microphone position changed randomly from utterance to utterance, it was allowed to use all the utterances from a single test condition and to perform full-batch processing such as environmental adaptation of the acoustic model (AM). This regulation was imposed to focus mainly on the effect of environmental adaptation rather than speaker adaptation.

15.3 Performance of Baseline and Top-Performing Systems

To give a rough idea of the degree of difficulty of the REVERB challenge data, this section summarizes the performance achieved by the baseline and some notable top-performing systems.

15.3.1 Benchmark Results with GMM-HMM and DNN-HMM Systems

First of all, let us introduce the performance that can be obtained with Gaussian-mixture-model–hidden-Markov-model (GMM-HMM) recognizers and deep-neural-network–hidden-Markov-model (DNN-HMM) recognizers without front-end processing. Table 15.1 shows the results of two versions of Kaldi-based baseline GMM-HMM recognizers [5], and two versions of simple DNN-HMM recognizers which were prepared by two different research institutes independently [2, 3]. The table shows that even a very complex GMM-HMM system (the second system in the table) is outperformed by simple fully connected DNN-HMM systems for both SimData and RealData, which clearly indicates the superiority of the DNN-based AM over the GMM-based AM. However, although these improvements are notable and may support a claim that DNNs are robust in adverse environments, the achieved performances are actually still very far from the WERs obtained with clean speech, which correspond to 3.5% for SimData and 6.1% for RealData. The goal of reverberation-robust ASR techniques is to close the performance gap between clean speech recognition and reverberant speech recognition. Note that the big gap between the SimData and RealData performance in Table 15.1 is partly due to the fact that many of the SimData settings were less reverberant than the RealData setting.

Table 15.1 Word error rate (WER) obtained by baseline GMM-HMM systems and simple DNN systems without front-end processing (Eval set) (%)

System	SimData WER (%)	RealData WER (%)
Baseline multicondition GMM-HMM system with bigram LM [5]	28.8	54.1
Baseline multicondition GMM-HMM system with MMI AM training, trigram LM, fMLLR, MBR decoding [5]	12.2	30.9
DNN system with fully connected seven hidden layers [2] with trigram LM	8.6	28.5
DNN system with fully connected five hidden layers [3] with trigram LM	8.9	28.2

LM language model

15.3.2 Top-Performing 1-ch and 8-ch Systems

Next, let us introduce the performance of the top-performing 1-ch systems to show how they are achieving their goal currently. In this subsection, for the sake of simplicity, we present only the results from utterance-based batch processing systems, which usually are suitable for online ASR applications. In addition, the results in this subsection are based on the baseline multicondition training dataset and the conventional trigram LM, excluding the effect of data augmentation and advanced techniques for LM. While there are a number of systems proposed to improve 1-ch ASR performance, among them, the systems proposed by [2, 3, 15] achieved good performances as shown in Table 15.2. Delcroix et al. [2] achieved 7.7% for SimData and 25.2% for RealData by employing linear-prediction-based dereverberation (introduced in Chap. 2) and a simple DNN-based AM. On the other hand, Giri et al. [3] achieved similar performance, i.e., 7.7% for SimData and 27.5% for RealData, by taking a completely different approach. They employed no front-end enhancement technique, but instead fully extended the capability of a DNN-based AM with multitask learning and an auxiliary input feature representing reverberation time [3]. Tachioka et al. [15] took a rather traditional approach, that is, spectral-subtraction-based dereverberation (introduced in Chap. 20) and system combination based on many GMM-SGMM AM-based systems and DNN AM-based systems, and achieved WERs of 8.5% for SimData and 23.7% for RealData.

Now, let us introduce the performance of the top-performing multichannel (here, 8-ch) systems. Tachioka et al. [15] achieved 6.7% for SimData and 18.6% for RealData by additionally employing an 8-ch delay–sum beamformer on top of their 1-ch system. Delcroix et al. [2] achieved 6.7% for SimData and 15.6% for RealData by employing 8-ch linear-prediction-based dereverberation (introduced in Chap. 2),

Table 15.2 WER obtained by top-performing 1-ch and 8-ch utterance-based batch processing systems (%)

System	SimData WER (%)	RealData WER (%)
1-ch		
Linear-prediction-based dereverb + DNN [2]	7.7	25.2
DNN with multitask learning and auxiliary reverb time information [3]	7.7	27.5
Spectral-subtraction dereverb + system combination of GMM and DNN recognizers [15]	8.5	23.7
8-ch		
Linear-prediction-based dereverb + MVDR beamformer + DNN [2]	6.7	15.6
Delay–sum beamformer + spectral-subtraction dereverb + system combination of GMM and DNN recognizers [15]	6.7	18.6

an 8-ch minimum variance distortionless response (MVDR) beamformer, and a simple DNN-based AM. These results clearly show superiority and importance of multichannel linear-filtering-based enhancement processing. Note that details of the other contributions presented in the REVERB challenge and their effectiveness are summarized in [7].

In summary, based on these results, we confirmed the importance of multichannel linear-filtering-based enhancement, an advanced DNN-based AM, and DNN-related techniques such as auxiliary input features. In addition, it was confirmed that, as in other ASR tasks, system combination provided consistently significant performance gains.

15.3.3   Current State-of-the-Art Performance

Table 15.3 serves as a reference for the current state-of-the-art performance obtained with the challenge data. All of these results were obtained with full-batch processing systems, which usually incorporate environmental adaptation and are generally suitable only for offline ASR applications. The major difference between these results and the ones in Table 15.2 lies in the back-end techniques. Specifically, the systems shown in Table 15.3 employ additionally (a) artificially augmented training data for AM training, (b) full-batch AM adaptation for environmental adaptation, i.e., additional back-propagation training using test data taken from a test condition, and (c) a state-of-the-art LM, i.e., a recurrent neural network (RNN) LM.

Table 15.3 Current state-of-the-art performance for 1-ch, 2-ch, and 8-ch scenarios (Eval set) (%)

System	SimData WER (%)	RealData WER (%)
1-ch		
1-ch linear-prediction-based dereverb + DNN-based AM + DNN adaptation + data augmentation + RNN LM [2]	5.0	15.9
2-ch		
2-ch linear-prediction-based dereverb + 2-ch MVDR beamformer + 1-ch model-based enhancement + DNN-based AM + DNN adaptation + data augmentation + RNN LM [2]	4.4	11.9
8-ch		
8-ch linear-prediction-based dereverb + 8-ch MVDR beamformer + 1-ch model-based enhancement + DNN-based AM + DNN adaptation + data augmentation + RNN LM [2]	4.1	9.1

15.4   Summary and Remaining Challenges for Reverberant Speech Recognition

This chapter introduced the scenario, data, and results for the REVERB challenge, which was a benchmark task carefully designed to evaluate reverberation-robust ASR techniques. As a result of the challenge, it was shown that notable improvement can be achieved by using algorithms such as linear-prediction-based dereverberation and DNN-based acoustic modeling. However, at the same time, it was found that there still remain a number of challenges in the field of reverberant speech recognition. For example the top performance currently obtained for the 1-ch scenario is still very far from that obtained for multichannel scenarios. This is partly due to the fact that there is no 1-ch enhancement technique that can greatly reduce WERs when used with a multicondition DNN AM. Finding a 1-ch enhancement algorithm which works effectively even in the DNN era is one of the key research directions to pursue. It is also important to note that there is still much room for improvement even for multichannel systems, especially for RealData.

The REVERB challenge was a benchmark task to evaluate technologies in reverberant environments where the amount of ambient noise is relatively moderate. However, if the remaining problems mentioned above are resolved in the future by further investigations, we should extend the scenario to include more noise in addition to reverberation, closely simulating more realistic distant speech recognition challenges.

References

1. Barker, J., Vincent, E., Ma, N., Christensen, C., Green, P.: The PASCAL CHiME speech separation and recognition challenge. Comput. Speech Lang. **27**(3), 621–633 (2013)
2. Delcroix, M., Yoshioka, T., Ogawa, A., Kubo, Y., Fujimoto, M., Nobutaka, I., Kinoshita, K., Espi, M., Araki, S., Hori, T., Nakatani, T.: Strategies for distant speech recognition in reverberant environments. Comput. Speech Lang. (2015). doi:10.1186/s13634-015-0245-7
3. Giri, R., Seltzer, M., Droppo, J., Yu, D.: Improving speech recognition in reverberation using a room-aware deep neural network and multi-task learning. In: Proceedings of International Conference on Acoustics, Speech and Signal Processing (ICASSP), pp. 5014–5018 (2015)
4. Huang, X., Acero, A., Hong, H.W.: Spoken Language Processing: A Guide to Theory, Algorithm and System Development. Prentice Hall, Upper Suddle River, NJ (2001)
5. Kaldi-based baseline system for REVERB challenge. https://github.com/kaldi-asr/kaldi/tree/master/egs/reverb
6. Kinoshita, K., Delcroix, M., Yoshioka, T., Nakatani, T., Habets, E., Haeb-Umbach, R., Leutnant, V., Sehr, A., Kellermann, W., Maas, R., Gannot, S., Raj, B.: The REVERB challenge: a common evaluation framework for dereverberation and recognition of reverberant speech. In: Proceedings of Workshop on Applications of Signal Processing to Audio and Acoustics (WASPAA) (2013)
7. Kinoshita, K., Delcroix, M., Gannot, S., Habets, E., Haeb-Umbach, R., Kellermann, W., Leutnant, V., Maas, R., Nakatani, T., Raj, B., Sehr, A., Yoshioka, T.: A summary of the REVERB challenge: state-of-the-art and remaining challenges in reverberant speech processing research. EURASIP J. Adv. Signal Process. (2016). doi:10.1186/s13634-016-0306-6

8. LDC: Multi-channel WSJ audio. https://catalog.ldc.upenn.edu/LDC2014S03
9. LDC: WSJCAMO Cambridge read news. https://catalog.ldc.upenn.edu/LDC95S24
10. Lincoln, M., McCowan, I., Vepa, J., Maganti, H.K.: The multi-channel Wall Street Journal audio visual corpus (MC-WSJ-AV): specification and initial experiments. In: Proceedings of IEEE Workshop on Automatic Speech Recognition and Understanding (ASRU), pp. 357–362 (2005)
11. Naylor, P.A., Gaubitch, N.D.: Speech Dereverberation. Springer, Berlin (2010)
12. Pearce, D., Hirsch, H.G.: The Aurora experimental framework for the performance evaluation of speech recognition systems under noisy conditions. In: Proceedings of International Conference on Spoken Language Processing (ICSLP), pp. 29–32 (2000)
13. REVERB Challenge. http://reverb2014.dereverberation.com/
14. Robinson, T., Fransen, J., Pye, D., Foote, J., Renals, S.: WSJCAM0: a British English speech corpus for large vocabulary continuous speech recognition. In: Proceedings of International Conference on Acoustics, Speech and Signal Processing (ICASSP), pp. 81–84 (1995)
15. Tachioka, Y., Narita, T., Weninger, F.J., Watanabe, S.: Dual system combination approach for various reverberant environments with dereverberation techniques. In: Proceedings of REVERB Challenge Workshop, p. 1.3 (2014)
16. Tashev, I.: Sound Capture and Processing. Wiley, Hoboken, NJ (2009)
17. Vincent, E., Araki, S., Theis, F.J., Nolte, G., Bofill, P., Sawada, H., Ozerov, A., Gowreesunker, B.V., Lutter, D.: The signal separation evaluation campaign (2007–2010): achievements and remaining challenges. Signal Process. **92**, 1928–1936 (2012)
18. Wölfel, M., McDonough, J.: Distant Speech Recognition. Wiley, Hoboken, NJ (2009)
19. Yoshioka, T., Sehr, A., Delcroix, M., Kinoshita, K., Maas, R., Nakatani, T., Kellermann, W.: Making machines understand us in reverberant rooms: robustness against reverberation for automatic speech recognition. IEEE Signal Process. Mag. **29**(6), 114–126 (2012)

Chapter 16
Distant Speech Recognition Experiments Using the AMI Corpus

Steve Renals and Pawel Swietojanski

Abstract This chapter reviews distant speech recognition experimentation using the AMI corpus of multiparty meetings. The chapter compares conventional approaches using microphone array beamforming followed by single-channel acoustic modelling with approaches which combine multichannel signal processing with acoustic modelling in the context of convolutional networks.

16.1 Introduction

Distant conversational speech recognition [30] poses many technical challenges such as multiple overlapping acoustic sources (including multiple talkers), reverberant acoustic environments and highly conversational speaking styles. Microphone-array-based approaches have been used to address the task since the early 1990s [3, 20, 29], and from about 2004 onwards there have been various evaluation frameworks for distant speech recognition, including the multichannel Wall Street Journal audio visual corpus (MC-WSJ-AV) [18], the NIST rich transcription (RT) series of evaluations [9], the REVERB challenge (Chap. 15) and the CHiME challenges (Chap. 14).

From 2004 to 2009, the NIST RT evaluations (http://www.itl.nist.gov/iad/mig//tests/rt) focused on the problem of meeting transcription, and enabled comparison between various automatic meeting transcription systems (e.g. [14, 26]). These evaluations of multiparty conversational speech recognition had a focus on meeting transcription. The acoustic data was classified by the recording condition: individual headset microphones (IHM), a single distant microphone (SDM), and multiple distant microphones (MDM). The MDM condition typically used tabletop microphone arrays, with the SDM condition choosing a single microphone from the array.

S. Renals (✉) • P. Swietojanski
Centre for Speech Technology Research, University of Edinburgh, Edinburgh, UK
e-mail: s.renals@ed.ac.uk; p.swietojanski@gmail.com

© Springer International Publishing AG 2017
S. Watanabe et al. (eds.), *New Era for Robust Speech Recognition*,
DOI 10.1007/978-3-319-64680-0_16

355

For MDM systems, microphone array processing was usually distinct from speech recognition. For instance, the AMIDA MDM system of Hain et al. [14] processed the multichannel microphone array data using a Wiener noise filter, followed by weighted filter–sum beamforming based on time-delay-of-arrival (TDOA) estimates, postprocessed using a Viterbi smoother. In practice, the beamformer tracked the direction of maximum energy, passing the beamformed signal on to a conventional automatic speech recognition (ASR) system—in the case of [14], a Gaussian mixture model/hidden Markov model (GMM/HMM) trained using the discriminative minimum phone error (MPE) criterion [21], speaker-adaptive training [4], and the use of bottleneck features [12] derived from a neural network trained as a phone classifier. The resulting system employed a complex multipass decoding scheme, including substantial cross-adaptation and model combination.

One of the main principles underpinning 'deep learning' is that systems for classification and regression can be constructed from multiple modules that are optimised using a common objective function [17]. In the context of distant speech recognition this can lead to approaches such as LIMABEAM [24, 25], in which the parameters of the microphone array beamformer are estimated so as to maximise the likelihood of the correct utterance model. Marino and Hain [19] explored removing the beamforming component entirely, and directly concatenating the feature vectors from the different microphones as the input features for an HMM/GMM speech recognition system. In contrast to the LIMABEAM approach, which retains explicit beamforming parameters but optimises them according to a criterion related to speech recognition accuracy, the concatenation approach makes the beamforming parameters implicit. More recently Xiao et al. (Chap. 4) introduced a neural network approach to optimise beamforming to maximise speech recognition performance, also allowing the beamforming and acoustic model to be optimised simultaneously, and Sainath et al introduced a multichannel neural network architecture operating on raw waveforms (Chap. 5).

This chapter is concerned with distant speech recognition of meeting recordings, based on experiments employing the AMI corpus (Sect. 16.2). We present experiments using beamformed microphone array features as a baseline (Sect. 16.3), comparing them with systems using concatenated features from multiple channels (Sect. 16.4) and systems using cross-channel convolutional networks (Sect. 16.5).

16.2   Meeting Corpora

Work on meeting transcription has been largely enabled by two corpora: the ICSI Meeting Corpus and the AMI corpus. The ICSI Meeting Corpus (http://www.icsi.berkeley.edu/Speech/mr/) contains about 75 h of recorded meetings with 3–15 participants, captured using IHMs, as well as an MDM condition comprising four boundary microphones placed about 1 m apart along the tabletop [15]. One limitation of this corpus was the fact that the distant microphones were widely spaced and not in known positions.

Fig. 16.1 AMI corpus recording setup

The AMI corpus (http://corpus.amiproject.org) comprises over 100 h of recordings of multiparty meetings. The meetings were recorded as part of the AMI/AMIDA projects (http://www.amiproject.org) using a common 'instrumented meeting room' (IMR) environment located at the University of Edinburgh, Idiap Research Institute and TNO Human Factors (Fig. 16.1). The corpus design, and the recording methodology, was driven by the multidisciplinary nature of the AMI/AMIDA projects, which included research in computer vision, multimodal processing, natural language processing, human computer interaction and social psychology, as well as speech recognition [6, 7]. The IMR recording environments each included at least six cameras (personal and room-view), MDMs configured as an eight-element circular microphone array placed on the meeting table, and IHM for each participant, as well as information capture using digital pens, smart whiteboards, shared laptop spaces, a data projector and videoconferencing if used. The different recorded streams were synchronised to a common timeline. In the initial recordings (2005), frame-level synchronisation was achieved using a hardware-based approach. Later meeting capture experiments used a high-resolution spherical digital video camera system and a 20-element microphone array with software synchronisation, as well as further experiments using digital MEMS microphone arrays [31]. The corpus also contains a verbatim word-level transcription synchronised to the same timeline. Additional annotations include dialogue acts, topic segmentation, extractive and abstractive summaries, named entities, limited forms of head and hand gestures, gaze direction, movement around the room, and head pose information. NXT (the NITE XML Toolkit (http://groups.inf.ed.ac.uk/nxt/) an XML-based open source software infrastructure for multimodal annotation [8]) was used to carry out and manage the annotations.

About two-thirds of the AMI corpus consists of 'scenario meetings' in which four participants play roles in a design across a set of four meetings, recorded in thirty replicas, ten in each of the IMRs. The remainder of the corpus comprises recordings of 'real' meetings which would have taken place irrespective of the recording. The use of scenario meetings had several advantages in the context of the interdisciplinary nature of the projects in which the corpus was produced: it allowed preferred meeting outcomes to be designed into the process, allowing the definition of group outcome and productivity measures; the knowledge and motivation of the participants were controlled, thus removing the confounding factors that would be present in a set of real meetings (for example the history of relationships between the participants, and the organisational context); and the meeting scenario could be replicated, enabling task-based evaluations. The main drawbacks of using scenario meetings are based around a reduction in diversity and naturalness. Although the recorded speech is spontaneous and conversational, the overall dialogue is less realistic. Furthermore, replicating the scenarios significantly reduces the linguistic variability across the corpus: for example, in 100 h of the AMI corpus there are about 8000 unique words, about half the number observed in that duration in other corpora such as the Wall Street Journal and Switchboard corpora.

16.3 Baseline Speech Recognition Experiments

In this paper, we focus on distant speech recognition using the AMI corpus. Unlike the NIST RT evaluations, where the AMI data was used together with other meeting corpora (e.g. [12, 14]), we carefully defined the training, development and test sets based on a 3-way partition of the AMI corpus, thus ensuring that our distant speech recognition experiments used identical microphone array configurations in the three different acoustic environments. The training, development and test sets all included a mix of scenario- and non-scenario-based meetings, and were designed such that no speaker appeared in more than one set. The definitions of these sets have also been made available on the AMI corpus website and are used in the associated Kaldi recipe (https://github.com/kaldi-asr/kaldi/tree/master/egs/ami/). We used the segmentation provided with the AMI corpus annotations (version 1.6.1). In this work, we considered all segments (including those with overlapping speech), and the speech recognition outputs were scored by the `asclite` tool [9] following the NIST RT recommendations for scoring simultaneous speech (http://www.itl.nist.gov/iad/mig/tests/rt/2009/).

- *IHM recordings.* Our baseline acoustic models used 13-dimension mel-frequency cepstral coefficients (MFCCs) (C0–C12), splicing together seven frames, projecting down from 91 to 40 dimensions using linear discriminant analysis (LDA) [13], and decorrelated using a single semi-tied covariance (STC) transform [10] (also referred to as a maximum likelihood linear transform, MLLT). These features are referred to as LDA/STC. Both GMM-HMM and artificial-neural-network (ANN)-HMM acoustic models were speaker-adaptively trained (SAT) on these LDA/STC features using a single constrained maximum likelihood linear regression (CMLLR) transform estimated per speaker. The GMM-HMM systems provided the state alignments for training the ANNs. Additionally, we also trained ANN systems on 40-dimension log mel filterbank (FBANK) features appended with first and second temporal derivatives. The state alignments obtained using the LDA/STC features were used for training the ANNs on FBANK features.

- *SDM/MDM recordings.* We used either a single element of the microphone array (SDM) or delay-sum beamforming on 2, 4 or 8 uniformly spaced array channels[1] using the BeamformIt toolkit [5] (MDM); the audio was then processed in a similar fashion to the IHM configuration. The major difference between the IHM and SDM/MDM configurations is that when audio is captured with distant microphones, it is not realistically possible to ascribe a speech segment to a particular speaker without having performed speaker diarisation. Hence we did not use any form of speaker adaptation or adaptive training in the SDM/MDM experiments, unless stated otherwise.

For all acoustic conditions, we trained (1) GMM-HMM systems using LDA/STC features (speaker-adapted in the IHM case) optimised according to the boosted maximum mutual information (BMMI) criterion, (2) ANN systems using LDA/STC features optimised according to the cross-entropy criterion, and (3) ANN systems using FBANK features optimised according to the cross-entropy criterion. We used a set of about 4000 tied states in each configuration, with about 80,000 Gaussians in each GMM-based system. The GMM-based systems were used to provide the state alignments for training the corresponding ANNs. The ANN systems were feedforward networks, each with six hidden layers of 2048 units, employing sigmoid transfer functions. The baseline experimental results are summarised in Table 16.1.

Table 16.1 Word error rates (%) for the GMM and ANN acoustic models for various microphone configurations

System	Microphone configuration				
	IHM	MDM8	MDM4	MDM2	SDM
AMI development set					
GMM BMMI on LDA/STC	30.2 (SAT)	54.8	56.5	58.0	62.3
ANN on LDA/STC	26.8 (SAT)	49.5	50.3	51.6	54.0
ANN on FBANK	26.8	49.2	–	50.1	53.1
AMI evaluation set					
GMM BMMI on LDA/STC	31.7 (SAT)	59.4	61.2	62.9	67.2
ANN on LDA/STC	28.1 (SAT)	52.4	52.6	52.8	59.0
ANN on FBANK	29.6	52.0	–	52.4	57.9

---

[1]Mics 1 and 5 were used in the 2-mic case; mics 1, 3, 5 and 7 in the 4-mic case.

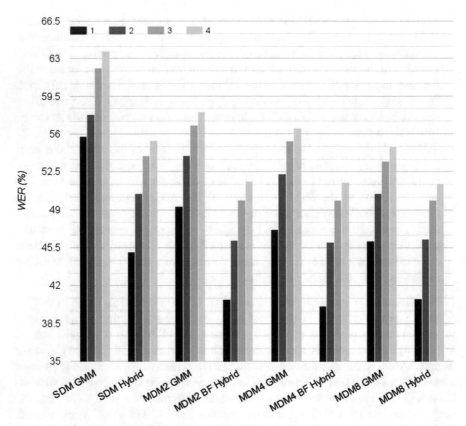

Fig. 16.2 Development set WERs for segments with 1, 2, 3 and 4 overlapping speakers. Acoustic models were trained on MFCC LDA/STC features. The figure comes originally from [27] and the results are not directly comparable to those reported in Table 16.1 because the latter benefit from later refinements in the recipe. The figure was included to visualise the overlapping-speakers issue across different systems

While Table 16.1 presents the word error rate (WER) for all segments, including those with overlapping speech, Fig. 16.2 shows the WERs for segments scored with different numbers of overlapping speakers. As expected, overlapping segments are harder to recognise. In fact, even if a beamformer can select the dominant source perfectly it still does not address the problem of recognising overlapping speech, which would require source separation and independent decoding for each identified source. Figure 16.2 presents results for different systems in terms of the number of overlapping speakers in the segment. There is an 8–12% reduction in WER when only considering segments with non-overlapping speech. One can also notice that the WER deteriorates relatively more in the presence of overlapping speech for ANNs. For example, in the SDM case a 12% relative drop in WER is observed for the GMM-HMM and over 19% relative for the ANN-HMM system. This may be because ANNs model non-overlapping segments more accurately, and part of this advantage diminishes for fragments containing overlapping speech. We do not further address the issue of overlapping speakers in this chapter, and to keep the exposition simple we report WERs for all segments as they are (including overlapping speakers).

16.4   Channel Concatenation Experiments

As an alternative to beamforming, it is possible to incorporate multiple channels into an ANN acoustic model by concatenating them, thus providing a sequence of higher-dimension acoustic vectors. We performed a set of experiments in order to evaluate the extent to which an ANN is able to learn to do front-end processing—both noise cancellation and beamforming—by providing the features extracted from multiple microphones as input to the networks (cf. [19]). In these experiments. the networks again had six hidden layers,[2] with a wider input layer of concatenated channels. There are some differences to the baseline experiments, since Wiener filtering and beamforming are time domain operations, whereas ANNs trained on concatenated features operate entirely in either the cepstral or the log-spectral domain. Nevertheless, the results offer an indication of the complementarity of the information from different channels. The results are tabulated in Table 16.2, and indicate that ANNs trained on concatenated inputs perform substantially better than the SDM case, achieving results approaching those obtained using beamforming. Since the ANNs trained on concatenated features do not use any knowledge of the array geometry, the technique is applicable to an arbitrary configuration of microphones.

To further understand the nature of the compensation being learned by the ANNs with multichannel inputs, we performed an additional control experiment. The input to the ANN was from a single channel, and at test time this was identical to the SDM case. However, during training, the data from other channels was also presented to

Table 16.2 WER for ANNs trained on multiple channels

Combining method	Recognition channel(s)	AMI dev set
SDM (no combination)	1	53.1
SDM (no combination)	2	52.9
Concatenate 1 + 5	3, 7	51.8
Concatenate 1 + 3 + 5 + 7	2, 4, 6, 8	51.7
Multistyle 1 + 3 + 5 + 7	1	51.8
Multistyle 1 + 3 + 5 + 7	2	51.7

SDM models were trained on channel 1

---

[2]However, since the networks were being tasked with additional processing, it may be that deeper architectures would be more suitable.

the network, although not at the same time. In other words, the ANN was presented with data drawn from multiple channels, while at test time it was only tested on a single channel. We call this multistyle training, and it is related to our work on low-resource acoustic modelling [11], where a similar concept was used to train ANNs in a multilingual fashion. From Table 16.2 we see that this approach performs similarly to the ANNs with concatenated input, without requiring multiple channels at the recognition stage. Recognition results on channel 2, which was not used in the multistyle training, show similar trends. These results strongly suggest that there is enough information in a single channel to enable accurate recognition. However, extraneous factors in the data may confound a learner trained only on data from a single channel. Being forced to classify data from multiple channels using the same shared representation (i.e. the hidden layers), the network learns how to ignore the channel-specific covariates. To the best of our knowledge, this is the first result to show that it is possible to improve recognition of audio captured with a single distant microphone by guiding the training using data from microphones at other spatial locations.

16.5 Convolutional Neural Networks

A channel concatenation network may be enriched by constraining one or more of the lower layers to have local connectivity and to share parameters—a convolutional neural network (CNN). CNNs have defined the state of the art on many vision tasks [17] and can reduce the speech recognition WER when applied to acoustic modelling [1, 23]. The major conceptual difference between recent CNN structures for speech modelling and previous trials in the form of both CNNs [17] and the closely related time-delay neural networks [16] lies in performing convolution and/or sharing parameters across frequency rather than time (see also Chap. 5).

The input to a CNN comprises FBANK features within an acoustic context window reordered such that each frequency band contains all the related static and dynamic coefficients. The hidden activations are then generated by a linear valid convolution[3] of a local frequency region. The same set of filters is then applied across different frequency regions to form a complete set of convolutional activations which can be subsampled, for instance by using the maxpooling operator, to further limit the variability across different frequencies.

Since the channels contain similar information (acoustic features shifted in time), we conjecture that the filter weights may be shared across different channels. Nevertheless, the formulation and implementation allow for different filter weights in each channel. Similarly, it is possible for each convolutional band to have a

---

[3]The convolution of two vectors of size X and Y may result either in a vector of size $X + Y - 1$ for a full convolution with zero-padding of non-overlapping regions, or a vector of size $X - Y + 1$ for a valid convolution where only the points which overlap completely are considered.

separate learnable bias parameter instead of the biases only being shared across bands [1, 23].

The complete set of convolutional layer activations is obtained by applying the (shared) set of filters across the whole (multichannel) input space (as depicted in the top part of Fig. 16.3). In this work the weights are tied across the input space; alternatively, the weights may be partially shared, tying only those weights spanning neighbouring frequency bands. Although limited weight sharing was reported to bring improvements for phone classification [1] and small-scale tasks [2], a recent study on larger tasks [23] suggests that full weight sharing with a sufficient number of filters can work equally well, while being easier to implement.

Multichannel convolution builds feature maps similarly to the LeNet-5 model [17], where each convolutional band is composed of filter activations

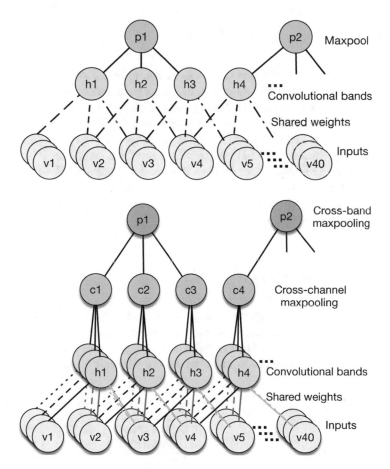

Fig. 16.3 Convolutional network layer with (*top*) cross-band maxpooling incorporating all channels, and (*bottom*) cross-channel maxpooling within each band, followed by cross-band maxpooling

spanning all input channels. We also constructed feature maps using maxpooling across channels, in which the activations are generated in channelwise fashion and then maxpooled to form a single cross-channel convolutional band. The resulting cross-channel activations can be further maxpooled along frequency (Fig. 16.3, bottom). Channelwise convolution may be viewed as a special case of two-dimensional convolution, where the effective pooling region is determined in frequency but varies in time depending on the actual time delays between the microphones. This CNN-based approach to multichannel speech recognition was first presented in [22, 28].

The CNN/ANN models in this section were trained on FBANK features appended with the first and the second time derivatives, which were presented in an 11-frame window.

16.5.1 SDM Recordings

The results of the single-channel CNN can be found in Table 16.3, with the first two lines presenting the GMM and ANN baselines from Table 16.1. The following three lines are results for the CNN using maxpool sizes of $R = N = 1, 2, 3$. By using CNNs, we were able to obtain a 3.4% relative reduction in WER with respect to the best ANN model and a 19% relative reduction in WER compared with a discriminatively trained GMM-HMM (baseline numbers taken from Table 16.1). The total number of parameters of the CNN models varied as $R = N$ while J was kept constant across the experiments. However, the best-performing model had neither the highest nor the lowest number of parameters, which suggests this is due to the optimal pooling setting.

16.5.2 MDM Recordings

For the MDM case, we compared a delay–sum beamformer with the direct use of multiple microphone channels as input to the network. For the beamforming experiments, we followed noise cancellation using a Wiener filter with delay–sum beamforming on eight uniformly spaced array channels using the BeamformIt

Table 16.3 Word error rates (%) on AMI-SDM, where R is the pool size

System	AMI dev set
BMMI GMM-HMM (LDA/STC)	63.2
ANN (FBANK)	53.1
CNN ($R = 3$)	51.4
CNN ($R = 2$)	51.3
CNN ($R = 1$)	52.5

Table 16.4 Word error rates
(%) on AMI-MDM

System	AMI dev set
MDM with beamforming (eight microphones)	
BMMI GMM-HMM	54.8
ANN	49.5
CNN	46.8
MDM without beamforming	
ANN 4 ch concatenated	51.2
CNN 2 ch conventional	50.5
CNN 4 ch conventional	50.4
CNN 2 ch channelwise	50.0
CNN 4 ch channelwise	49.4

toolkit [5]. The results are summarised in Table 16.4. The first block of Table 16.4 presents the results for the case in which the models were trained on a beamformed signal from eight microphones. The first two rows show the WER for the baseline GMM and ANN acoustic models as reported in Table 16.1. The following row contains the CNN model trained on eight beamformed channels, obtaining 2.7% absolute improvement (5.5% relative) over the ANN. The configuration of the MDM CNN is the same as the best SDM CNN ($R = N = 2$).

The second part of Table 16.4 shows WERs for the models directly utilising multichannel features. The first row is a baseline ANN variant trained on four concatenated channels from Table 16.2. Then we present the CNN models with MDM input convolution performed as in Fig. 16.3 (top) and a pooling size of 2, which was optimal for the SDM experiments. This scenario decreases WER by 1.6% relative to the ANN structure with concatenated channels (this approach can be seen as a channel concatenation for CNN models). Applying channelwise convolution with two-way pooling (Fig. 16.3, bottom) brings further gains of 3.5% WER relative. Furthermore, channelwise pooling works better for more input channels: conventional convolution on four channels achieves 50.4% WER, practically the same as the 2-channel network, while channelwise convolution with four channels achieves 49.5% WER, compared to 50.0% for the 2-channel case. These results indicate that picking the best information (selecting the feature receptors with maximum activations) within the channels is crucial when doing model-based combination of multiple microphones.

16.5.3 IHM Recordings

We observe similar relative WER improvements between the ANN and CNN for close-talking speech recordings (Table 16.5) as were observed for the MDM and SDM experiments. The CNN achieves 3.6% WER reduction relative to the ANN

Table 16.5 Word error rates
(%) on AMI Dev set—IHM

System	WER (%)
BMMI GMM-HMM (SAT)	29.4
ANN	26.6
CNN	25.6

model. Both ANN and CNN systems outperform a BMMI-GMM system trained in a speaker-adaptive fashion by 9.4% and 12.9% relative WER, respectively. We did not see any improvements by increasing the pooling size. Sainath et al. [23] previously suggested that pooling may be task dependent.

16.6 Discussion and Conclusions

In this chapter we have presented some baseline experiments for distant speech recognition of multiparty meetings using the AMI corpus. ANN-based systems provide WER reductions compared with GMM-based systems, and further reductions are obtained using convolutional hidden layers followed by maxpooling. We presented a number of experiments exploring the effect of replacing microphone array beamforming with ANN and CNN architectures that take multichannel input. Although multichannel CNNs do not outperform beamforming approaches on the AMI corpus, our results indicate that these CNN architectures are able to learn from multichannel signals. We have applied these approaches to the ICSI corpus, in which the microphone array is less calibrated, and our results indicate that cross-channel CNN architectures outperform beamforming by a small amount [22].

Our current experiments do not explicitly attempt to optimise the acoustic model for overlapping talkers, or for reverberation. The promising results using raw multiple-channel input features in place of beamforming open up possibilities of learning representations taking into account aspects such as overlapping speech. One interesting research direction is the use raw-waveform features in a multichannel context, as discussed in Chap. 5.

References

1. Abdel-Hamid, O., Mohamed, A.R., Hui, J., Penn, G.: Applying convolutional neural networks concepts to hybrid NN–HMM model for speech recognition. In: Proceedings of the IEEE ICASSP, pp. 4277–4280 (2012)
2. Abdel-Hamid, O., Deng, L., Yu, D.: Exploring convolutional neural network structures and optimization techniques for speech recognition. In: Proceedings of the ICSA Interspeech (2013)
3. Adcock, J., Gotoh, Y., Mashao, D., Silverman, H.: Microphone-array speech recognition via incremental MAP training. In: Proceedings of the IEEE ICASSP, pp. 897–900 (1996)

4. Anastasakos, T., McDonough, J., Schwartz, R., Makhoul, J.: A compact model for speaker-adaptive training. In: Proceedings of the ICSLP, pp. 1137–1140 (1996)
5. Anguera, X., Wooters, C., Hernando, J.: Acoustic beamforming for speaker diarization of meetings. IEEE Trans. Audio Speech Lang. Process. **15**, 2011–2021 (2007)
6. Carletta, J., Lincoln, M.: Data collection. In: Renals, S., Bourlard, H., Carletta, J., Popescu-Belis, A. (cds.) Multimodal Signal Processing: Human Interactions in Meetings, chap. 2, pp. 11–27. Cambridge University Press, Cambridge (2012)
7. Carletta, J., Ashby, S., Bourban, S., Flynn, M., Guillemot, M., Hain, T., Kadlec, J., Karaiskos, V., Kraaij, W., Kronenthal, M., Lathoud, G., Lincoln, M., Lisowska, A., McCowan, I., Post, W., Reidsma, D., Wellner, P.: The AMI meeting corpus: a pre-announcement. In: Proceedings of the Machine Learning for Multimodal Interaction (MLMI), pp. 28–39 (2005)
8. Carletta, J., Evert, S., Heid, U., Kilgour, J.: The NITE XML toolkit: data model and query language. Lang. Resour. Eval. **39**, 313–334 (2005)
9. Fiscus, J., Ajot, J., Radde, N., Laprun, C.: Multiple dimension Levenshtein edit distance calculations for evaluating ASR systems during simultaneous speech. In: Proceedings of the LREC (2006)
10. Gales, M.: Semi-tied covariance matrices for hidden Markov models. IEEE Trans. Speech Audio Process. **7**(3), 272–281 (1999)
11. Ghoshal, A., Swietojanski, P., Renals, S.: Multilingual training of deep neural networks. In: Proceedings of the IEEE ICASSP (2013)
12. Grezl, F., Karafiat, M., Kontar, S., Cernocky, J.: Probabilistic and bottle-neck features for LVCSR of meetings. In: Proceedings of IEEE ICASSP, pp. IV-757–IV-760 (2007)
13. Haeb-Umbach, R., Ney, H.: Linear discriminant analysis for improved large vocabulary continuous speech recognition. In: Proceedings of the IEEE ICASSP, pp. 13–16 (1992). http://dl.acm.org/citation.cfm?id=1895550.1895555
14. Hain, T., Burget, L., Dines, J., Garner, P., Grézl, F., El Hannani, A., Karafiat, M., Lincoln, M., Wan, V.: Transcribing meetings with the AMIDA systems. IEEE Trans. Audio Speech Lang. Process. **20**, 486–498 (2012)
15. Janin, A., Baron, D., Edwards, J., Ellis, D., Gelbart, D., Morgan, N., Peskin, B., Pfau, T., Shriberg, E., Stolcke, A., Wooters, C.: The ICSI meeting corpus. In: Proceedings of the IEEE ICASSP, pp. I-364–I-367 (2003)
16. Lang, K., Waibel, A., Hinton, G.: A time-delay neural network architecture for isolated word recognition. Neural Netw. **3**, 23–43 (1990)
17. LeCun, Y., Bottou, L., Bengio, Y., Haffner, P.: Gradient-based learning applied to document recognition. Proc. IEEE **86**, 2278–2324 (1998)
18. Lincoln, M., McCowan, I., Vepa, J., Maganti, H.: The multi-channel Wall Street Journal audio visual corpus (MC-WSJ-AV): specification and initial experiments. In: Proceedings of the IEEE ASRU (2005)
19. Marino, D., Hain, T.: An analysis of automatic speech recognition with multiple microphones. In: Proceedings of the Interspeech, pp. 1281–1284 (2011)
20. Omologo, M., Matassoni, M., Svaizer, P., Giuliani, D.: Microphone array based speech recognition with different talker-array positions. In: Proceedings of the IEEE ICASSP, pp. 227–230 (1997)
21. Povey, D., Woodland, P.: Minimum phone error and I-smoothing for improved discriminative training. In: Proceedings of the IEEE ICASSP, pp. 105–108 (2002)
22. Renals, S., Swietojanski, P.: Neural networks for distant speech recognition. In: Proceedings of the HSCMA (2014)
23. Sainath, T., Kingsbury, B., Mohamed, A., Dahl, G., Saon, G., Soltau, H., Beran, T., Aravkin, A., Ramabhadran, B.: Improvements to deep convolutional neural networks for LVCSR. In: Proceedings of the IEEE ASRU (2013)
24. Seltzer, M., Stern, R.: Subband likelihood-maximizing beamforming for speech recognition in reverberant environments. IEEE Trans. Audio Speech Lang. Process. **14**, 2109–2121 (2006)
25. Seltzer, M., Raj, B., Stern, R.: Likelihood-maximizing beamforming for robust hands-free speech recognition. IEEE Trans. Speech Audio Process. **12**, 489–498 (2004)

26. Stolcke, A., Anguera, X., Boakye, K., Cetin, O., Janin, A., Magimai-Doss, M., Wooters, C., Zheng, J.: The SRI-ICSI spring 2007 meeting and lecture recognition system. In: Stiefelhagen, R., Bowers, R., Fiscus, J. (eds.) Multimodal Technologies for Perception of Humans. Lecture Notes in Computer Science, vol. 4625, pp. 373–389. Springer, New York (2008)
27. Swietojanski, P., Ghoshal, A., Renals, S.: Hybrid acoustic models for distant and multi-channel large vocabulary speech recognition. In: Proceedings of the IEEE ASRU (2013). doi:10.1109/ASRU.2013.6707744
28. Swietojanski, P., Ghoshal, A., Renals, S.: Convolutional neural networks for distant speech recognition. IEEE Signal Process. Lett. **21**, 1120–1124 (2014)
29. Van Compernolle, D., Ma, W., Xie, F., Van Diest, M.: Speech recognition in noisy environments with the aid of microphone arrays. Speech Commun. **9**, 433–442 (1990)
30. Wölfel, M., McDonough, J.: Distant Speech Recognition. Wiley, Chichester (2009)
31. Zwyssig, E., Lincoln, M., Renals, S.: A digital microphone array for distant speech recognition. In: Proceedings of the IEEE ICASSP, pp. 5106–5109 (2010). doi:10.1109/ICASSP.2010.5495040

Chapter 17
Toolkits for Robust Speech Processing

Shinji Watanabe, Takaaki Hori, Yajie Miao, Marc Delcroix, Florian Metze, and John R. Hershey

Abstract Recent robust automatic speech recognition (ASR) techniques have been developed rapidly due to the demand placed on ASR applications in real environments, with the help of publicly available tools developed in the community. This chapter overviews major toolkits available for robust ASR, covering general ASR toolkits, language model toolkits, speech enhancement/microphone array front-end toolkits, deep learning toolkits, and emergent end-to-end ASR toolkits. The aim of this chapter is to provide information about functionalities (features, functions, platform, and language), license, and source location so that readers can easily access such tools to build their own robust ASR systems. Some of the toolkits have actually been used to build state-of-the-art ASR systems for various challenging tasks. The references in this chapter also includes the URLs of the resource webpages.

17.1 Introduction

Speech recognition technology consists of a lot of different components such as speech enhancement, feature extraction, acoustic modeling, language modeling, and decoding, which are all indispensable to accomplish its functionality with sufficient accuracy. Therefore, creating a new automatic speech recognition (ASR) system is quite time-consuming and requires specific technical knowledge about each component. This complication makes it difficult to incorporate new ideas in state-of-the-art ASR technology and validate their efficacy. Furthermore, this also

S. Watanabe (✉) · T. Hori · J.R. Hershey
Mitsubishi Electric Research Laboratories (MERL), Cambridge, MA, USA
e-mail: shinjiw@ieee.org

Y. Miao · F. Metze
Carnegie Mellon University, 5000 Forbes Ave, Pittsburgh, PA, USA

M. Delcroix
NTT Corporation, 2-4, Hikaridai, Seika-cho, Kyoto, Japan

© Springer International Publishing AG 2017
S. Watanabe et al. (eds.), *New Era for Robust Speech Recognition*,
DOI 10.1007/978-3-319-64680-0_17

369

prevents people from rapid development of new applications with cutting-edge speech technology.

Recently, publicly available toolkits have been playing an important role in overcoming the above problems. Although these kinds of toolkits have long been contributing to the progress of ASR technology, currently they are getting much easier to use and develop by many people and are accelerating the research and development in the community.

In the following sections, we describe publicly available toolkits for robust speech recognition. Generally, a toolkit for ASR includes feature extraction, acoustic-model training, and decoding. Signal processing for speech enhancement and language modeling are usually separate toolkits since they are designed for more generic purposes, for example, language model toolkits are available for other applications such as statistical machine translation and optical character recognition. Also, some toolkits use OpenFst [3] to build a decoding graph, which is not included in this chapter.

17.2 General Speech Recognition Toolkits

Table 17.1 shows the major publicly available ASR toolkits with their functionalities. These tools usually include feature extraction, an acoustic-model trainer, and a large-vocabulary continuous speech recognition (LVCSR) decoder:

HTK [38]: a portable toolkit for building and manipulating hidden Markov models. HTK is primarily used for speech recognition research, although it has been used for numerous other applications, including research into speech synthesis, character recognition, and DNA sequencing.

Julius [22]: high-performance, small-footprint LVCSR decoder software.

Kaldi [30]: a toolkit for speech recognition written in C++ and licensed under the Apache License v2.0. Kaldi is intended for use by speech recognition researchers.

RASR [31]: a software package containing a speech recognition decoder together with tools for the development of acoustic models, for use in speech recognition systems.

Sphinx [21]: a toolkit containing speech recognition algorithms for efficient speech recognition. It includes various decoders (full, lightweight, and adjustable/modifiable versions) with acoustic-model trainers.

HTK is the most well-known toolkit for its long history of supporting the full set of ASR tools, including feature extraction, an acoustic-model trainer, and an LVCSR decoder. In fact, several formats developed in HTK are used as the de facto standard, especially feature file format (HTK format), which can be supported by many other tools listed in the above. CMU Sphinx also has a long history, and has a unique property of supporting various platforms with various implementations, including a lightweight version (PocketSphinx), a normal version, and adjustable/modifiable

Table 17.1 ASR toolkits

Toolkit	Affiliation	Function	Interface	Platform	Implementation	GPU	License	Example	Open source
HTK	Cambridge	AM trainer + decoder	C++	Unix	C++/Cuda	Yes	Own license	RM	N/A
Julius	Nitech	Decoder	C	Unix/Windows	C	N/A	Own license	JNAS	GitHub
Kaldi	JHU	AM trainer + decoder	C++	Unix/Windows	C++/Cuda	Yes	Apache v2.0	>40 recipes	GitHub
RASR	RWTH	AM trainer + decoder	C++	Unix	C++/Cuda	Yes	Own license	AN4	N/A
Sphinx	CMU	AM trainer + decoder	C/C++/ Java/Python	Unix/Windows	C/C++/ Java/Python	Yes	Multiple licenses	AN4	GitHub

versions (Sphinx 4). Julius specializes its function to an LVCSR decoder, and can use acoustic models trained by the other toolkits, including the HTK-format acoustic model. Compared with the above toolkits, RASR and Kaldi are relatively new. Note that all toolkits except for Julius now support graphics processing unit (GPU) computing, and provide fast deep network training functions in their acoustic-model trainers.

Among them, Kaldi is getting more popular in the ASR community by making full use of open source benefits. That is, compared with the other toolkits that are mainly maintained by one research group, there are many contributors to Kaldi from many research groups, which enables Kaldi to actively implement new technologies in the main source repository. Also, Kaldi has many ASR examples (called recipes) and can provide end-to-end system construction for many ASR benchmark tasks of data preprocessing, acoustic and language modeling, decoding, and scoring. For researchers, it is important to reproduce exactly the same results with state-of-the-art ASR techniques, and the Kaldi recipes greatly contribute to the community in terms of the reproducibility and diffusion of cutting-edge ASR techniques.

17.3 Language Model Toolkits

Similarly to the speech recognition toolkits, Table 17.2 shows the major publicly available language model (LM) toolkits with their functionalities. We have referred to the features of each toolkit from its webpage, as follows:

CSLM [32]: a continuous-space LM toolkit including a neural network language model.

CUED-RNNLM [9]: efficient training and evaluation of recurrent neural network language models (RNNLMs).

IRSTLM [13]: suitable for estimating, storing, and accessing very large n-gram LMs.

KENLM [16]: efficient n-gram language model queries, reducing both time and memory costs.

MITLM [17]: training and evaluation of n-gram LMs with efficient data structure and algorithms.

RNNLM TOOLKIT [27]: a recurrent-neural-network-based language-modeling toolkit.

RWTHLM [34]: training feedforward, recurrent, and long short-term memory (LSTM) LMs.

SRILM [33]: training and evaluation of n-gram LMs with many extensions.

Compared with the speech recognition toolkits in Table 17.1, these LM toolkits are based on major free software licenses, many of them permissive or weakly protective ones, and these toolkits are widely used for various applications and are not restricted to ASR. We can roughly categorize these LM toolkits based

Table 17.2 Language-modeling toolkits

Tool	Affiliation	Interface	Platform	Implementation	GPU	License	Example	Open source
CSLM	LIUM	C++	Unix/Windows	C++(+Cuda)	Yes	LGPLv3	Training and rescoring scripts	Yes
CUED-RNNLM	Cambridge	Shell/C++	Unix	C++(+Cuda)	Yes	BSD3	Training and rescoring scripts	Yes
IRSTLM	FBK		Unix	C++	No	LGPLv2	Command-level examples	GitHub
KENLM	CMU	C++/Java/Python	Unix/Windows	C++	No	LGPLv2	Command-level examples	GitHub
MITLM	MIT	Shell/C++	Unix	C++	No	MIT	A test script	GitHub
RNNLM TOOLKIT	BUT	Shell/C++	Unix	C++	No	BSD3	Training and rescoring scripts	Yes
RWTHLM	RWTH	Shell/C++	Unix	C++ (+OpenMP)	No	Own license	Command-level examples	Yes
SRILM	SRI	Shell/C++	Unix/Windows	C++	No	Own license	Command-level examples	Yes

on n-gram LMs (IRSTLM, KENLM, MITLM, SRILM) or neural network LMs (CSLM, CUED-RNNLM, RNNLM TOOLKIT, RWTHLM). If using n-gram-based toolkits, the language model format is unified with the ARPA format or its variants. Since all speech recognition toolkits support the ARPA format LM, the LMs produced by them can basically be applied to all speech recognition toolkits. Among n-gram-based toolkits, SRILM is often used for ASR experiments due to its many functionalities (including most of the major n-gram smoothing techniques and n-gram pruning) and various examples of ASR applications, including lattice rescoring.

On the other hand, neural network LMs have different model structures depending on the network architecture, and they often have different model formats from each other. For example, the most well-known toolkit is RNNLM TOOLKIT, which was developed by T. Mikolov. It supports a recurrent neural network architecture. CUED-RNNLM is also based on the same RNN as it is an extension of RNNLM TOOLKIT, and supports GPU-based parallel computation. On the other hand, CSLM and RWTHLM are based on a feedforward neural network and an LSTM, respectively, and have a different model structure from RNNLM. Therefore, compared with n-gram LMs, neural-network-based LM toolkits do not have a unified model format, and their models cannot be integrated with various LVCSR decoders easily, unlike n-gram LMs. Instead, each toolkit provides lattice-rescoring scripts for major lattice formats in speech recognition toolkits. Note that some generic deep learning toolkits discussed in Sect. 17.5 (e.g., CHAINER, CNTK, THEANO, TENSORFLOW, and TORCH) also include RNNLM/LSTMLM functions as an example of their toolkits.

17.4 Speech Enhancement Toolkits

We list the following speech enhancement software:

BeamformIT [5]:   an acoustic beamforming tool that accepts a variable amount of input channels and computes an output via a filter and sum beamforming technique.

BTK [20]:   C++ and Python libraries that implement speech processing and microphone array techniques.

FASST [29]:   a flexible audio source separation toolbox with the purpose of speeding up the conception and automating the implementation of new model-based audio source separation algorithms.

HARK [28]:   open-source robot audition software consisting of sound source localization modules, sound source separation modules, and automatic speech recognition modules for separated speech signals that work on any robot with any microphone configuration.

ManyEars [15]:   real-time microphone array processing to perform sound source localization, tracking, and separation. It was designed for mobile robot audition in dynamic environments.
WPE [37]:   a speech dereverberation tool for single- and multichannel recordings.

Many tools deal with acoustic beamforming by using multichannel inputs, while WPE deals with dereverberation, which is described in Chap. 2. These tools have proved their effectiveness at several distant speech recognition challenges. Although these tools are publicly available, compared with the other toolkits, several speech enhancement toolkits have been developed with in a closed community. This is because speech enhancement techniques do not require large and complex programs, and the development can be done without open source communities. Also, MATLAB[1] (with a signal-processing toolbox) provides a strong research platform. However, as noise-robust speech recognition becomes important, these enhancement toolkits are gathering attention from ASR communities. Therefore, open source activities for speech enhancement toolkits has been more active with the integration of ASR toolkits or other speech-related applications.

17.5   Deep Learning Toolkits

Deep learning is now playing an important part in most ASR systems, which directly yields a big improvement of the recognition accuracy. Recently, many research groups have released deep learning toolkits to deal with generic machine learning problems. Among them, this section lists the following toolkits, which are used for robust speech processing, with their functionalities, and Table 17.3 provides details about these tools.

Caffe [19]:   a deep learning framework made with expression, speed, and modularity.
Chainer [35]:   a flexible, intuitive, and powerful architecture.
CNTK [2]:   a unified deep learning toolkit that describes neural networks as a series of computational steps via a directed graph.
CURRENNT [36]:   a machine learning library for RNN/LSTM with GPUs.
Mxnet [8]:   lightweight, portable, flexible distributed/mobile deep learning with a dynamic, mutation-aware dataflow dep scheduler.
TensorFlow [1]:   a library for numerical computation using data flow graphs.
Theano [7]:   symbolic scripting in Python, dynamic C-code generation.
TORCH [10]:   supporting neural network, and energy-based models, etc.

All of the toolkits have a set of built-in functions used in deep learning, including affine transformation, sigmoid and softmax operations, and cross-entropy and mean

---

[1]The MathWorks Inc., http://www.mathworks.com/products/matlab/.

Table 17.3 Deep learning toolkits

Tool	Author/affiliation	Application	Interface	Platform	Implementation	GPU	License
Caffe	UC Berkeley	Computer vision	C++	Unix/Win	C++/Cuda	Yes	BSD 2-clause
Chainer	Preferred Networks, Inc.	General	Python	Unix/Win	Python based on CuPy	Yes	MIT License
CNTK	Microsoft	General	C++	Unix/Win	C++/Cuda	Yes	Microsoft custom
CURRENNT	TUM	Sequence recog/conv	C++	Unix/Win	C++/Cuda	Yes	GPLv3
MxNet	Tianqi Chen et al.	General	Python/MATLAB/ R/etc.	Unix/Win	C++/Cuda	Yes	Apache v2.0
TensorFlow	Google Brain	General	Python/C++	Unix	C++/Cuda	Yes	Apache v2.0
Theano	Univ. of Montreal	General	Python	Unix/Mac/Win	Python/C++/Cuda	Yes	BSD 3-clause
Torch	Facebook, Google+DeepMind	General	C + LuaJIT	Almost any	C++/Cuda	Yes	MIT open source

square error cost functions. Therefore, users can implement their own networks by combining these functions. In addition, many of these toolkits support GPU computing, and are scalable to a large amount of data; thus the toolkits are widely used for robust speech processing.

Furthermore, these toolkits are useful for applying new techniques of deep learning to speech recognition, because most toolkit developers are trying to implement novel techniques and utilize new GPU functionalities in their own toolkit as soon as possible, and example codes written by many users also become available through source-code-sharing sites.

17.6   End-to-End Speech Recognition Toolkits

Recently, a new area of ASR research has emerged, which is called *end-to-end speech recognition*. The goal of this approach is to build an ASR system without any internal knowledge about speech such as a phoneme set and a pronunciation lexicon. Although the technology is still in the early stages of research, there are already several toolkits publicly available for end-to-end ASR. Table 17.4 summarizes end-to-end speech recognition software, which aims to build ASR without complicated pipelines:

ATTENTION-LVCSR [6]:   end-to-end attention-based large-vocabulary speech recognition.
EESEN [26]:   end-to-end speech recognition using deep RNN models and WFST-based decoding.
stanford-ctc [24]:   Neural net code for lexicon-free speech recognition with connectionist temporal classification (CTC).
warp-ctc [4]:   A fast parallel implementation of CTC, on both CPU and GPU.

Most tools can train ASR models that deal with direct mapping from input feature to output character sequences without a hidden-Markov-model state-tying module and lexicon and language models. This significantly simplifies their codes compared with those of the conventional ASR toolkits. In the decoding stage, by using the language model, the performance approaches to that of state-of-the-art ASR with a very complicated pipeline. ATTENTION-LVCSR uses an attention mechanism, while the other toolkits use CTC [14].

EESEN [26] leverages a lot of the Kaldi infrastructure, and presents a simple yet powerful acoustic model based on CTC for end-to-end speech recognition. It is presented in Sect. 13.3.

Table 17.4 End-to-end speech recognition toolkits

Tool	Author/affiliation	Interface	Platform	Implementation	GPU	License	Example/recipe	Open source
ATTENTION-LVCSR	Dzmitry Bahdanau	Python	Unix/Windows	Python	Yes	MIT	WSJ/TIMIT	GitHub
EESEN	Yajie Miao, CMU	C++	Unix	C++/Cuda	Yes	Apache v2.0	Four recipes	GitHub
stanford-ctc	Andrew Maas/Stanford	Python	Unix/Windows	Python	Yes	Apache v2.0	Three recipes	GitHub
warp-ctc	Baidu Research	C	Unix	C/Lua	Yes	Apache v2.0		GitHub

17.7 Other Resources for Speech Technology

Other than the toolkits mentioned above, there is another type of activity to help the progress of research, development, and education of speech technology by providing a collaborative repository.

COVAREP [11] is an open-source repository of advanced speech-processing algorithms, stored in a GitHub project where researchers in speech processing can store original implementations of published algorithms. This framework accelerates reproducible research and enables other researchers to perform fair comparisons without reimplementation of research.

Speech Recognition Virtual Kitchen [25] provides an environment to promote community sharing of research techniques, foster innovative experimentation, and provide solid reference systems as a tool for education, research, and evaluation. The feature of this site is that it hosts virtual machines (VMs) that provide a consistent environment for experimentation without the need to install other software or data, and cope with their incompatibilities and peculiarities.

Bob [18] is a free signal-processing and machine learning toolbox composed of a reasonably large number of packages that implement tools for image, audio, and video processing, machine learning, and pattern recognition.

On the other hand, speech and language corpora are also crucial in research and development for ASR technology. A large amount of speech data with correct annotations is required to train high-accuracy models and perform effective evaluation. However, collecting and annotating speech data are very costly. Therefore, some institutions undertake to host individual corpora generated in different research projects, and license and provide each corpus to people who need the data. The Linguistic Data Consortium (LDC) [23] is a major institution that provides well-known corpora such as WSJ, TIMIT, ATIS, and Switchboard. The European Language Resources Association (ELRA) [12] is another institution, in Europe, which provides speech and language corpora in different European languages such as AURORA, CHIL, and TC-STAR.

17.8 Conclusion

We have summarized the toolkits used in robust ASR, which cover general ASR toolkits, language model toolkits, speech enhancement/microphone array front-end toolkits, deep learning toolkits, and end-to-end ASR toolkits. We have also introduced some collaborative repositories and major institutions that manage speech and language corpora. Many of these toolkits are actually used in the other chapters to realize state-of-the-art systems including Kaldi, SRILM, WPE, BeamformIt, Theano, CNTK, and EESEN. We hope this chapter will help readers to accelerate their own research and development with the help of the existing toolkits listed in the chapter.

References

1. Abadi, M., Agarwal, A., Barham, P., Brevdo, E., Chen, Z., Citro, C., Corrado, G.S., Davis, A., Dean, J., Devin, M., et al.: TensorFlow: large-scale machine learning on heterogeneous distributed systems (2016). arXiv preprint arXiv:1603.04467. https://www.tensorflow.org/
2. Agarwal, A., Akchurin, E., Basoglu, C., Chen, G., Cyphers, S., Droppo, J., Eversole, A., Guenter, B., Hillebrand, M., Hoens, T.R., et al.: An introduction to computational networks and the computational network toolkit. Microsoft Technical Report MSR-TR-2014-112 (2014). https://github.com/Microsoft/CNTK
3. Allauzen, C., Riley, M., Schalkwyk, J., Skut, W., Mohri, M.: OpenFst: a general and efficient weighted finite-state transducer library. In: International Conference on Implementation and Application of Automata, pp. 11–23. Springer, New York (2007). http://www.openfst.org/
4. Amodei, D., Anubhai, R., Battenberg, E., Case, C., Casper, J., Catanzaro, B., Chen, J., Chrzanowski, M., Coates, A., Diamos, G., et al.: Deep speech 2: end-to-end speech recognition in English and Mandarin (2015). arXiv preprint arXiv:1512.02595. https://github.com/baidu-research/warp-ctc
5. Anguera, X., Wooters, C., Hernando, J.: Acoustic beamforming for speaker diarization of meetings. IEEE Trans. Audio Speech Lang. Process. 15(7), 2011–2022 (2007). http://www.xavieranguera.com/beamformit/
6. Bahdanau, D., Chorowski, J., Serdyuk, D., Brakel, P., Bengio, Y.: End-to-end attention-based large vocabulary speech recognition. In: 2016 IEEE International Conference on Acoustics, Speech and Signal Processing (ICASSP), pp. 4945–4949 (2016). https://github.com/rizar/attention-lvcsr
7. Bergstra, J., Breuleux, O., Bastien, F., Lamblin, P., Pascanu, R., Desjardins, G., Turian, J., Warde-Farley, D., Bengio, Y.: Theano: A CPU and GPU math compiler in Python. In: Proceedings of the 9th Python in Science Conference, pp. 1–7 (2010). http://deeplearning.net/software/theano/
8. Chen, T., Li, M., Li, Y., Lin, M., Wang, N., Wang, M., Xiao, T., Xu, B., Zhang, C., Zhang, Z.: Mxnet: a flexible and efficient machine learning library for heterogeneous distributed systems. In: Proceedings of Workshop on Machine Learning Systems (LearningSys) in 29th Annual Conference on Neural Information Processing Systems (NIPS) (2015). http://mxnet-mli.readthedocs.io/en/latest/
9. Chen, X., Liu, X., Qian, Y., Gales, M., Woodland, P.: CUED-RNNLM: an open-source toolkit for efficient training and evaluation of recurrent neural network language models. In: IEEE International Conference on Acoustics, Speech and Signal Processing (ICASSP), pp. 6000–6004. IEEE, New York (2016). http://mi.eng.cam.ac.uk/projects/cued-rnnlm/
10. Collobert, R., Kavukcuoglu, K., Farabet, C.: Torch7: a MATLAB-like environment for machine learning. In: BigLearn, NIPS Workshop, EPFL-CONF-192376 (2011). http://torch.ch/
11. Degottex, G., Kane, J., Drugman, T., Raitio, T., Scherer, S.: COVAREP: a collaborative voice analysis repository for speech technologies. In: 2014 IEEE International Conference on Acoustics, Speech and Signal Processing (ICASSP), pp. 960–964. IEEE, New York (2014). http://covarep.github.io/covarep/
12. ELRA: ELDA Portal. http://www.elra.info/en/
13. Federico, M., Bertoldi, N., Cettolo, M.: IRSTLM: An open source toolkit for handling large scale language models. In: Interspeech, pp. 1618–1621 (2008). http://hlt-mt.fbk.eu/technologies/irstlm
14. Graves, A., Jaitly, N.: Towards end-to-end speech recognition with recurrent neural networks. In: ICML, vol. 14, pp. 1764–1772 (2014)
15. Grondin, F., Létourneau, D., Ferland, F., Rousseau, V., Michaud, F.: The ManyEars open framework. Auton. Robot. 34(3), 217–232 (2013). https://sourceforge.net/projects/manyears/
16. Heafield, K.: KenLM: faster and smaller language model queries. In: Proceedings of the Sixth Workshop on Statistical Machine Translation, pp. 187–197. Association for Computational Linguistics (2011). http://kheafield.com/code/kenlm/

17. Hsu, B.J.P., Glass, J.R.: Iterative language model estimation: efficient data structure & algorithms. In: INTERSPEECH, pp. 841–844 (2008). https://github.com/mitlm/mitlm
18. Idiap Research Institute: Bob 2.4.0 documentation. https://pythonhosted.org/bob/
19. Jia, Y., Shelhamer, E., Donahue, J., Karayev, S., Long, J., Girshick, R., Guadarrama, S., Darrell, T.: Caffe: convolutional architecture for fast feature embedding. In: Proceedings of the 22nd ACM International Conference on Multimedia, pp. 675–678. ACM (2014). http://caffe.berkeleyvision.org/
20. Kumatani, K., McDonough, J., Schacht, S., Klakow, D., Garner, P.N., Li, W.: Filter bank design based on minimization of individual aliasing terms for minimum mutual information subband adaptive beamforming. In: IEEE International Conference on Acoustics, Speech and Signal Processing (ICASSP), pp. 1609–1612. IEEE (2008). http://distantspeechrecognition.sourceforge.net/
21. Lee, K.F., Hon, H.W., Reddy, R.: An overview of the SPHINX speech recognition system. IEEE Trans. Acoust. Speech Signal Process. **38**(1), 35–45 (1990). http://cmusphinx.sourceforge.net/
22. Lee, A., Kawahara, T., Shikano, K.: Julius – an open source real-time large vocabulary recognition engine. In: Interspeech, pp. 1691–1694 (2001). http://julius.osdn.jp/en_index.php
23. Linguistic Data Consortium: https://www.ldc.upenn.edu/
24. Maas, A.L., Xie, Z., Jurafsky, D., Ng, A.Y.: Lexicon-free conversational speech recognition with neural networks. In: Proceedings of the North American Chapter of the Association for Computational Linguistics (NAACL) (2015). https://github.com/amaas/stanford-ctc
25. Metze, F., Fosler-Lussier, E.: The speech recognition virtual kitchen: An initial prototype. In: Interspeech, pp. 1872–1873 (2012). http://speechkitchen.org/
26. Miao, Y., Gowayyed, M., Metze, F.: EESEN: end-to-end speech recognition using deep RNN models and WFST-based decoding. In: 2015 IEEE Workshop on Automatic Speech Recognition and Understanding (ASRU), pp. 167–174 (2015). https://github.com/srvk/eesen
27. Mikolov, T., Karafiát, M., Burget, L., Cernocký, J., Khudanpur, S.: Recurrent neural network based language model. In: Interspeech, pp. 1045–1048 (2010). http://www.rnnlm.org/
28. Nakadai, K., Takahashi, T., Okuno, H.G., Nakajima, H., Hasegawa, Y., Tsujino, H.: Design and implementation of robot audition system "hark" open source software for listening to three simultaneous speakers. Adv. Robot. **24**(5–6), 739–761 (2010). http://www.hark.jp/
29. Ozerov, A., Vincent, E., Bimbot, F.: A general flexible framework for the handling of prior information in audio source separation. IEEE Trans. Audio Speech Lang. Process. **20**(4), 1118–1133 (2012). http://bass-db.gforge.inria.fr/fasst/
30. Povey, D., Ghoshal, A., Boulianne, G., Burget, L., Glembek, O., Goel, N., Hannemann, M., Motlicek, P., Qian, Y., Schwarz, P., Silovsky, J., Stemmer, G., Vesely, K.: The Kaldi speech recognition toolkit. In: IEEE 2011 Workshop on Automatic Speech Recognition and Understanding (2011). http://kaldi-asr.org/
31. Rybach, D., Gollan, C., Heigold, G., Hoffmeister, B., Lööf, J., Schlüter, R., Ney, H.: The RWTH Aachen University open source speech recognition system. In: Interspeech, pp. 2111–2114 (2009). https://www-i6.informatik.rwth-aachen.de/rwth-asr/
32. Schwenk, H.: CSLM – a modular open-source continuous space language modeling toolkit. In: INTERSPEECH, pp. 1198–1202 (2013). http://www-lium.univ-lemans.fr/cslm/
33. Stolcke, A., et al.: SRILM – an extensible language modeling toolkit. In: Interspeech, vol. 2002, pp. 901–904 (2002). http://www.speech.sri.com/projects/srilm/
34. Sundermeyer, M., Schlüter, R., Ney, H.: RWTHLM – the RWTH Aachen University neural network language modeling toolkit. In: INTERSPEECH, pp. 2093–2097 (2014). https://www-i6.informatik.rwth-aachen.de/web/Software/rwthlm.php
35. Tokui, S., Oono, K., Hido, S., Clayton, J.: Chainer: a next-generation open source framework for deep learning. In: Proceedings of Workshop on Machine Learning Systems (LearningSys) in 29th Annual Conference on Neural Information Processing Systems (NIPS) (2015). http://chainer.org/

36. Weninger, F., Bergmann, J., Schuller, B.: Introducing CURRENNT – the Munich open-source CUDA RecurREnt neural network toolkit. J. Mach. Learn. Res. **16**(3), 547–551 (2015). https://sourceforge.net/projects/currennt/

37. Yoshioka, T., Nakatani, T., Miyoshi, M., Okuno, H.G.: Blind separation and dereverberation of speech mixtures by joint optimization. IEEE Trans. Audio Speech Lang. Process. **19**(1), 69–84 (2011). http://www.kecl.ntt.co.jp/icl/signal/wpe/

38. Young, S., Evermann, G., Gales, M., Hain, T., Kershaw, D., Liu, X., Moore, G., Odell, J., Ollason, D., Povey, D., et al.: *The HTK Book*, vol. 3, p. 175. Cambridge University Engineering Department (2002). http://htk.eng.cam.ac.uk/

Part IV
Applications

Chapter 18
Speech Research at Google to Enable Universal Speech Interfaces

Michiel Bacchiani, Françoise Beaufays, Alexander Gruenstein, Pedro Moreno, Johan Schalkwyk, Trevor Strohman, and Heiga Zen

Abstract Since the wide adoption of smartphones, speech as an input modality has developed from a science fiction dream to a widely accepted technology. The quality demand on this technology that allowed fueling this adoption is high and has been a continuous focus of research activities at Google. Early adoption of large neural network model deployments and training of such models on large datasets has significantly improved core recognition accuracy. Adoption of novel approaches like long short-term memory models and connectionist temporal classification have further improved accuracy and reduced latency. In addition, algorithms that allow adaptive language modeling improve accuracy based on the context of the speech input. Focus on expanding coverage of the user population in terms of languages and speaker characteristics (e.g., child speech) has lead to novel algorithms that further pushed the universal speech input vision. Continuing this trend, our most recent investigations have been on noise and far-field robustness. Tackling speech processing in those environments will enable applications of in-car, wearable, and in-the-home scenarios and as such be another step towards true universal speech input. This chapter will briefly describe the algorithmic developments at Google over the past decade that have brought speech processing to where it is today.

18.1 Early Development

Although there was some speech activity in Google, it wasn't until 2005 that speech development got a more serious emphasis. A decision was made at that point that speech is a key technology and that Google should obtain its own implementation

M. Bacchiani (✉) • P. Moreno • J. Schalkwyk
Google Inc., 76 Ninth Ave, New York, NY 10011, USA
e-mail: michiel@google.com

F. Beaufays • A. Gruenstein • T. Strohman
Google Inc., 1600 Amphitheatre Parkway, Mountain View, CA 94043, USA

H. Zen
Google Inc., 1-13 Saint Giles High Street, London WC2H 8AG, USA

© Springer International Publishing AG 2017   385
S. Watanabe et al. (eds.), *New Era for Robust Speech Recognition*,
DOI 10.1007/978-3-319-64680-0_18

of that technology to support a speech modality in upcoming projects. Although this might sound logical at this point, it is important to realize the state of technology in general at that time. Google as a search engine company was very well known and had a lot of use, but the bulk of that use was from desktop computers where users would engage by typing into a search box. Cell phones were ubiquitous as well, but they would function as phones alone; the smartphones as we know them now had not yet emerged. As a result, an investment in building speech infrastructure was exploratory, with an agreement on potential, but not with obvious use.

In terms of applications, we focused extensively on an application named GOOG411. This application was an automation of the directory assistance service common in the United States at that time. When out and armed with only a cell phone, contacting a business would require knowing the phone number. Since the smartphone model had not yet emerged, the user would reach out to a telephone service to obtain the number of the business of interest. The GOOG411 service automated this, showing the feasibility of large-vocabulary speech recognition. The fairly reliable service this provided, together with a monetary incentive (the service was free), made this a fairly popular service. The substantial flow of data this service generated and the application-specific modeling challenges were interesting as an application as well as a powerful model to help build up our infrastructure. Details of the early development are described in detail in [10].

Other early investigations were in transcription applications of speech recognition. Two areas of application received a fair amount of attention, one in voicemail transcription, a service still alive today in Google Voice, the other in the automatic subtitling of YouTube video content. The latter started as an exploration into a niche domain around an election cycle [3].

Algorithmically, the early work for building up our infrastructure was not very innovative, as it was a large effort to "catch up." That said, the engineering was interesting as we based our implementation on using core Google infrastructure, particularly MapReduce [18], which allowed us to scale to very large databases. The other area where we invested heavily was to build a foundation using weighted finite state transducers (WFSTS) [7]. To establish the notation used in the rest of this chapter, our models generally employ an acceptor WFST G which encodes the language model, a lexicon transducer L which maps from word to phone strings, and a transducer C to implement the phonetic context dependency model. The optimized composed graph is generally referred to as CLG. That infrastructure with the popular GOOG411 application allowed us to quickly grow the ability of our core platform. The transcription efforts received less acceptance as an application but offered interesting research challenges like language identification [2] and punctuation restoration [23, 60]. The latter topic received even more attention later on in light of the transcription systems described in Sect. 18.4.

These early efforts in speech development at Google had traction but none like the adoption of our voice search application that became a major focus of our efforts with the emergence of smartphones, and we describe this in more detail in Sect. 18.2.

18.2   Voice Search

Using our voice to access information has been part of science fiction ever since the days of Captain Kirk talking to the *Star Trek* computer. With the advent of web-enabled smartphones, information access became a ubiquitous aspect of our daily life. With this came a significant shift in users expectations, and the nature of the services they expected—e.g., new types of up-to-the-minute information (where's the closest parking spot?) or communications (e.g., "update my Facebook status to 'seeking chocolate'").

There is also the growing expectation of ubiquitous availability. Users increasingly expect to have constant access to the information and services of the web. Over the years these expectations have evolved to many new devices. Today you can speak to your phone, your car, your watch, your TV, your home, and many more to come. You can use these devices to navigate, to listen to music, to ask what's the weather like, to change channel, to call your wife, to remind you to pick up milk on the way home, to book an Uber, and many more. These devices have become part of our daily lives, assisting us in our daily needs.

Given the nature of delivery devices and the increased range of usage scenarios, speech technology has taken on new importance in accommodating user needs for ubiquitous access—any time, any place, any scenario. A goal at Google is to make spoken access ubiquitously available. Users should be able to take it for granted that you can always just speak and state your need by voice. Achieving ubiquity requires two things: availability (i.e., built into every possible interaction where speech input or output can make sense), and performance (i.e., works so well that the modality adds no friction to the interaction).

Performance has two major aspects that form the core of our algorithmic investments. The obvious is core recognition quality: are we transcribing every word heard correctly? However, a second and equally important aspect is latency. The interaction needs to be really fast. This is another major aspect of making the interaction frictionless. In [55] we describe the various technical challenges and algorithmic solutions we developed. This work shows the benefit of using very large amounts of training data to build accurate models. But it also focuses on voice-search-specific challenges. It describes the unique challenges in text normalization, corpus recency, user interface design around a multimodel application, and error handling, to mention a few of the topics.

18.3   Text to Speech

Google used text-to-speech (TTS) systems provided by third parties in its early services such as GOOG411. However, as the speech-based modality was getting more important within Google, having its own implementation of TTS became reasonable. In 2010, Google acquired Phonetic Arts, which was a start-up company in the UK providing TTS, and started developing its own TTS systems based on the acquired technology. Google has released TTS in more than 30 languages. It has been used in various Google services such as Google Maps, Google Translate, and Android.

A typical TTS system consists of text analysis and speech synthesis modules. The text analysis module includes a number of natural language processing (NLP) sub-modules, e.g., sentence segmentation, word segmentation, part-of-speech tagging, text normalization, and grapheme-to-phoneme (G2P) predictions. Google developed a flexible text-normalization system called "Kestrel" [20]. At the core of Kestrel, text-normalization grammars are compiled into libraries of WFSTs. This tool has been open-sourced [62].

There are two main approaches to implementing the speech synthesis module; concatenative unit-selection and statistical parametric approaches. The former approach concatenates real speech units (e.g., diphones) from a speech database to synthesize speech given a text. The latter approach uses a statistical acoustic model to predict a sequence of acoustic features given a text, then speech is reconstructed from the predicted acoustic features using a speech analysis/synthesis technique (a.k.a. vocoder). The concatenative approach can synthesize speech with high segmental quality but it requires large disk space and memory. On the other hand, the statistical parametric approach is compact but its segmental quality is bounded by that of the vocoder. Google uses the concatenative approach for the TTS services on servers [22] and the statistical parametric approach for the TTS services on mobile devices [73].

Google has been actively developing new technologies in the speech synthesis area. For example, Google is one of the pioneers in utilizing deep learning in the statistical parametric approach, e.g., acoustic modeling based on deep neural networks [72], mixture density networks [71], and long short-term memory recurrent neural networks [70]. These neural-network-based TTS systems have been deployed to production [73]. The progress of acoustic modeling in speech synthesis can be found in [69]. As the segmental quality obtained from the statistical parametric approach is bounded by that of the vocoder, improving the vocoder itself is also important. Google has also developed new speech analysis [31] and synthesis [1] techniques. Integrating acoustic modeling and the vocoder into a unified framework to eliminate the bound has also been explored [63, 64].

18.4 Dictation/IME/Transcription

Shortly after launching voice search, we realized that our users would want to use speech recognition not only to talk to a machine, but also to dictate messages to their friends. This put new constraints on the technology: if you type "imags of elefnats" into google.com, it won't mind and will return the desired results, but if the recognizer spells your voice message this way, your correspondent may be surprised. Also, for longer messages, capitalization and punctuation inference are needed to improve readability. More fundamentally, longer voice inputs have a higher probability of being recognized with some mistake and to be deemed imperfect, which increases the demands on raw accuracy. We tackled all these issues.

The formatting of speech recognition results can be handled in different ways. For example, the language model can be all lower-cased, without explicit support for entity rendering, e.g., "seventy six ninth avenue new york." Rule-based grammars compiled as finite state transducers (FSTs) can then be used to postprocess the string into "76 9th Ave, New York" [23, 60]. Another approach we later investigated consisted in formatting the training corpus used to build the system language model, and relying on the relatively higher order of the n-gram language model to maintain capitalization information [13]. Here, instead of relying on rules, we learned capitalization patterns from an auxiliary training set that was well-formatted, e.g., typed documents.

We later extended our formatting efforts to also cover entities, such as "seventy six" or "ninth avenue." The proposed solution was to insert a "verbalizer" FST in the CLG decoder graph composition to bridge the "spoken form" pronunciation of a phrase "seventy six" in the L transducer to its "written form" in the G transducer, "76" [49]. This approach drastically improved our rendition of alphanumerical entities as the language model context was now involved in disambiguation choices such as "7-Eleven" vs. "Boeing 711" vs. "7:11."

Improving the raw accuracy of long-form transcription is complicated. Text messages dictated while driving a car do not sound like voicemails left from home, and even less like a YouTube video uploaded by a private user or a news agency. Yet we aim at recognizing all of these with great accuracy. Bayesian interpolation technology helps us leverage different data sources in our models [6], but when the raw data sources are themselves speech recognition results, our best-matched data, they may be corrupted by previous recognition errors such as those resulting from bad pronunciations in our lexicon.

We developed sophisticated techniques to learn word pronunciations from audio data (see, e.g., [39, 42]), but also learned how to iteratively re-recognize our data to progressively erase errors. This can be done with a model trained on cleaner data, as in our previously mentioned capitalization work and as detailed in [12], through semisupervised learning exploiting a concept of "islands of confidence", i.e., training only on data that is likely to be error-free [34], or in some cases through mostly unsupervised techniques. In our work on Google Voice, such techniques along with improvements in acoustic modeling technologies allowed us to halve our word error rate (WER) compared to a previously state-of-the-art baseline [11].

18.5   Internationalization

Google started its expansion into more languages in 2008, initially with the launch of different dialectal versions of English, first with British English, followed by Australian and then Indian English. In 2009 Mandarin voice search was made publicly available, and since then Google voice search has been available in more than 50 languages.

Our approach to developing new voice search systems is based on fast data collection as described in [30], followed by quick iterations once real data flows into our servers. We typically "refresh" acoustic models, lexicons, and language models within months of launch.

Our core research in speech recognition is conducted on US English, and once new techniques are sufficiently validated we transfer them to all our fleet of languages. While we strive for automation across all languages, each new language presents different challenges not encountered in English. Some examples of issue that arise are text segmentation (e.g., Mandarin), tone modeling (e.g., Thai), vocabulary size control in highly inflected languages (e.g., Russian), and ambiguity in the orthography (e.g., Arabic, where short vowels are omitted).

Each of these problems has required specific solutions; however, we have always tried to turn these solutions into generic language-processing modules that can be applied to other languages that exhibit similar phenomena. Our goal in general is to build data-driven generic solutions for each of these problems and then deploy them repeatedly. For example, we make extensive use of conditional random fields (CRFs) to build segmenters, which we have applied verbatim over several languages. Similarly, tone modeling is done by expanding the phonetic inventory to cover all the vowel–tone pairs, and this simple idea has been successfully applied in Mandarin, Cantonese, and Thai [59].

We split our 55 languages into different tiers of interest, giving more research effort to some languages (e.g., French, German, Korean, Japanese, Mandarin) than others. For some of the languages, we make extensive use of unsupervised or semisupervised data extraction techniques to improve acoustic models and lexicons.

Since 2009, we have launched more and more languages and, by our own estimates, in order to reach 99% of the world population we will need to reach a total of 200 languages and dialects. As we reach languages with smaller numbers of speakers, or languages with smaller textual sources on the web, the challenge of finding data to build acoustic models, lexicons, and language models increases dramatically. Also, many of these populations often speak more than one language, so automatically identifying the spoken language becomes quite useful. Hence we have investigated and deployed in production language identification techniques [21].

Despite these investments in technologies like language ID, some phenomena such as code switching are still difficult to solve and will require further investments in research.

In summary, while our multilingual efforts started in 2009, we still face plenty of challenges to deliver on our ultimate goal of seamless voice search in 200 languages.

18.6   Neural-Network-Based Acoustic Modeling

For decades, acoustic modeling for speech recognition has been dominated by triphone-state Gaussian mixture models (GMMs) whose likelihood estimates were fed into a hidden-Markov-model (HMM) backbone. These simple models had many advantages, including their mathematical elegance, which allowed researchers to propose principled solutions to practical problems such as speaker or task adaptation. GMM training lent itself well to parallelization, since each model was trained to model data likelihood under a given state assumption, and the models were also fast to evaluate at run time.

Around 1990, the idea of training triphone states discriminatively picked up momentum, with new, hybrid neural network architectures [14, 68], including even recurrent neural networks (RNNs) [41]. These architectures replaced thousands of independent GMMs with a single neural network that evaluated posterior probabilities over the entire state space. Such models, however, were computationally expensive and remained for 20 years in the realm of academic research.

Neural network acoustic models really came to life in 2012 [28, 65], when more powerful computers and engineering prowess made them fast enough to serve real traffic. The promised accuracy gains were finally realized, ranging from 20 to 30% relative WER reduction, across languages. But this was also the beginning of a technology revolution.

At Google a new training infrastructure was developed, DistBelief, to facilitate the development of deep neural networks (DNNs). DistBelief implements the Hogwild! algorithm [40] where asynchronous gradient descent (ASGD) optimization is performed using a parameter server that aggregates and distributes gradient updates, and model replicas that operate on subsets of the training data [19]. GMMs, originally retained to initialize DNN models and triphone-state definitions, were progressively retired [9, 57]. Interestingly, the long-used front-end processing to create cepstrum vectors was also abandoned in favor of simple log-filtered energies and later direct waveform processing [29, 46].

DNNs were first trained with a frame-level cross-entropy (CE) optimization. Soon after, sentence-level sequence discriminative criteria such as maximum mutual information (MMI) and state-level minimum Bayes risk (sMBR) were implemented in DistBelief, leading to WER gains of 15% over models trained with CE only [25].

The same year, feedforward neural networks were outpaced by RNNs. Regular RNNs, whose training was known to be delicate due to the instability of gradient propagation, were replaced with long short-term memory (LSTM) RNNs that, much like transistor-based electronic circuits, introduce gates to store, refresh, and propagate inner signals. Scaling LSTMs to large output layers (tens of thousands of states) was made possible by the introduction of recurrent and nonrecurrent projection layers. With these, CE-trained LSTMs outperformed CE-trained DNNs [50], and sequence-trained LSTMs outperformed sequence-trained DNNs by10% [51].

Within less than a year, a cascade of convolutional, LSTM, and DNN layers was shown to outperform LSTMs by another 5% WER relative [44]. In the convolutional long short-term memory deep neural network (CLDNN), a vector of input frames is first processed by two convolutional layers with maxpooling to reduce the frequency variance in the signal. A linear layer then reduces the dimensionality of the signal prior to processing it with a couple of LSTM layers, followed by a few DNN layers.

Meanwhile, it was recognized that phone state modeling in HMMs was a vestigial artifact of long-gone GMMs: LSTMs have all the memory power needed to keep temporal context, and context-dependent whole-phone models replaced triphone states with no loss of accuracy [58]. This was perhaps a first step in a new direction: connectionist temporal classification (CTC).

CTC, first introduced in [24], allows asynchronous sequence-to-sequence modeling by adding a "blank" label to the list of desired output classification labels. At run time, the model can predict for every frame of input data either a real label, a context-dependent phone in this case, or the blank label. This encourages the model to provide spiky outputs: nothing (blank) for many frames, and then a phone label when it has accumulated enough evidence. The asynchronism between acoustic events in the input stream and output labels means that no alignment is needed to bootstrap the models and no HMM is needed to orchestrate the output sequencing.

CTC proved difficult to bend to production constraints, though. Early results showed success with bidirectional monophone and models were encouraging [54], but would not allow streaming recognition. Context-dependent unidirectional models offered better accuracies, but were rather unstable, a problem that was resolved by stacking several frames and sampling the input stream at lower frame rates [53]. This also brought a considerable reduction in computation and runtime latency. Overfitting was solved by scaling the training-set multistyle training (MTR), which also increased its robustness to noise. All in all, CTC proved equally accurate as CLDNN HMM models, but halved recognition latency [52].

Looking back, the move to neural networks for acoustic modeling was a decisive turning point in the history of speech recognition, one that allowed massive improvements in the field and accelerated the adoption of speech as a key input modality for mobile devices.

18.7 Adaptive Language Modeling

The traditional approach to language modeling for the most part has been based on building static language models. In this approach multiple sources of text, such as previous search queries, web documents, newspapers, etc., are mined and then n-gram-based language models are built and deployed to production. These language models are refreshed on a weekly or biweekly schedule, but this is still far from language models that are truly dynamic and contextual.

In the last few years we have started to radically change the nature of our production language models by making them more responsive to user's contextual changes. We have accomplished this using two basic techniques, dynamic classes and language model (LM) biasing (or twiddling).

The idea of dynamic classes is well known in the language-modeling community. Simply put, it consists of introducing nonterminal tags in n-gram-based LMs and at run time dynamically replacing those nonterminals with a small LM that contains a class-based element. Many of these new elements can be new words not previously known in the language model vocabulary. We currently do this to improve the performance of name recognition in Google voice search [4]. In our research we have demonstrated that mining the user contacts from the user's contacts list can yield significant reductions in WER.

Our experience in contact recognition motivated us to explore other ways to dynamically adjust the LM. In [5] we introduced the concept of biasing the language model to specific n-grams for which we have very strong evidence that they might be spoken. We call this technique language model biasing or twiddling. This allows us to leverage contextual information that has the potential to carry very accurate information about what the user might say next. Examples of such context cues are the dialog state, previous queries from the user, or text currently displayed on the phone screen.

At the same time, contextual information can be more or less reliable; for example, if the user is in a confirmation state where Google voice asks for "yes/no/cancel", this knowledge can be encoded in the language model to strongly "bias" the LM to increase the probability of yes/no/cancel-related n-grams in the LM. Alternatively, the user way see in their screen some text string that might influence the recognizer a little bit, but perhaps not as strongly as in the previous example.

The implementation of this biasing is by encoding all contextual information into a WFST, which is sent to the server and then interpolated with the static server LM. This approach allows us to control the strength of the bias and in addition allows the introduction of new vocabulary.

Our ultimate goal is to transform Google voice search language models into dynamic and contextual data structures that accurately model the user-expected behavior with significant WER reductions.

18.8 Mobile-Device-Specific Technology

Most speech recognition research focuses on speech recognition "in the cloud", accessed over the network by mobile devices. However, there are key use cases for speech recognition technology that runs a the mobile device itself. Many tasks can be performed on mobile devices without a network connection, or with a slow or flaky one: dictating an SMS message, or setting alarms and timers. In addition, algorithms such as keyword spotting must run on the device itself, and must consume extremely small amounts of resources if they are to be used to wake a device prior to performing a command or search, as in "Ok Google. Set a timer for 5 minutes."

In [32, 35] we give an overview of large-vocabulary automatic speech recognition (ASR) systems we have developed that are capable of running in real time on most Android phones, have memory footprints of around 20–30 MB, and yet are competitive with "cloud" ASR accuracy for dictation and voice commands. We have explored GMM, DNN, and LSTM acoustic models, achieving best results with an LSTM trained with CTC to directly predict phoneme targets. Using joint factorization of the current and interlayer weight matrices, we are able to compress the LSTM to one third of its size, while maintaining the same accuracy [38]. Quantization of the weight matrices from floating-point to 8 bit integers further reduces the model size by a factor of 4 and significantly reduces computation. Accuracy loss is minimal when the training is made quantization aware [8].

We employ a number of techniques to reduce memory usage during our FST-based decoding. First, we limit the vocabulary size to 64K so that arc labels can be stored as 16-bit integers. Second, in the case of the CTC LSTM with context-independent phoneme targets, we need only decode with an *LG*, instead of a *CLG*, greatly reducing size of the FST. Third, we employ on-the-fly rescoring during decoding: the *LG* is built with an aggressively pruned language model, while a larger language model is used to rescore. The larger language model is compressed using LOUDS [61]. Furthermore, we can combine the domains of dictation and spoken commands into a single language model using Bayesian interpolation [6]. Personalization is supported by adaptation as described in Sect. 18.7.

In addition to large-vocabulary ASR, we have also developed an embedded keyword spotter that requires an order of magnitude less memory and computational power, with a total memory footprint of under 500 KB. [15] describes the overall architecture, in which a DNN is trained to slide over a window of log-filterbank energies and identify individual words. A simple algorithm smooths the posteriors for each word over time, which are combined to form an overall confidence score. Accuracy can be improved, particularly in noise, by using convolutional neural networks [43] as well as multistyle training [37]. We have found that rank-constrained neural networks are an effective way to reduce the size of the DNN by 75%, without accuracy loss [36]. Finally, we have presented a novel approach to using LSTMs for query-by-example keyword spotting [16].

While the keyword spotter is speaker independent, we have built an embedded speaker verifier of a similar size that can be used to verify that a known user spoke the keyword. Here, again, we pioneered the use of DNNs for this task, developing a d-vector-based algorithm that outperforms the traditional i-vector approach to the task [66]. We train a neural net to predict speaker labels, and then discard the final layer—using the penultimate layer for feature extraction to produce a "d-vector" for each utterance. In [17], it was shown that using locally connected and convolutional networks can further improve accuracy. [26] describes an end-to-end approach to training the algorithm, and demonstrates that it improves both DNNs and LSTMs being used for this task.

18.9   Robustness

Given the recent success of speech recognition for smartphone input and its mass adoption by the public, a natural progression of this input modality is towards allowing this interface in noisy and far-field conditions. This requirement comes up in applications like assistants and wearables or in the car. The technical complexity that this shift of context creates is significant, yet it is perceived by the user as insignificant and hence it gives rise to an expectation that they can enjoy the same experience they have grown accustomed to using their phone.

Many years of research in speech enhancement have resulted in various algorithms to allow ASR in far-field and/or noisy conditions. Still, even with the application of advanced algorithms, alleviating the issues raised by the environment remains challenging. Focusing particularly on multimicrophone (or multichannel) systems, they generally apply speech enhancement techniques to multichannel input to transform it into a single-channel signal. The objective in this transformation is to reduce the negative impact of reverberation and noise on recognition accuracy. That enhancement process can be characterized by three stages: localization, beamforming, and postfiltering. The beamforming implements spatial filtering, amplifying the signal from a particular direction, and suppressing the input from other directions. This requires a localization model, i.e., one that estimates the directions the spatial filtering should emphasize or de-emphasize. The filter design commonly uses objectives like minimum variance distortionless response or multi-channel Wiener filtering to define a figure of merit for the enhanced signal resulting from this processing.

Practical application of multichannel processing is challenging in real-world environments. If the localization estimate has an error, the subsequent beamforming will actively degrade the performance as it will enhance the noise and suppress the speech. Another challenge is in optimization. The localization, spatial filtering, and postfiltering are optimized with proxy figures of merit. The end objective is to improve the recognition accuracy, but the objectives used in the optimization of the subparts are distinct from that objective. As a result, the joint system might not benefit even if the individual subparts are successful in optimizing their objectives.

Joint optimization has received some attention where enhancement and recognition models are generative, e.g. in [56]. However, novel ASR systems based on neural networks are generally trained by gradient descent and, as such, the joint optimization of such a system with generative enhancement models is at best very complex (e.g., see [27]). To retain the joint-optimization paradigm and make this compatible with our neural network-based models, we have extended our neural-architectures. First, independently of multichannel processing, we showed that we can incorporate the front-end processing of the recognition system directly into the neural network architecture. In [29, 46] we showed that, by use of a convolutional input layer, we can recognize speech by processing directly from the waveform signal. That base architecture is directly amenable to multichannel processing by replicating the input layer for the multiple channels and, as such, integrating the enhancement processing with the recognition model directly. As we showed in [45], this joint model is very successful in implicitly doing the localization and beamforming in the network. And since it is implemented as a single joint network, the enhancement and recognition models are optimized jointly with the same objective. Factoring the input layer to more specifically allow look direction optimization further enhances the performance of this model [48]. Given the success of this approach, we further investigated modeling choices that fit within this framework. In [33] we showed that instead of factoring for look directions, an adaptive approach, one where the beamforming network parameters are computed at inference time, is successful as well. Finally, we showed in [47, 67] that when we make use of the duality of time and (complex) frequency domain processing, we are presented with yet another option for the joint-processing model. This spectrum of implementation options is all more or less equally effective in getting enhancement to provide gains for recognition, but the options differ in terms of computational cost and other modeling nuances. Many more details of these joint-modeling approaches are described in our contribution on this topic in Chap.5.

References

1. Agiomyrgiannakis, Y.: Vocaine the vocoder and applications in speech synthesis. In: Proceedings of ICASSP (2015)
2. Alberti, C., Bacchiani, M.: Discriminative features for language identification. In: Proceeding of Interspeech (2011)
3. Alberti, C., Bacchiani, M., Bezman, A., Chelba, C., Drofa, A., Liao, H., Moreno, P., Power, T., Sahuguet, A., Shugrina, M., Siohan, O.: An audio indexing system for election video material. In: Proceedings of ICASSP (2009)
4. Aleksic, P., Allauzen, C., Elson, D., Kracun, A., Casado, D.M., Moreno, P.J.: Improved recognition of contact names in voice commands. In: Proceedings of ICASSP, pp. 4441–4444 (2015)
5. Aleksic, P., Ghodsi, M., Michaely, A., Allauzen, C., Hall, K., Roark, B., Rybach, D., Moreno, P.: Bringing contextual information to Google speech recognition. In: Proceedings of Interspeech (2015)
6. Allauzen, C., Riley, M.: Bayesian language model interpolation for mobile speech input. In: Proceedings of Interspeech, pp. 1429–1432 (2011)
7. Allauzen, C., Riley, M., Schalkwyk, J., Skut, W., Mohri, M.: OpenFst: a general and efficient weighted finite-state transducer library. In: Proceedings of the 12th International Conference on Implementation and Application of Automata (CIAA) (2007)
8. Alvarez, R., Prabhavalkar, R., Bakhtin, A.: On the efficient representation and execution of deep acoustic models. In: Proceedings of Interspeech (2016)
9. Bacchiani, M., Rybach, D.: Context dependent state tying for speech recognition using deep neural network acoustic models. In: Proceedings of ICASSP (2014)
10. Bacchiani, M., Beaufays, F., Schalkwyk, J., Schuster, M., Strope, B.: Deploying GOOG-411: early lessons in data, measurement, and testing. In: Proceedings of ICASSP (2008)
11. Beaufays, F.: The neural networks behind Google voice transcription. In: Google Research blog (2015). https://research.googleblog.com/2015/08/the-neural-networks-behind-google-voice.html

12. Beaufays, F.: How the dream of speech recognition became a reality. In: Google (2016). www.google.com/about/careers/stories/how-one-team-turned-the-dream-of-speech-recognition-into-a-reality
13. Beaufays, F., Strope, B.: Language modeling capitalization. In: Proceedings of ICASSP (2013)
14. Bourlard, H., Morgan, N.: Connectionist Speech Recognition: A Hybrid Approach. Kluwer Academic, Dordrecht (1993)
15. Chen, G., Parada, C., Heigold, G.: Small-footprint keyword spotting using deep neural networks. In: Proceedings of ICASSP (2014)
16. Chen, G., Parada, C., Sainath, T.N.: Query-by-example keyword spotting using long short-term memory networks. In: Proceedings of ICASSP (2015)
17. Chen, Y.H., Lopez-Moreno, I., Sainath, T., Visontai, M., Alvarez, R., Parada, C.: Locally-connected and convolutional neural networks for small footprint speaker recognition. In: Proceedings of Interspeech (2015)
18. Dean, J., Ghemawat, S.: MapReduce: simplified data processing on large Clusters. In: OSDI'04, Sixth Symposium on Operating System Design and Implementation (2004)
19. Dean, J., Corrado, G.S., Monga, R., Chen, K., Devin, M., Le, Q.V., Mao, M.Z., Ranzato, M., Senior, A., Tucker, P., Yang, K., Ng, A.Y.: Large scale distributed deep networks. In: Proceedings of Neural Information Processing Systems (NIPS) (2012)
20. Ebden, P., Sproat, R.: The Kestrel TTS text normalization system. J. Nat. Lang. Eng. **21**(3), 333–353 (2014)
21. Gonzalez-Dominguez, J., Lopez-Moreno, I., Moreno, P.J., Gonzalez-Rodriguez, J.: Frame by frame language identification in short utterances using deep neural networks. In: Neural Networks, Special Issue: Neural Network Learning in Big Data, pp. 49–58 (2014)
22. Gonzalvo, X., Tazari, S., Chan, C.A., Becker, M., Gutkin, A., Silen, H.: Recent advances in Google real-time HMM-driven unit selection synthesizer. In: Proceeding of Interspeech (2016)
23. Gravano, A., Jansche, M., Bacchiani, M.: Restoring punctuation and capitalization in transcribed speech. In: Proceedings of ICASSP (2009)
24. Graves, A.: Supervised Sequence Labelling with Recurrent Neural Networks. Studies in Computational Intelligence, vol. 385. Springer, New York (2012)
25. Heigold, G., McDermott, E., Vanhoucke, V., Senior, A., Bacchiani, M.: Asynchronous stochastic optimization for sequence training of deep neural networks. In: Proceedings of ICASSP (2014)
26. Heigold, G., Moreno, I., Bengio, S., Shazeer, N.M.: End-to-end text-dependent speaker verification. In: Proceedings of ICASSP (2016)
27. Hershey, J.R., Roux, J.L., Weninger, F.: Deep unfolding: model-based inspiration of novel deep architectures. CoRR abs/1409.2574 (2014)
28. Hinton, G., Deng, L., Yu, D., Dahl, G., Rahman Mohamed, A., Jaitly, N., Senior, A., Vanhoucke, V., Nguyen, P., Sainath, T., Kingsbury, B.: Deep neural networks for acoustic modeling in speech recognition. Signal Process. Mag. **29**(6), 82–97 (2012)
29. Hoshen, Y., Weiss, R.J., Wilson, K.W.: Speech acoustic modeling from raw multichannel waveforms. In: Proceedings of ICASSP (2015)
30. Hughes, T., Nakajima, K., Ha, L., Vasu, A., Moreno, P., LeBeau, M.: Building transcribed speech corpora quickly and cheaply for many languages. In: Proceedings of Interspeech (2010)
31. Kawahara, H., Agiomyrgiannakis, Y., Zen, H.: Using instantaneous frequency and aperiodicity detection to estimate f0 for high-quality speech synthesis. In: ISCA SSW9 (2016)
32. Lei, X., Senior, A., Gruenstein, A., Sorensen, J.: Accurate and compact large vocabulary speech recognition on mobile devices. In: Proceedings of Interspeech (2013)
33. Li, B., Sainath, T.N., Weiss, R.J., Wilson, K.W., Bacchiani, M.: Neural network adaptive beamforming for robust multichannel speech recognition. In: Interspeech (2016)
34. Liao, H., McDermott, E., Senior, A.: Large scale deep neural network acoustic modeling with semi-supervised training data for YouTube video transcription. In: Proceedings of IEEE Workshop on Automatic Speech Recognition and Understanding (ASRU) (2013)

35. McGraw, I., Prabhavalkar, R., Alvarez, R., Arenas, M.G., Rao, K., Rybach, D., Alsharif, O., Sak, H., Gruenstein, A., Beaufays, F., Parada, C.: Personalized speech recognition on mobile devices. In: Proceedings of ICASSP (2016)
36. Nakkiran, P., Alvarez, R., Prabhavalkar, R., Parada, C.: Compressing deep neural networks using a rank-constrained topology. In: Proceedings of Interspeech, pp. 1473–1477 (2015)
37. Prabhavalkar, R., Alvarez, R., Parada, C., Nakkiran, P., Sainath, T.: Automatic gain control and multi-style training for robust small-footprint keyword spotting with deep neural networks. In: Proceedings of ICASSP, pp. 4704–4708 (2015)
38. Prabhavalkar, R., Alsharif, O., Bruguier, A., McGraw, I.: On the compression of recurrent neural networks with an application to LVCSR acoustic modeling for embedded speech recognition. In: Proceedings of ICASSP (2016)
39. Rao, K., Peng, F., Beaufays, F.: Grapheme-to-phoneme conversion using long short-term memory recurrent neural networks. In: Proceedings of ICASSP (2016)
40. Recht, B., Re, C., Wright, S., Feng, N.: Hogwild: a lock-free approach to parallelizing stochastic gradient descent. In: Shawe-Taylor, J., Zemel, R.S., Bartlett, P.L., Pereira, F., Weinberger, K.Q. (eds.) Advances in Neural Information Processing Systems, vol. 24, pp. 693–701. Curran Associates, Red Hook (2011)
41. Robinson, T., Hochberg, M., Renals, S.: The Use of Recurrent Neural Networks in Continuous Speech Recognition. Springer, New York (1995)
42. Rutherford, A., Peng, F., Beaufays, F.: Pronunciation learning for named-entities through crowd-sourcing. In: Proceedings of Interspeech (2014)
43. Sainath, T., Parada, C.: Convolutional neural networks for small-footprint keyword spotting. In: Proceedings of Interspeech (2015)
44. Sainath, T., Vinyals, O., Senior, A., Sak, H.: Convolutional, long short-term memory, fully connected deep neural networks. In: Proceedings of ICASSP (2015)
45. Sainath, T.N., Weiss, R.J., Wilson, K.W., Narayanan, A., Bacchiani, M., Senior, A.: Speaker localization and microphone spacing invariant acoustic modeling from raw multichannel waveforms. In: Proceedings of IEEE Workshop on Automatic Speech Recognition and Understanding (ASRU) (2015)
46. Sainath, T.N., Weiss, R.J., Wilson, K.W., Senior, A., Vinyals, O.: Learning the speech front-end with raw waveform CLDNNS. In: Proceedings of Interspeech (2015)
47. Sainath, T.N., Narayanan, A., Weiss, R.J., Wilson, K.W., Bacchiani, M., Shafran, I.: Improvements to factorized neural network multichannel models. In: Interspeech (2016)
48. Sainath, T.N., Weiss, R.J., Wilson, K.W., Narayanan, A., Bacchiani, M.: Factored spatial and spectral multichannel raw waveform CLDNNS. In: Proceedings of ICASSP (2016)
49. Sak, H., Sung, Y., Beaufays, F., Allauzen, C.: Written-domain language modeling for automatic speech recognition. In: Proceedings of Interspeech (2013)
50. Sak, H., Senior, A.W., Beaufays, F.: Long short-term memory recurrent neural network architectures for large scale acoustic modeling. In: Proceedings of Interspeech, pp. 338–342 (2014)
51. Sak, H., Vinyals, O., Heigold, G., Senior, A., McDermott, E., Monga, R., Mao, M.: Sequence discriminative distributed training of long short-term memory recurrent neural networks. In: Proceedings of Interspeech (2014)
52. Sak, H., Senior, A., Rao, K., Beaufays, F., Schalkwyk, J.: Google voice search: faster and more accurate. In: Google Research blog (2015). https://research.googleblog.com/2015/09/google-voice-search-faster-and-more.html
53. Sak, H., Senior, A.W., Rao, K., Beaufays, F.: Fast and accurate recurrent neural network acoustic models for speech recognition. CoRR abs/1507.06947 (2015)
54. Sak, H., Senior, A.W., Rao, K., Irsoy, O., Graves, A., Beaufays, F., Schalkwyk, J.: Learning acoustic frame labeling for speech recognition with recurrent neural networks. In: Proceedings of ICASSP, pp. 4280–4284 (2015)
55. Schalkwyk, J., Beeferman, D., Beaufays, F., Byrne, B., Chelba, C., Cohen, M., Garrett, M., Strope, B.: Google Search by Voice: A Case Study. Springer, New York (2010)

56. Seltzer, M., Raj, B., Stern, R.M.: Likelihood-maximizing beamforming for robust handsfree speech recognition. IEEE Trans. Audio Speech Lang. Process. **12**(5), 489–498 (2004)
57. Senior, A., Heigold, G., Bacchiani, M., Liao, H.: GMM-free DNN training. In: Proceedings of ICASSP (2014)
58. Senior, A.W., Sak, H., Shafran, I.: Context dependent phone models for LSTM RNN acoustic modelling. In: Proceedings of ICASSP, pp. 4585–4589 (2015)
59. Shan, J., Wu, G., Hu, Z., Tang, X., Jansche, M., Moreno, P.J.: Search by voice in Mandarin Chinese. In: Proceedings of Interspeech, pp. 354–357 (2010)
60. Shugrina, M.: Formatting time-aligned ASR transcripts for readability. In: 2010 Annual Conference of the North American Chapter of the Association for Computational Linguistics (2010)
61. Sorensen, J., Allauzen, C.: Unary data structures for language models. In: Proceedings of Interspeech (2011)
62. Sparrowhawk. https://github.com/google/sparrowhawk (2016)
63. Tokuda, K., Zen, H.: Directly modeling speech waveforms by neural networks for statistical parametric speech synthesis. In: Proceedings of ICASSP, pp. 4215–4219 (2015)
64. Tokuda, K., Zen, H.: Directly modeling voiced and unvoiced components in speech waveforms by neural networks. In: Proceedings of ICASSP (2015)
65. Vanhoucke, V.: Speech recognition and deep learning. In: Google Research blog (2012). https://research.googleblog.com/2012/08/speech-recognition-and-deep-learning.html
66. Variani, E., Lei, X., McDermott, E., Moreno, I.L., Gonzalez-Dominguez, J.: Deep neural networks for small footprint text-dependent speaker verification. In: Proceedings of ICASSP (2014)
67. Variani, E., Sainath, T.N., Shafran, I., Bacchiani, M.: Complex Linear Prediction (CLP): a discriminative approach to joint feature extraction and acoustic modeling. In: Proceedings of Interspeech (2016)
68. Waibel, A., Hanazawa, T., Hinton, G., Shikano, K., Lang, K.: Phoneme recognition using time-delay neural networks. In: Proceedings of ICASSP, vol. 37, pp. 328–339 (1989)
69. Zen, H.: Acoustic modeling for speech synthesis – from HMM to RNN. Invited Talk. In: ASRU (2015)
70. Zen, H., Sak, H.: Unidirectional long short-term memory recurrent neural network with recurrent output layer for low-latency speech synthesis. In: Proceedings of ICASSP, pp. 4470–4474 (2015)
71. Zen, H., Senior, A.: Deep mixture density networks for acoustic modeling in statistical parametric speech synthesis. In: Proceedings of ICASSP, pp. 3872–3876 (2014)
72. Zen, H., Senior, A., Schuster, M.: Statistical parametric speech synthesis using deep neural networks. In: Proceedings of ICASSP, pp. 7962–7966 (2013)
73. Zen, H., Agiomyrgiannakis, Y., Egberts, N., Henderson, F., Szczepaniak, P.: Fast, compact, and high quality LSTM-RNN based statistical parametric speech synthesizers for mobile devices. In: Interspeech (2016)

Chapter 19
Challenges in and Solutions to Deep Learning Network Acoustic Modeling in Speech Recognition Products at Microsoft

Yifan Gong, Yan Huang, Kshitiz Kumar, Jinyu Li, Chaojun Liu, Guoli Ye, Shixiong Zhang, Yong Zhao, and Rui Zhao

Abstract Deep learning (DL) network acoustic modeling has been widely deployed in real-world speech recognition products and services that benefit millions of users. In addition to the general modeling research that academics work on, there are special constraints and challenges that the industry has to face, e.g., the run-time constraint on system deployment, robustness to variations such as the acoustic environment, accents, lack of manual transcription, etc. For large-scale automatic speech recognition applications, this chapter briefly describes selected developments and investigations at Microsoft to make deep learning networks more effective in a production environment, including reducing run-time cost with singular-value-decomposition-based training, improving the accuracy of small-size deep neural networks (DNNs) with teacher–student training, the use of a small amount of parameters for speaker adaptation of acoustic models, improving the robustness to the acoustic environment with variable-component DNN modeling, improving the robustness to accent/dialect with model adaptation and accent-dependent modeling, introducing time and frequency invariance with time–frequency long short-term memory recurrent neural networks, exploring the generalization capability to unseen data with maximum margin sequence training, the use of unsupervised data to improve speech recognition accuracy, and increasing language capability by reusing speech-training material across languages. The outcome has enabled the deployment of DL acoustic models across Microsoft server and client product lines including Windows 10 desktop/laptop/phones, XBOX, and skype speech-to-speech translation.

Y. Gong (✉) • Y. Huang • K. Kumar • J. Li • C. Liu • G. Ye • S. Zhang • Y. Zhao • R. Zhao
Microsoft, One Microsoft Way, Redmond, WA 98052, USA
e-mail: ygong@microsoft.com

© Springer International Publishing AG 2017
S. Watanabe et al. (eds.), *New Era for Robust Speech Recognition*,
DOI 10.1007/978-3-319-64680-0_19

19.1   Introduction

Deep learning (DL) has been the mainstream of the speech recognition community in recent years, both in industry and in academia. While sharing research topics with academia, at Microsoft we pursue many specialized topics to deliver fast, scalable, and accurate automatic speech recognition (ASR) systems under practical constraints and requirements.

One requirement on efficient DL models is that, when switching the acoustic models from Gaussian mixture models (GMMs) to deep neural networks (DNNs), the user perceives the ASR system to be both faster and more accurate. Acoustic-model personalization benefits the individual user experience, and speaker adaptation with limited user data is being intensively studied at Microsoft as we can afford adapting only a small amount of parameters given limited user data. The on-device acoustic model typically is given much less parameters than the server-based acoustic model, due to footprint constraints. How to improve the accuracy of device-based ASR under limited modeling capability is thus a key to speech recognition (SR) products.

The robustness [12, 15] of ASR systems is always a very important topic. Given that mobile devices are used in all kinds of environments and users with different accents and dialects, how to make the recognition accuracy invariant to noisy environments and speakers' accents and dialects is a major challenge to the industry. Also, we hope the ASR performance can be invariant to adverse time and frequency variation, and be robust to unseen data.

Last but not least, live SR service provides unlimited untranscribed data, compared to only thousands of hours of transcribed data. Effectively leveraging unsupervised data improves ASR accuracy and development speed. We always need to develop a new language in a new scenario with a small amount of training data, while for other languages a large amount of data is available. How to leverage the resource-rich languages to develop high-quality ASR for resource-limited languages is an interesting topic.

In the following, we elaborate on the technologies developed at Microsoft to address the above challenges.

19.2   Effective and Efficient DL Modeling

With much more parameters than the traditional GMM–hidden-Markov-model (HMM) framework, the great performance of the CD-DNN-HMM comes with huge computational cost. The significant run-time cost increase is very challenging to the service, as the users may feel the system is running slower although more accurate. When deploying to mobile devices, this issue is even more challenging as, due to power consumption, mobile devices can only afford a very small footprint and CPU cost. Furthermore, the large number of DNN parameters also limits the use of speaker personalization due to the huge storage cost in large-scale deployments. This section addresses these challenges with effective and efficient deep learning model technologies proposed at Microsoft.

19.2.1   Reducing Run-Time Cost with SVD-Based Training

To reduce run-time cost, we proposed singular-value-decomposition (SVD) [24]-based model restructuring. The original full-rank DNN model is converted to a much smaller low-rank DNN model without loss of accuracy.

An $m \times n$ weight matrix $\mathbf{A}$ in a DNN can be approximated as the product of two low-rank matrices by applying

$$\mathbf{A}_{m \times n} \approx \mathbf{U}_{m \times k} \mathbf{N}_{k \times n}. \tag{19.1}$$

If $\mathbf{A}_{m \times n}$ is a low-rank matrix, k will be much smaller than n and therefore the number of parameters of the matrices $\mathbf{U}_{m \times k}$ and $\mathbf{N}_{k \times n}$ is much smaller than that of matrix $\mathbf{A}_{m \times n}$. Applying this decomposition to the DNN model, it acts as adding a linear bottleneck layer with fewer units between the original layers. If the number of parameters is reduced too aggressively, a stochastic-gradient-descent-based fine tuning can be used to recover accuracy. With such an SVD-based model-restructuring method, we can reduce the model size and the run-time CPU cost by 75% without losing any accuracy. SVD-based DNN modeling is now used in all Microsoft's SR products.

19.2.2   Speaker Adaptation on Small Amount of Parameters

Speaker adaptation is an established field [9, 10, 25], which seeks a speaker-dependent personalization of one of the speaker-independent (SI) ASR components, such as the acoustic model (AM). Typically the SI models are trained on a large dataset with an objective to work best for all speakers. While working well on average, it leaves out substantial opportunities to best account for different accents, speech content, speaking rate, etc. Personalization methods adapt the SI model to best perform for target speakers.

We focus on AM adaptation in production scenarios for millions of speakers. We typically have limited adaptation data and, given the prohibitive transcription cost, we use unsupervised data from production logs. Since the adapted models are speaker dependent (SD), the size of the SD parameters is a critical challenge when we scale to millions of speakers. We also provide solutions to minimize the SD model parameters while retaining adaptation benefits.

19.2.2.1 SVD Bottleneck Adaptation

We proposed the SVD bottleneck adaptation in [25] to produce low-footprint SD models by making use of the SVD-restructured topology. The linear transformation is applied to each of the bottleneck layers by adding an additional layer of k units. We have

$$\mathbf{A}_{s,m \times n} = \mathbf{U}_{m \times k} \mathbf{S}_{s,k \times k} \mathbf{N}_{k \times n}, \tag{19.2}$$

where $\mathbf{S}_{s,k \times k}$ is the transformation matrix for speaker s and is initialized to be an identity matrix $I_{k \times k}$. The advantage of this approach is that only a couple of small matrices need to be updated for each speaker. Consider $k = 256$ and $m = n = 2048$ as an instance. Directly adapting the initial matrix $\mathbf{A}$ needs to update $2048 \times 2048 = 4M$ parameters, while adapting $\mathbf{S}$ only updates $256 \times 256 = 64K$ parameters, reducing the footprint to 1.6%. This dramatically reduces the deployment cost for speaker personalization while producing a more reliable estimate of the adaptation model [25].

Given the constraints and challenges in speaker personalization, our previous work [9, 25] proposed an intermediate-layer adaptation scheme. This significantly reduces the number parameters to be adapted, and acts as regularization.

We extended this work by additionally constraining the inserted layer coefficients to be nonnegative. Nonnegativity constraints have been applied to a number of speech applications like speech factorization, dereverberation, etc. [12, 15]. This results in a sparse and robust set of model parameters as nonessential parameters are retained as 0. The DNN layer coefficients aren't necessarily nonnegative; still, we motivate nonnegativity [10] for the inserted layer to constrain the nonzero coefficients in the SD parameters.

The SVD-based work was applied to the Windows Phone en-US server task with 1000 h of transcribed SI training data. The baseline had 66-dim dynamic log mel features, along with a context of 11 frames. We considered 50–300 speakers, 50 untranscribed utts. (4–5 min) for adaptation and 50 test utts. per speaker. The SI DNN had five nonlinear hidden layers each with 2k nodes, the SVD layer had 200–300 nodes, and the output layer had 6k units. We adapted a layer inserted in the fourth SVD layer. The word error rate (WER) for the SI baseline was 14.15%, and that for adaptation without a nonnegativity constraint was 12.55%, thus 11.3% WER relative reduction (WERR). With nonnegativity constraints on the SD parameters, we can retain only 72.1% of the original parameters with 11.17% WERR. Building on this idea, we then chose a small positive value as a threshold by truncating values smaller than the threshold to 0. For a threshold of 0.0005, we retained only 13.8% of the original parameters for 10.46% WERR, demonstrating that the SD parameters can be reduced by about 86% with minimal loss in adaptation benefits.

19.2.2.2 DNN Adaptation Through Activation Function

The aforementioned adaptation methods either adapt or add transformation matrices to characterize the target speaker. The DNN model can also be adapted by adjusting the node activation functions [32]. We modify the sigmoid function in a general form

$$\tilde{\sigma}(v) = 1/(1 + e^{-(\alpha v+\beta)}), \tag{19.3}$$

where α is the slope and β is the bias, initialized to 1 and 0, respectively, and updated for each speaker. The main advantage of adapting through activation functions is that the total number of adaptation parameters is small, twice the total number of hidden units.

It can be shown that adapting the slopes and biases through the activation functions amounts to adding a linear layer right before the activation functions with a one-to-one correspondence.

19.2.2.3 Low-Rank Plus Diagonal (LRPD) Adaptation

To make the model scalable, it is desired that the model footprint can be adjusted according to the amount of available adaptation data. We proposed the LRPD adaptation method to control the number of adaptation parameters in a flexible way while maintaining modeling accuracy [33].

One simple heuristic is to reapply the SVD to decompose the adaptation matrix S_s in (19.2). It is observed that the adaptation matrices are very close to an identity matrix, which is expected as the adapted model should not deviate too far from the SI model given the limited amount of adaptation data. Because S_s is close to an identity matrix, singular values of S_s center around 1 and decrease slowly. With such a high-rank matrix, SVD would not yield a high compression rate. In contrast, the singular values of $(S_s - I)$ steadily decrease and approach zero, suggesting that we should apply SVD to $(S_s - I)$ to reduce the speaker-dependent footprint.

Given an adaptation matrix $S_{s,k\times k}$, we approximate it as a superposition of a diagonal matrix $D_{s,k\times k}$ and a product of two smaller matrices $P_{s,k\times c}$ and $Q_{s,c\times k}$. Hence

$$S_{s,k\times k} \approx D_{s,k\times k} + P_{s,k\times c}Q_{s,c\times k} \tag{19.4}$$

The number of elements in the LRPD decomposition is $k(2c+1)$, while the original S_s has k^2 elements. If $c \ll k$, this can significantly reduce the adaptation model footprint.

The LRPD bridges the gap between the full and diagonal transformation matrices. When $c = 0$, the LRPD is reduced to adaptation with a diagonal matrix. Specifically, if we apply the diagonal transforms before or after all nonlinear layers, we may achieve the sigmoid adaptation as described in Sect. 19.2.2.2 [32], or LHUC [22] adaptation.

19.2.3 Improving the Accuracy of Small-Size DNNs with Teacher–Student Training

The SVD-based DNN modeling method enables us to deploy accurate and fast DNN server models. However, this is far from enough when we deploy DNNs on mobile devices which have very limited computational and storage resources. A common practice is to train a DNN with a small number of hidden nodes using the standard training process, leading to significant accuracy loss. In [14], we proposed a new DNN training criterion to better address this issue by utilizing the DNN output distribution as shown in Fig. 19.1. To enable a small-size (student) DNN to copy the output distribution of a larger one (teacher), we minimize the Kullback–Leibler

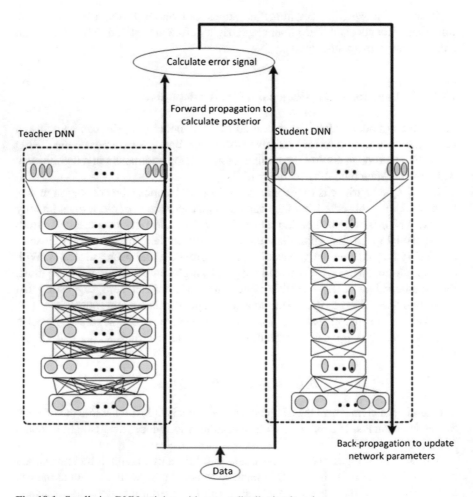

Fig. 19.1 Small-size DNN training with output distribution learning

divergence between the output distributions of the small-size DNN and a standard large-size DNN by utilizing a large amount of untranscribed data.

This teacher–student learning method leverages the large amount of untranscribed data available in the industry: the more data we use, the closer the student DNN can approach the accuracy of the teacher DNN. With such a teacher–student learning method, we can reduce 75% of the accuracy gap between the small-size and large-size DNN. Together with lossless SVD model compression, we can reduce the number of DNN model parameters from 30 million to 2 million, which enables the deployment of DNN on devices with high accuracy.

19.3   Invariance Modeling

A good model should be invariant to all the disturbing factors not related to the phoneme-modeling goal. These disturbing factors can be the accent or dialect of speakers, noisy environments, time and frequency warping from different speakers, and even unseen samples during testing time. This section focuses on developing methods to make models more invariant to those disturbing factors.

19.3.1   Improving the Robustness to Accent/Dialect with Model Adaptation

Foreign-accented speech, characterized by systematic segmental and/or suprasegmental deviations from native speech, degrades the intelligibility and may result in poor speech recognition performance. There exists a large speech recognition accuracy gap between native and foreign-accented speech. For example, we observe a nearly doubled word error rate for accented speech in a real speech service system.

To achieve better user experience, we proposed a modularized multiaccent DNN acoustic model [7], where a set of accent-specific subneural-network modules are trained to model the accent-specific patterns. The rest of the DNN is shared across the native and the accented speech to allow maximum knowledge transfer and data sharing. For the accent-specific module, its size is determined by the amount of accent training data and its placement is based upon the effectiveness of accent modularization at different network locations. Our study shows that accent modularization at the top neural network layers performs better than at the bottom layers. We thus believe that the accent-specific patterns are primarily captured by the abstract speech features at the top neural network layers.

The accent-specific module can be optimized using Kullback–Leibler-divergence (KLD)-regularized model adaptation [4, 21]. In this approach, the KLD ($\mathscr{F}_{\text{KLD}}$) between the baseline and the adapted model is added to the standard cross-entropy objective ($\mathscr{F}_{\text{CE}}$) or subtracted from the maximum mutual information (MMI) objective ($\mathscr{F}_{\text{MMI}}$) to form the new regularized objectives ($\hat{\mathscr{F}}_{\text{CE}}$ and $\hat{\mathscr{F}}_{\text{MMI}}$):

$$\begin{cases} \hat{\mathscr{F}}_{\text{CE}} = (1-\rho)\mathscr{F}_{\text{CE}} + \rho\mathscr{F}_{\text{KLD}}, \\ \hat{\mathscr{F}}_{\text{MMI}} = (1-\rho)\mathscr{F}_{\text{MMI}} - \rho\mathscr{F}_{\text{KLD}}. \end{cases} \tag{19.5}$$

Here ρ is the weight of the KLD regularization. As the MMI objective is to be maximized, the KLD regularization term is introduced with a negative sign in (19.6). The new regularized objectives can be optimized through standard back-propagation. f-Smoothing is a frame-level regularization used in the MMI sequence-training implementation in [27]. We further derived the KLD-regularized MMI adaptation with f-smoothing ($\hat{\mathscr{F}}_{\text{MMI}}^{(f)}$) as a general formulation [4]:

$$\begin{aligned} \hat{\mathscr{F}}_{\text{MMI}}^{(f)} &= (1-\rho)\mathscr{F}_{\text{MMI}}^{(f)} + \rho\mathscr{F}_{\text{KLD}} \\ &= (1-\rho)[(1-\rho_F)\mathscr{F}_{\text{MMI}} + (-\rho_F\mathscr{F}_{\text{CE}})] + \rho\mathscr{F}_{\text{KLD}}, \end{aligned} \tag{19.6}$$

where ρ_F is the weight of f-smoothing. When ρ_F equals 1, $\hat{\mathscr{F}}_{\text{MMI}}^{(f)}$ becomes the KLD-regularized cross-entropy adaptation; when ρ_F equals 0, it becomes the KLD-regularized MMI adaptation.

On a mobile short message dictation task, with 1K, 10K, and 100K British or Indian accent adaptation utterances, the KLD-regularized adaptation achieves 18.1%, 26.0%, and 28.5% or 16.1%, 25.4%, and 30.6% WER reduction, respectively, against a 400 h baseline EN-US native DNN. Comparable performance has been achieved from a baseline model trained an 2000 h EN-US native speech.

The accent modeling relies on obtaining reliable accent signals. One way to get the accent signal is to ask the users to specify their accents when using the application, which requires cooperation from the users. A more feasible way is to automatically identify the users' accent [1], which introduces additional run-time cost and the result is not always correct. Also, to train a robust identification module requires costly accent-labeled data.

Inspired by the fact that users from the same geolocation region likely share a similar accent, we proposed to directly utilize the user's province (state) as the accent signal [26]. The user's current province inferred from GPS is used as the accent signal. Since a GPS signal is available from most commercial speech services, this method is zero-cost in both run time and accent labeling.

The training data is first partitioned into groups based on each utterance's province label. One DNN for each province is then trained. At run time, the user's current province inferred from GPS is used as the signal to choose the right DNN to recognize the user's voice.

19.3.2 Improving the Robustness to Acoustic Environment with Variable-Component DNN Modeling

To improve the robustness to the acoustic environment, we can either adapt the DNN model to the new environment or make the DNN model itself robust to the new environment. In [13], we proposed a novel factorized adaptation method to adapt a DNN with only a limited number of parameters by taking into account the underlying distortion factors. To improve the performance on the type of data unavailable at training time, it is desirable that DNN components can be modeled as a function of a continuous environment-dependent variable. At recognition time, a set of DNN components specific to the given value of the environment variable is instantiated and used for recognition. Even if the test environment is not seen in the training, the values of the DNN components can still be predicted by the environment-dependent variable. Variable-component DNN (VCDNN) [30, 31] has been proposed for this purpose.

In the VCDNN method, any component in the DNN can be modeled as a set of polynomial functions of an environment variable. For example, the weight matrix and bias of the DNN can be environment-variable dependent as in Fig. 19.2, and we call such a DNN a variable-parameter DNN (VPDNN), in which the weight matrix $\mathbf{A}$ and bias $\mathbf{b}$ of layer l are modeled as a function of the environment variable u (e.g., the signal-to-noise ratio):

$$\mathbf{A}^l = \sum_{j=0}^{J} \mathbf{H}_j^l v^j \qquad 0 < l \le L, \tag{19.7}$$

$$\mathbf{b}^l = \sum_{j=0}^{J} \mathbf{p}_j^l v^j \qquad 0 < l \le L. \tag{19.8}$$

J is the polynomial function order. $\mathbf{H}_j^l$ is a matrix with the same dimensions as $\mathbf{A}^l$, and $\mathbf{p}_j^l$ is a vector with the same dimension as $\mathbf{b}^l$. Zhao et al. [30, 31] show the advantage of VCDNNs, which achieved 3.8% and 8.5% relative WER

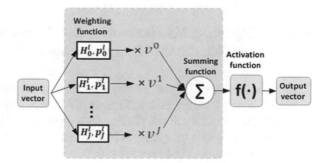

Fig. 19.2 The layer l of a variable-parameter DNN. Its weight matrix and bias are polynomial functions of weight matrices $\mathbf{H}_j^l$ and biases $\mathbf{p}_j^l$, respectively. v is the environment variable

reduction from the standard DNN under seen conditions and unseen conditions, respectively. This indicates that a standard DNN has a strong ability to model the various environments it is trained in, and has room for improvement for unseen environments.

19.3.3 Improving the Time and Frequency Invariance with Time–Frequency Long Short-Term Memory RNNs

DNNs only consider information in a fixed-length sliding window of frames and thus cannot exploit long-range correlations in the signal. Recurrent neural networks (RNNs), on the other hand, can encode sequence history in their internal state, and thus have the potential to predict phonemes based on all the speech features observed up to the current frame. To address the gradient-vanishing issue in RNNs, long short-term memory (LSTM) RNNs [3] were developed and have been shown to outperform DNNs on a variety of ASR tasks [2, 20]. In [18], we reduced the run-time cost of an LSTM with model simplification and frame-skipping evaluation. All previously proposed LSTMs use a recurrence along the time axis to model the temporal patterns of speech signals, and we call them T-LSTMs in this chapter.

Typically, log-filterbank features are often used as the input to the neural-network-based acoustic model [11, 19]. Switching two filterbank bins will not affect the performance of the DNN or LSTM. However, this is not the case when a human reads a spectrogram: a human relies on patterns that evolve in both time and frequency to predict phonemes. This inspired us to propose a 2-D, time–frequency (TF) LSTM [16, 17] as in Fig. 19.3 which jointly scans the input over the time and frequency axes to model spectro-temporal warping, and then uses the output activations as the input to a T-LSTM. The joint time–frequency modeling better normalizes the features for the upper-layer T-LSTMs. Evaluated on a 375-h short message dictation task, the proposed TF-LSTM obtained a 3.4% relative WERR over the best T-LSTM. The invariance property achieved by joint time–frequency analysis was demonstrated on a mismatched test set, where the TF-LSTM achieved a 14.2% relative WERR over the best T-LSTM.

19.3.4 Exploring the Generalization Capability to Unseen Data with Maximum Margin Sequence Training

Traditional DNNs/RNNs use multinomial logistic regression (softmax activation) at the top layer for classification. The new DNN/RNN instead uses an SVM at the top layer (see Fig. 19.4). The SVM has several prominent features. First, it has been proven that maximizing the margin is equivalent to minimizing an upper bound of generalization errors [23]. Second, the optimization problem of the SVM is convex,

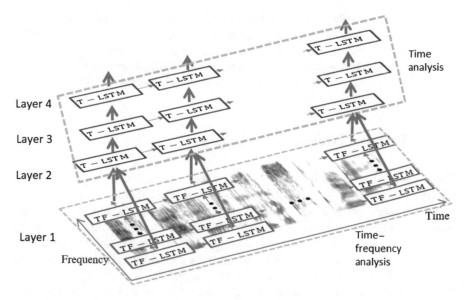

Fig. 19.3 An example of a time–frequency LSTM-RNN which scans both the time and frequency axes at the bottom layer using TF-LSTM, and then scans the time axis at the upper layers using T-LSTM

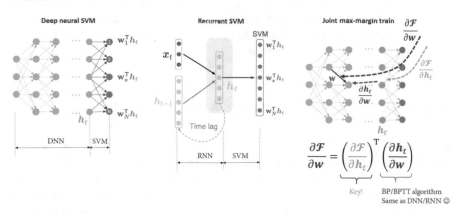

Fig. 19.4 The architecture of deep neural SVMs and recurrent SVMs. The parameters of the SVM and DNN/RNN are jointly trained

which is guaranteed to produce a global optimal solution. Third, the size of the SVM model is determined by the number of support vectors [23], which is learned from training data, instead of a fixed design in DNNs/RNNs. The resulting models are named neural SVMs and recurrent SVMs, respectively [28, 29].

The algorithm iterates on two steps. The first estimates the parameters of the SVM in the last layer, keeping the parameters of the DNN/RNN in the previous layers fixed. This step is equivalent to the training of structured SVMs with DNN bottleneck features [28, 29]. The second updates the parameters of the DNN/RNN in all previous layers, keeping the parameters of the SVM fixed. This joint training uses the DNN/RNN to learn the feature space, while using the SVM (in the last layer) for sequence classification. We have verified its effectiveness on the Windows Phone task for large-vocabulary continuous speech recognition [28, 29].

19.4   Effective Training-Data Usage

Untranscribed data from live speech service traffic is unlimited and literally free. Using untranscribed data to improve acoustic-model accuracy is an ideal and economic model development strategy. This becomes even more important in the new types of deep learning acoustic model with ever-enlarged model capacity. Furthermore, due to the common practice of frequent model updates with fresh data to improve accuracy or simply for legal reasons, developing technologies to make use of untranscribed data is immensely appealing and extremely valuable. It is also a common requirement to deploy an ASR system for a new, resource-limited language, for which we don't have sufficient live data to train an accurate acoustic model. On the other hand, we may have large amount of training data for a resource-rich language, such as US English. One challenge is how to leverage the training data from the resource-rich language to generate a good acoustic model for resource-limited languages. In this section, we focus on how to utilize untranscribed data and training data across languages.

19.4.1   Use of Unsupervised Data to Improve SR Accuracy

High-quality transcription inference, effective importance data sampling, and transcription-error-robust model training are the keys to the success of using untranscribed data for acoustic-model training. For transcription inference, we use a multiview learning-based system combination and confidence recalibration to generate accurately inferred transcriptions and reject erroneous transcriptions [6]. User click and correction information is further used to improve the transcription quality. As untranscribed data is literally unlimited, effective importance data sampling optimizes the accuracy gain per added datum and helps to control the model-training cost. Lastly, since machine-inferred transcription is never perfect, it is necessary to develop transcription-error-robust semisupervised training.

We studied semisupervised training for a fully connected DNN, unfolded RNN, and LSTM-RNN with respect to the transcription quality, the importance-based data sampling, and the training data amount [8]. We found that the DNN, unfolded RNN, and LSTM-RNN were increasingly more sensitive to labeling errors as shown in Fig. 19.5 (left). For example, with the training transcription simulated at 5%, 10%, or 15% WER level, the semisupervised DNN yields 2.37%, 4.84%, or 7.46%

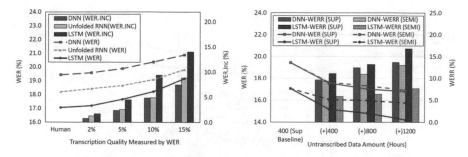

Fig. 19.5 *Left*: Performance comparison of semisupervised DNN, unfolded RNN, and LSTM-RNN. These models were trained using 400 h speech with transcription simulated at different WER levels. *Right*: Supervised and semisupervised DNN and LSTM-RNN with increased training data

relative WER increase against the baseline trained with human transcription. In comparison, the corresponding WER increase is 2.53%, 4.89%, or 8.85% for an unfolded RNN and 4.47%, 9.38%, or 14.01% for an LSTM-RNN. We further found that the importance-based sampling has similar impact on all three models with 2–3% relative WERR compared to random sampling. Lastly, we compared the model capability with increased training data in Fig. 19.5 (right). Experimental results suggested that the LSTM-RNN benefits more from an increased training data amount under supervised training. On a mobile speech recognition task, a semisupervised LSTM-RNN using 2600 h transcribed and 10,000 h untranscribed data yields 6.56% relative WERR against the supervised baseline.

19.4.2   Expanded Language Capability by Reusing Speech-Training Material Across Languages

To reuse speech-training materials across languages, we proposed [5] a shared-hidden-layer multilingual DNN (SHL-MDNN) architecture in which the input and hidden layers are shared across multiple languages and serve as a universal feature transformation, as shown in Fig. 19.6. In this architecture, the input and hidden layers are shared across all the languages and can be considered as a universal feature transform. In contrast, each language has its own output layer, used to estimate the posteriors of the senones specific to that language.

The shared hidden layers are jointly trained with data from multiple source languages. As such, they carry rich information to distinguish phonetic classes in multiple languages and can be carried over to distinguish phones in new languages. With such a structure, we can first train the SHL-MDNN with data from resource-rich languages, and then do a cross-lingual model transfer by adding a new softmax layer on top of the shared hidden layers, specific to the resource-limited language. The limited training data from this new language is used to train the top language-

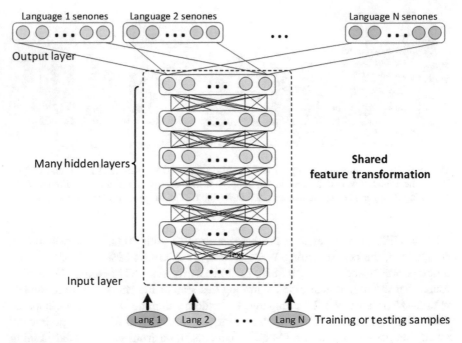

Fig. 19.6 The architecture of a shared-hidden-layer multilingual DNN

specific softmax layer only. In [5], we demonstrate that the SHL-MDNN can reduce errors by relatively 3–5% over the monolingual DNNs trained using only the language-specific data. Further, we show that the learned hidden layers shared across languages can be transferred to improve the recognition accuracy of new languages, with relative error reductions ranging from 6% to 28% against DNNs trained without exploiting the transferred hidden layers. Hence, this SHL-MDNN has been proven to be effective and has been widely adopted by many research sites.

19.5 Conclusion

In this chapter, we presented selected deep learning technologies developed at Microsoft to address the challenges in the deployment of SR products and services:

- To meet the run-time computational requirement, we proposed SVD-based training, which significantly reduces the run-time cost of a DNN and maintains the same accuracy as the full-size DNN.
- To do better personalization with limited adaptation data, we have developed a series of methods such as SVD-based adaptation, adaptation through an

activation function, and LRPD, which only need to adapt a very small amount of parameters, compared to the huge amount of parameters in DNNs.

- To improve DL modeling power, we designed a teacher–student learning strategy to use a potentially unlimited amount of untranscribed data, and enable a device-based DNN to approximate asymptotically the accuracy of a server-based DNN.
- We developed a time–frequency LSTM to extract the time–frequency patterns in the spectrogram to improve the representation of time–frequency invariance.
- We designed modularized multiaccent DNNs where the accent-specific module can be optimized using KLD regularization. We also utilized geolocation information to determine accent clusters at zero cost to the runtime.
- To address the noise-robustness issue, we proposed a variable-component DNN which instantiates at run-time DNN components by using a continuous function of environment variables such that the estimated DNN components can work well even if the test environment is unseen at the training time.
- To increase the effectiveness of using untranscribed data, we designed a process that works well with both DNNs and LSTM-RNNs. With shared-hidden-layer transfer learning, we can leverage resource-rich languages to build high-quality models for resource-limited languages.
- To improve the generalization capability to unseen data, we investigated replacing the multinomial logistic regression in the top layer of DNNs/RNNs with an SVM, and optimized it with maximum margin sequence training.

The aforementioned technologies enable us to ship high-quality models for all Microsoft SR products, on both servers and devices, and for both resource-rich and resource-limited languages.

References

1. Chen, T., Huang, C., Chang, E., Wang, J.: Automatic accent identification using Gaussian mixture models. In: Proceedings of the Workshop on Automatic Speech Recognition and Understanding (2001)
2. Graves, A., Mohamed, A., Hinton, G.: Speech recognition with deep recurrent neural networks. In: Proceedings of the IEEE International Conference on Acoustics, Speech and Signal Processing, pp. 6645–6649 (2013)
3. Hochreiter, S., Schmidhuber, J.: Long short-term memory. Neural Comput. 9(8), 1735–1780 (1997)
4. Huang, Y., Gong, Y.: Regularized sequence-level deep neural network model adaptation. In: Proceedings of the Interspeech (2015)
5. Huang, J.T., Li, J., Yu, D., Deng, L., Gong, Y.: Cross-language knowledge transfer using multilingual deep neural network with shared hidden layers. In: Proceedings of the IEEE International Conference on Acoustics, Speech and Signal Processing, pp. 7304–7308 (2013)
6. Huang, Y., Yu, D., Gong, Y., Liu, C.: Semi-supervised GMM and DNN acoustic model training with multi-system combination and confidence re-calibration. In: Proceedings of the Interspeech (2013)
7. Huang, Y., Yu, D., Liu, C., Gong, Y.: Multi-accent deep neural network acoustic model with accent-specific top layer using the KLD-regularized model adaptation. In: Proceedings of the Interspeech (2014)

8. Huang, Y., Wang, Y., Gong, Y.: Semi-supervised training in deep learning acoustic models. In: Proceedings of the Interspeech (2016)
9. Kumar, K., Liu, C., Yao, K., Gong, Y.: Intermediate-layer DNN adaptation for offline and session-based iterative speaker adaptation. In: Sixteenth Annual Conference of the International Speech Communication Association (2015)
10. Kumar, K., Liu, C., Gong, Y.: Non-negative intermediate-layer DNN adaptation for a 10-kb speaker adaptation profile. In: Proceedings of the IEEE International Conference on Acoustics, Speech and Signal Processing (ICASSP) (2016)
11. Li, J., Yu, D., Huang, J.T., Gong, Y.: Improving wideband speech recognition using mixed-bandwidth training data in CD-DNN-HMM. In: Proceedings of the IEEE Spoken Language Technology Workshop, pp. 131–136 (2012)
12. Li, J., Deng, L., Gong, Y., Haeb-Umbach, R.: An overview of noise-robust automatic speech recognition. IEEE/ACM Trans. Audio Speech Lang. Process. 22(4), 745–777 (2014)
13. Li, J., Huang, J.T., Gong, Y.: Factorized adaptation for deep neural network. In: Proceedings of the IEEE International Conference on Acoustics, Speech and Signal Processing (2014)
14. Li, J., Zhao, R., Huang, J.T., Gong, Y.: Learning small-size DNN with output-distribution-based criteria. In: Proceedings of the Interspeech (2014)
15. Li, J., Deng, L., Haeb-Umbach, R., Gong, Y.: Robust Automatic Speech Recognition: A Bridge to Practical Applications. Academic, London (2015)
16. Li, J., Mohamed, A., Zweig, G., Gong, Y.: LSTM time and frequency recurrence for automatic speech recognition. In: Proceedings of the Workshop on Automatic Speech Recognition and Understanding (2015)
17. Li, J., Mohamed, A., Zweig, G., Gong, Y.: Exploring multidimensional LSTMs for large vocabulary ASR. In: Proceedings of the IEEE International Conference on Acoustics, Speech and Signal Processing (2016)
18. Miao, Y., Li, J., Wang, Y., Zhang, S., Gong, Y.: Simplifying long short-term memory acoustic models for fast training and decoding. In: Proceedings of the IEEE International Conference on Acoustics, Speech and Signal Processing (2016)
19. Mohamed, A., Hinton, G., Penn, G.: Understanding how deep belief networks perform acoustic modelling. In: Proceedings of the IEEE International Conference on Acoustics, Speech and Signal Processing, pp. 4273–4276 (2012)
20. Sak, H., Senior, A., Beaufays, F.: Long short-term memory recurrent neural network architectures for large scale acoustic modeling. In: INTERSPEECH, pp. 338–342 (2014)
21. Su, H., Li, G., Yu, D., Seide, F.: Error back propagation for sequence training of context-dependent deep networks for conversational speech transcription. In: Proceedings of the IEEE International Conference on Acoustics, Speech and Signal Processing (2013)
22. Swietojanski, P., Renals, S.: Learning hidden unit contributions for unsupervised speaker adaptation of neural network acoustic models. In: Proceedings of the SLT, pp. 171–176 (2014)
23. Vapnik, V.N.: The Nature of Statistical Learning Theory. Springer, New York (1995)
24. Xue, J., Li, J., Gong, Y.: Restructuring of deep neural network acoustic models with singular value decomposition. In: Proceedings of the Interspeech, pp. 2365–2369 (2013)
25. Xue, J., Li, J., Yu, D., Seltzer, M., Gong, Y.: Singular value decomposition based low-footprint speaker adaptation and personalization for deep neural network. In: Proceedings of the IEEE International Conference on Acoustics, Speech and Signal Processing, pp. 6359–6363 (2014)
26. Ye, G., Liu, C., Gong, Y.: Geo-location dependent deep neural network acoustic model for speech recognition. In: Proceedings of the IEEE International Conference on Acoustics, Speech and Signal Processing, pp. 5870–5874 (2016)
27. Yu, D., Yao, K., Su, H., Li, G., Seide, F.: KL-divergence regularized deep neural network adaptation for improved large vocabulary speech recognition. In: Proceedings of the IEEE International Conference on Acoustics, Speech and Signal Processing (2013)
28. Zhang, S.X., Liu, C., Yao, K., Gong, Y.: Deep neural support vector machines for speech recognition. In: ICASSP, pp. 4275–4279. IEEE, New York (2015)
29. Zhang, S.X., Zhao, R., Liu, C., Li, J., Gong, Y.: Recurrent support vector machines for speech recognition. In: ICASSP. IEEE, New York (2016)

30. Zhao, R., Li, J., Gong, Y.: Variable-activation and variable-input deep neural network for robust speech recognition. In: Proceedings of the IEEE Spoken Language Technology Workshop (2014)
31. Zhao, R., Li, J., Gong, Y.: Variable-component deep neural network for robust speech recognition. In: Proceedings of the Interspeech (2014)
32. Zhao, Y., Li, J., Xue, J., Gong, Y.: Investigating online low-footprint speaker adaptation using generalized linear regression and click-through data. In: Proceedings of the ICASSP, pp. 4310–4314 (2015)
33. Zhao, Y., Li, J., Gong, Y.: Low-rank plus diagonal adaptation for deep neural networks. In: Proceedings of the ICASSP (2016)

Chapter 20
Advanced ASR Technologies for Mitsubishi Electric Speech Applications

Yuuki Tachioka, Toshiyuki Hanazawa, Tomohiro Narita, and Jun Ishii

Abstract Mitsubishi Electric Corporation has been developing speech applications for 20 years. Our main targets are car navigation systems, elevator-controlling systems, and other industrial devices. This chapter deals with automatic speech recognition technologies which were developed for these applications. To realize real-time processing with small resources, syllable N-gram-based text search is proposed. To deal with reverberant environments in elevators, spectral-subtraction-based dereverberation techniques with reverberation time estimation are used. In addition, discriminative methods for acoustic and language models are developed.

20.1 Introduction

First, we describe the problems of far-field noisy and reverberant automatic speech recognition (ASR) for our applications. Many applications are used in a remote scenario. Our main targets here are car navigation and hands-free elevator ASR systems. In cars, it is necessary to perform ASR with limited computational resources because server-based ASR systems cannot be used without a connection to the Internet. However, the target vocabulary is huge because there are many points of interest (POIs). Section 20.2 introduces a method that can efficiently conduct large-vocabulary ASR with limited resources. In elevators, noise is relatively small but reverberation degrades ASR performance. For limited computational resources, a simple dereverberation method is proposed in Sect. 20.3. We have developed several discriminative methods for acoustic models and language models (LMs). These advanced technologies are described in Sect. 20.4.

Y. Tachioka (✉) • T. Hanazawa • T. Narita • J. Ishii
Information Technology R&D Center, Mitsubishi Electric Corporation, Kamakura, Kanagawa, Japan
e-mail: ytachioka@d-itlab.co.jp; Hanazawa.Toshiyuki@ab.MitsubishiElectric.co.jp; Narita.Tomohiro@ct.MitsubishiElectric.co.jp; Ishii.Jun@ab.MitsubishiElectric.co.jp

© Springer International Publishing AG 2017
S. Watanabe et al. (eds.), *New Era for Robust Speech Recognition*,
DOI 10.1007/978-3-319-64680-0_20

20.2 ASR for Car Navigation Systems

This section introduces flexible POI search for car navigation systems, which use statistical ASR and postprocessing such as syllable N-gram-based text search.

20.2.1 Introduction

Mitsubishi Electric is striving to improve ASR capabilities for car navigation systems. One of the key problems encountered is when the user utters a word that is not in the recognizable vocabulary. To solve this problem, the ASR system was equipped with a smart POI search function [5], which automatically generates variations of each word in the recognizable vocabulary. However, along with the expanded vocabulary, both the cost of updating the word variation list and the amount of computation increase. In addition, since the words for operation commands are limited to those listed in the command lists, no variations can be recognized. Consequently, we divided the system into two parts [4]: an ASR part that converts an input speech into a character string, and a postprocessing part that performs the text match processing. We developed a new ASR system that avoids an increase in computational resources when the number of words increases. This section describes the newly developed ASR and postprocessing technologies, as well as the voice interface installed in commercial products.

20.2.2 ASR and Postprocessing Technologies

20.2.2.1 ASR Using Statistical LM

Figure 20.1 shows a new POI name search system that consists of the ASR and text match processes. An LM is a set of data that contains bigrams or trigrams on the vocabulary to be recognized. The output from the ASR process is a sequence of

Fig. 20.1 Overview of POI name search system

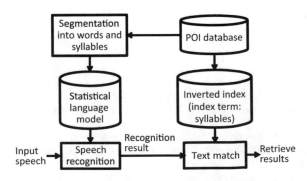

words chosen from the vocabulary registered in the LM so that the word string has the highest likelihood (recognition score) of matching the input speech. Therefore, if the LM contains only the exact name of a POI, e.g., "Tokyo National Museum of Modern Art," "Tokyo Museum of Modern Art," which omits "National," would not be recognized. Therefore, in the above-mentioned smart POI search, the exact name of the POI is divided into separate words: "Tokyo," "National," "Museum," "Modern," and "Art," and then these words along with their conjunction rule are registered in the LM. For example, by registering the rule that "National" may be omitted, the inexact name "Tokyo Museum of Modern Art" would now be recognized. The drawback of this method is the difficulty related to constructing rules that cover a huge number of variations of POI names, and the resulting high cost. In contrast, the newly developed ASR system employs a statistical LM, where the likelihood of a bigram is expressed as a numerical value referred to as a language score. In the case of the above-mentioned POI name, e.g., language scores for the word combinations of "Tokyo" followed by "National" and by "Modern" are registered as numerical values, and these language scores are summed to determine the recognition score of the candidate recognition output. The construction effort of the LM is much smaller than that of hand-made rules because the language score can be automatically calculated from a large number of POI names. However, practically, when the POI name list has several hundred thousand entries, the size of the vocabulary for constructing POI names exceeds a hundred thousand, which is prohibitive for a resource-restricted car navigation system to perform the recognition. Therefore, each word is segmented into smaller element words. For example, "Bijutsu-Kan" (art museum) is expressed by the combination of element words, "Bijutsu" (art) and "Kan" (museum). The element words are limited to about the top 5000 words most frequently appearing in POI names. If a word cannot be expressed by a word N-gram, it is expressed by a syllable N-gram. For example, "Kindai" (modern) is expressed by syllable N-grams "ki," "n," "da," and "i." Thus, by using the syllable as the base unit of the LM, the vocabulary size can be limited to below a certain level even as the number of POIs to be recognized increases.

20.2.2.2   POI Name Search Using High-Speed Text Search Technique

As shown in Fig. 20.1, the new POI name search system consists of the ASR and text match processes. In the ASR part, as described in Sect. 20.2.2.1, each word to be searched is recognized as a sequence of segmented words and syllables. The result of recognition may not have a complete match in the registered vocabulary. In such a case, the text match process searches for a POI name that most closely matches the syllable string given by the recognition process. An additional method has also been developed to generate an approximate language score for a word junction containing a word not included in the exact subject name, which enables the recognition of inexact names as well. The score for matching is the number of matched syllable N-grams. An advantage of using the syllable N-gram as the base unit is that the system is more robust to ASR errors compared to systems using a word or word

N-gram. For example, if "Bijutsukan" (art museum) is erroneously recognized as "Bujutsukan" (martial arts gymnasium), the word-based matching process gives no match results or no score, whereas the syllable-based matching process finds three matching syllable N-grams, "ju-tsu," "tsu-ka," and "ka-n," which are contained in the POI names including "Bijutsukan," and thus contribute to the score. This score can be calculated at high speed by referring to a predetermined inverted index.

20.2.2.3 Application to Commercial Car Navigation System

The POI name search described above has been implemented in a commercial car navigation system. The system enables fast input of any POI name and greatly enhances user convenience compared with character-by-character manual input using a kana keypad.

Operation of this system is as follows. For example, a user utters "Sky Building" by touching the "Input by voice" button, the search system is activated. The ASR result "Sky Building" is displayed (simultaneously, speech output is also provided). And 48 facilities that contain "Sky Building" in their names are retrieved from the POIs and some of the 48 facilities are listed on the screen, e.g., "Kiji Sky Building," "Sky Building Parking Lot," etc. At this stage, the user can choose the requested POI either manually or by uttering "Next" or "Previous"; in addition, the user can utter an additional POI name to narrow down the candidates.

20.3 Dereverberation for Hands-Free Elevator

20.3.1 Introduction

Elevators are one of the most important products of our company. To enable people with a disability to control elevators, ASR systems for elevators have been developed. Since elevators are rectangular in shape with rigid walls, they are highly reverberant. In such environments, reverberant components of speech degrade ASR performance. Some researchers have proposed dereverberation methods with a low computational load based on a statistical model of reverberation [9]. The key to using statistical models for dereverberation is to limit the number of parameters and to estimate them robustly. Lebart et al. proposed a dereverberation method [6] using Polack's statistical model [9], whose parameter is the reverberation time (RT). This method is effective and its computational load is relatively low; however, its performance is unstable because it estimates RT only from the end of an utterance.

We propose a dereverberation method in which spectral subtraction (SS) is used [1]. We also use Polack's statistical model and propose an RT estimation method by utilizing the decay characteristics of not only the end of utterances but also whole utterances in a frequency bin.

20.3.2 A Dereverberation Method Using SS

When the reverberation time T_r is much longer than the frame size, an observed power spectrum $|\mathbf{x}|^2$ is modeled as a weighted sum of the source's power spectrum $|\hat{\mathbf{y}}|^2$ to be estimated with a stationary noise power spectrum $|\mathbf{n}|^2$ as

$$|\mathbf{x}_t|^2 = \sum_{\mu=0}^{t} w_\mu |\hat{\mathbf{y}}_{t-\mu}|^2 + |\mathbf{n}|^2, \tag{20.1}$$

where μ and w are the delay frame and the weight coefficient, respectively. The source's power spectrum $|\hat{\mathbf{y}}|^2$ is related to $|\mathbf{x}|^2$ as

$$|\hat{\mathbf{y}}_{t-\mu}|^2 = \eta(T_r)|\mathbf{x}_{t-\mu}|^2 - |\mathbf{n}|^2, \tag{20.2}$$

where η is the ratio of the direct sound components to the sum of the direct and reflected sound components, which is a decreasing function of T_r because for longer T_r, the energy of the reflected sound components increases. Assuming that w_0 is unity, (20.3) can be derived from the above relations:

$$|\hat{\mathbf{y}}_t|^2 = |\mathbf{x}_t|^2 - \sum_{\mu=1}^{t} w_\mu \left[\eta(T_r)|\mathbf{x}_{t-\mu}|^2 - |\mathbf{n}|^2 \right] - |\mathbf{n}|^2. \tag{20.3}$$

Reverberation is divided into two stages: early reverberation and late reverberation. The threshold between them, after arrival of the direct sound, is denoted by D (in frames). Early reverberation is complex but can be ignored because the ASR performance is mainly degraded by the late reverberation. The proposed method focuses on the late reverberation, where the sound-energy density decays exponentially with time according to Polack's statistical model [9]. Hence, w is determined as

$$w_\mu = \begin{cases} 0 & (1 \le \mu \le D), \\ \dfrac{\alpha_s}{\eta(T_r)} e^{-2\Delta\varphi\mu} & (D < \mu), \end{cases} \tag{20.4}$$

where φ is the frame shift and α_s is the subtraction parameter to be set. The upper condition and lower condition correspond to early and late reverberation, respectively. Assuming η is constant, (20.3) is a process similar to spectral subtraction [1]. If the subtracted power spectrum $|\hat{\mathbf{y}}|^2$ is less than $\beta|\mathbf{x}|^2$, it is substituted by $\beta|\mathbf{x}|^2$, where β is a flooring parameter. We define the floored ratio r as the ratio of the number of floored time–frequency bins to the total number of bins.

Two observations are exploited to estimate T_r from floored ratios r. First, when some arbitrary reverberation times (T_a) are assumed, r increases monotonically with T_a. This is modeled as linear with inclination Δ_r. Second, r increases with T_r at the same T_a. Since the actual $\eta(T_r)$ decreases with T_r, the power spectrum after dereverberation assuming constant η is more likely to be floored for a longer T_r because the second term of (20.3) is larger than the actual one in the condition with longer T_r. Therefore, T_r has a positive correlation with Δ_r and we have modeled this as $T_r = a\Delta_r - b$ with two constants a and b. The estimation process of T_r is summarized as follows: Calculate r and the inclination Δ_r by least-squares regression for some values of T_a, and estimate T_r.

20.3.3   Experiments

We evaluated the word recognition rate using JEIDA-JCSD (B-set) and CENSREC-4 [8], where eight different reverberant environments were prepared. We compared the performance of the proposed method with that of Lebart's method [6]. Figure 20.2 shows the recognition rate in terms of RT. The proposed method improves the recognition rate for all cases, and significantly in three environments whose RTs are over 0.5 s. In these three environments, the proposed method improves the recognition rate by 9.9, 11.0, and 13.7%, respectively, whereas the improvements obtained by Lebart's method are 7.5, 7.1, and 7.3%, respectively. The proposed method improves the average recognition rate by 5.0%, whereas Lebart's method improves that by 3.6%. The recognition rate given by the proposed method is better than that given by Lebart's method in almost all cases. The proposed method and Lebart's method are equivalent in computational time.

Fig. 20.2 Recognition rate of reverberant speech by the proposed and Lebart methods

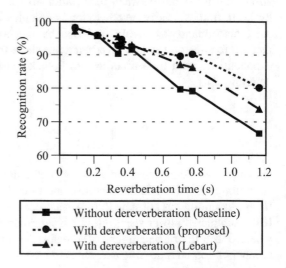

20.4   Discriminative Methods

This section introduces our advanced technologies, especially focused on discriminative methods for the acoustic model (AM) and the LM. Discriminative training of AMs is described in Sect. 20.4.2, in addition, discriminative training of recurrent neural network LMS (RNN-LMs) is described in Sect. 20.4.3.

20.4.1   Introduction

Many researchers have pointed out that combining different systems effectively improves performance (e.g., Recognizer Output Voting Error Reduction (ROVER) [3]) even if the performance of the complementary systems is lower than that of the base system. Because effective system combination relies on a combination of hypotheses with different trends, system combinations do not necessarily improve the performance when the hypotheses of the complementary systems have similar trends or yield too many errors [10, 12]. Classical system combination approaches require trial-and-error attempts because they do not rely on a general theoretical background such as an objective function in discriminative training.

To address this problem, a complementary system-training algorithm for acoustic models for system combination based on the minimum phone error (MPE) criterion has been proposed [2]. This lattice-based approach provides a theoretical background for training complementary systems and is promising because conventional discriminative training methods can be easily applied. We propose a general framework for sequential discriminative training for system combinations, encompassing various model-training methods for AMs. Our method generalizes the objective function of discriminative training in order to balance the objective function given by the correct labels and that given by the hypotheses of the base systems. The advantages of our proposed method are the fact it leads to a simple extension of conventional lattice-based discriminative training and its clear resemblance to a discriminative training method. In addition, because the formulation of our proposed method includes margin-based discriminative training, one can adjust the degree of deviation of the complementary systems' outputs with respect to those of the base systems.

In addition to discriminative training of AMs, discriminative training of RNN-LMs has been proposed. Neural networks have been recently introduced and used for language processing. Among them, RNN-LM has become popular due to its high performance [7]. An RNN is a neural network that contains one or more hidden layers with recursive inputs. Although their computational costs are high, an RNN-LM greatly improves ASR performance. The greatest difference between the RNN-LM and conventional n-gram models is the available word context length. A long context provides much information but the simple use of a long context by a conventional n-gram LM encounters data sparsity problems. To address these problems, the RNN-LM first maps a high-dimensional 1-of-N representation of a target word to a low-dimensional continuous space in a hidden layer and directly estimates the posterior probability of the target word. The hidden-layer units from the previous frame are then connected to the input vector in the next frame. These recursive inputs collect the history of words in the low-dimensional hidden-layer units. The RNN-LM is typically used for postprocessing such as N-best or lattice rescoring.

However, the training criteria of the RNN-LM are based on cross-entropy (CE) between predicted and reference words. That is, the CE criterion does not explicitly consider discriminative criteria calculated from ASR hypotheses and references. The RNN-LM CE criterion is discriminative in the sense of considering the posterior distribution of a target word given the history, but a discriminative criterion of the RNN-LM that considers ASR hypotheses can further correct ASR errors. Our proposed method is based on an RNN-LM framework, and can consider a long context with consideration of ASR hypotheses.

20.4.2 Discriminative Training for AMs

We describe a discriminative method that constructs complementary systems for a appropriate system combination [10, 12]. Complementary systems are constructed by discriminatively training a model starting from an initial model. Assuming Q base systems have already been constructed, the discriminative training objective function $\mathscr{F}$ is generalized to the following proposed objective function $\mathscr{F}^c$, which subtracts from the original objective function involving the correct labels s_r the objective functions involving the 1-best hypotheses (lattice) $s_{q,1}$ of the qth base systems:

$$\mathscr{F}^c(\varphi, s_r) = (1 + \alpha_c)\mathscr{F}(\varphi, s_r) - \frac{\alpha_c}{Q} \sum_{q=1}^{Q} \mathscr{F}(\varphi, s_{q,1}), \qquad (20.5)$$

where φ is the set of model parameters of a complementary system to be optimized and α_c is a scaling factor. The discriminative criterion $\mathscr{F}$ is selected as the maximum mutual information (MMI) or MPE. If α_c equals zero, this objective function matches the original $\mathscr{F}$. The first term in (20.5) promotes good performance according to the discriminative training criterion, whereas the second term makes the target system generate hypotheses that have a different tendency from the original base models.

We evaluated the performance improvement provided by these system combination techniques on the second CHiME challenge, Track 2 [13]. Although the database provided two-channel data, we used noise-suppressed single-channel data obtained by prior-based binary masking [11]. mel-frequency cepstral coefficients (MFCCs) were used as acoustic features with linear discriminant analysis (LDA), maximum likelihood linear transformation (MLLT), speaker adaptive training (SAT), and feature-space maximum likelihood linear regression (fMLLR). We used ROVER for combining output hypotheses from multiple systems. The baseline word error rate (WER) was 29.46% (evaluation set) by using an MMI Gaussian mixture model. The proposed complementary system improved the WER by 0.66% (28.80%).

20.4.3   Discriminative Training for RNN-LM

To introduce discriminative training into an RNN-LM, we start from the word-level likelihood ratio objective function $\mathscr{F}^{\mathrm{LR}}$:

$$\mathscr{F}^{\mathrm{LR}}(C, H) = -\sum_t \log \frac{y_t(c_t)}{y_t(h_t)^\beta}, \tag{20.6}$$

where h_t is an index of the tth word of the 1-best ASR hypothesis aligned with the reference sequence $C = \{c_t | t = 1, \ldots, T\}$, and $H = \{h_t | t = 1, \ldots, T\}$ denotes the 1-best ASR sequence. β is a scaling factor.

Equation (20.6) can also be rewritten as

$$\mathscr{F}^{\mathrm{LR}}(C, H) = -\sum_n \sum_t \delta(n, c_t) \log y_t(n) - \beta \delta(n, h_t) \log y_t(n)$$

$$= \mathscr{F}^{\mathrm{CE}}(C) - \beta \mathscr{F}^{\mathrm{CE}}(H), \tag{20.7}$$

where n is an index of elements in the output layer. Therefore, (20.6) can be interpreted as a weighted difference of the CE for the correct label and the ASR hypothesis.

For our proposed model, the update rule is derived from differentiation of (20.7) such that

$$\frac{\partial \mathscr{F}^{\mathrm{LR}}(C, H)}{\partial a_t(n)} = -[\delta(n, c_t) - \beta \delta(n, h_t) - (1 - \beta) y_t(n)], \tag{20.8}$$

where a_t is an activation of the nth word. In our implementation, we assume $(1 - \beta) y_t(n)$ as $y_t(n)$ for simplicity; thus we obtain

$$\frac{\partial \mathscr{F}^{\mathrm{LR}}(C, H)}{\partial a_t(n)} \approx -[\delta(n, c_t) - \beta \delta(n, h_t) - y_t(n)]. \tag{20.9}$$

First, alignments of correct word sequences and ASR hypotheses are fixed using dynamic programming. Second, the weight for the correct label is discounted (i.e., $1-\beta$) and the model is retrained with these discounted weights. Note that we assume

Table 20.1 WER (%) on CSJ using a deep-neural-network acoustic model with a conventional n-gram, rescoring with RNN-LM, and rescoring with the proposed dRNN-LM

	E1	E2	E3	Avg.
Baseline	12.81	10.64	11.13	11.53
+RNN-LM	11.97	10.18	10.51	10.89
+dRNN-LM	11.84	10.02	10.39	10.75

that $\delta(n, c_t) - \beta\delta(n, h_t) = 0$ when $\delta(n, c_t) - \beta\delta(n, h_t) < 0$ to avoid the possibility that the value of the target reference word becomes negative. Finally, the weights of the RNN-LM models W are smoothed by an interpolation of the weights of the proposed discriminative method W^{LR} with those of the original CE model W such that

$$\{U, V\} \leftarrow \tau\{U^{CE}, V^{CE}\} + (1 - \tau)\{U^{LR}, V^{LR}\}, \tag{20.10}$$

where τ is a smoothing factor. This avoids overtraining.

We evaluated the observed performance improvement on the Corpus of Spontaneous Japanese (CSJ), which is one of the most widely used large-vocabulary continuous speech recognition (LVCSR) tasks used to build Japanese ASR systems. The vocabulary size is about 70k. Although the size of the original LM was 70k, the vocabulary size of the RNN-LM was limited to 10k, which corresponds to the number of input layer dimensions. The number of hidden-layer units was 30. The LM score was obtained by linear interpolation of the RNN-LM score and the original n-gram model score. The weight of interpolation was 0.5, and the 100 best hypotheses for each utterance were used for rescoring. We used three types of test sets, where each set consisted of lecture-style examples from ten speakers. Test sets E1, E2, and E3 contained 22,682, 23,226, and 14,896 words, respectively. We used a DNN and a hidden Markov model for the AM.

Table 20.1 shows the WER. The RNN-LM improved the WER of the baseline for all cases. In addition, the proposed discriminative RNN-LM (dRNN-LM) improved the WER further.

20.5 Conclusion

Our main targets are used in noisy and reverberant environments with limited computation resources. We have developed a search system and dereverberation method for small-computational-resource systems such as embedded systems. Advanced technologies related to discriminative training for AMs and LMs were also introduced. These methods improved the robustness of ASR in noisy and reverberant environments.

References

1. Boll, S.: Suppression of acoustic noise in speech using spectral subtraction. IEEE Trans. Acoust. Speech Signal Process. **27**(2), 113–120 (1979)
2. Diehl, F., Woodland, P.: Complementary phone error training. In: Proceedings of INTERSPEECH (2012)
3. Fiscus, J.: A post-processing system to yield reduced error word rates: recognizer output voting error reduction (ROVER). In: Proceedings of ASRU, pp. 347–354 (1997)
4. Hanazawa, T., Okato, Y., Iwasaki, T.: Speech recognition using statistical language model and text match based large vocabulary search by voice. In: Proceedings of 2009 Autumn Meeting of the Acoustical Society of Japan, pp. 61–62 (2009)
5. Iwasaki, T., Kosaka, M., Nanba, T., Narita, T.: Voice interface of car navigation system – current technologies and the future. In: Mitsubishi Denki Giho, pp. 51–54 (2004)
6. Lebart, K., Boucher, J.M., Denbigh, P.N.: A new method based on spectral subtraction for speech dereverberation. Acta Acustica **87**, 359–366 (2001)
7. Mikolov, T., Karafiát, M., Burget, L., Černocký, J., Khudanpur, S.: Recurrent neural network based language model. In: Proceedings of INTERSPEECH, pp. 1045–1048 (2010)
8. Nakayama, M., Nishiura, T., Denda, Y., Kitaoka, N., Yamamoto, K., Yamada, T., Tsuge, S., Miyajima, C., Fujimoto, M., Takiguchi, T., Tamura, S., Ogawa, T., Matsuda, S., Kuroiwa, S., Takeda, K., Nakamura, S.: CENSREC-4: development of evaluation framework for distant-talking speech recognition under reverberant environments. In: Proceedings of Interspeech, pp. 968–971 (2008)
9. Naylor, P., Gaubitch, N.: Speech Dereverberation. Springer, New York (2010)
10. Tachioka, Y., Watanabe, S.: Discriminative training of acoustic models for system combination. In: Proceedings of INTERSPEECH, pp. 2355–2359 (2013)
11. Tachioka, Y., Watanabe, S., Hershey, J.: Effectiveness of discriminative training and feature transformation for reverberated and noisy speech. In: Proceedings of ICASSP, pp. 6935–6939 (2013)
12. Tachioka, Y., Watanabe, S., Le Roux, J., Hershey, J.: A generalized framework of discriminative training for system combination. In: Proceedings of ASRU, pp. 43–48 (2013)
13. Vincent, E., Barker, J., Watanabe, S., Le Roux, J., Nesta, F., Matassoni, M.: The second "CHiME" speech separation and recognition challenge: datasets, tasks and baselines. In: Proceedings of ICASSP, pp. 126–130 (2013)

Index

Printed in the United States
By Bookmasters